TC 3-21.76

HANDBOOK
Not for the weak or fainthearted

*"Let the enemy come till he's almost close enough to touch.
Then let him have it and jump out and finish him with your hatchet."*
Major Robert Rogers, 1759.

APRIL 2017

DISTRIBUTION RESTRICTION: Approved for public release; distribution is unlimited.

Headquarters, Department of the Army

Training Circular		Headquarters
No. 3-21.76		Department of the Army
		Washington, DC, 26 April 2017

Ranger Handbook

Contents

		Page
	PREFACE	xvi
	RANGER HISTORY	xvii
	MEDAL OF HONOR RECIPIENTS	xxi
Chapter 1	LEADERSHIP	1-1
	Principles	1-1
	Assumption of Command	1-7
Chapter 2	OPERATIONS	2-1
	Troop Leading Procedures	2-1
	Combat Intelligence	2-5

DISTRIBUTION RESTRICTION: Approved for public release; distribution is unlimited.

TC 3-21.76

Contents

	Operation Order	2-12
	Fragmentary Order	2-18
	Annexes	2-21
	Coordination Checklists	2-31
	Terrain Model	2-41
Chapter 3	**FIRE SUPPORT**	**3-1**
	Basic Tasks and Targeting	3-1
	Interdiction	3-1
	Risk Estimate Distance	3-4
	Close Air Support	3-13
	Army Attack Aviation	3-18
Chapter 4	**COMMUNICATIONS**	**4-1**
	Equipment	4-1
	Antennas	4-6
Chapter 5	**DEMOLITIONS**	**5-1**
	Initiating (Priming) Systems	5-2
	Detonation (Firing) Systems	5-4
	Expedient Explosives	5-5
	Charges	5-11
Chapter 6	**MOVEMENT**	**6-1**
	Formations	6-1
	Movement Techniques	6-2
	Danger Areas	6-8
Chapter 7	**PATROLS**	**7-1**
	Principles	7-1
	Planning	7-1

Contents

	Reconnaissance Patrols 7-4
	Combat Patrols 7-8
	Performing a Raid 7-16
	Supporting Tasks.. 7-18
	Movement to Contact 7-25
	Task Standards 7-26
Chapter 8	**BATTLE DRILLS**... 8-1
	React to Direct Fire Contact (07-3-D9501) 8-1
	Conduct a Platoon Assault (07-3-D9514)......................... 8-5
	React to Ambush (Near) (07-3-D9502) 8-14
	Enter and Clear a Room (07-4-D9509) 8-18
	React to Indirect Fire (07-3-D9504) 8-21
Chapter 9	**MILITARY MOUNTAINEERING** 9-1
	Training and Planning 9-1
	Dismounted Mobility 9-2
	Task Organization 9-3
	Rescue Equipment 9-3
	Mountaineering Equipment 9-4
	Rope Installations 9-24
Chapter 10	**MACHINE GUN EMPLOYMENT** 10-1
	Specifications... 10-1
	Classes of Automatic Weapons Fire 10-4
	Offense 10-9
	Defense.. 10-13
	Control of Machine Guns 10-14

26 April 2017   TC 3-21.76   iii

Contents

Chapter 11	URBAN OPERATIONS	11-1
	Planning	11-1
	Preparation	11-2
	Analyzing the Urban Environment	11-3
	Close Quarters Combat	11-5
Chapter 12	WATERBORNE OPERATIONS	12-1
	Rope Bridge	12-1
	Poncho Raft	12-7
	Watercraft	12-8
	Preparation, Personnel, and Equipment	12-9
	Conduct Capsize Procedures	12-14
	River Movement, Navigation, and Formations	12-15
Chapter 13	MOUNTED PATROL OPERATIONS	13-1
	Planning	13-1
	Forced Stops	13-8
Chapter 14	AVIATION..	14-1
	Reverse Planning Sequence..	14-1
	Air Assault Formations	14-4
	Pickup Zone Operations	14-10
	Rotary Wing Aircraft Specifications..	14-18
Chapter 15	FIRST AID	15-1
	LifeSaving Steps and Care Under Fire...................................	15-1
	Treating Injuries...	15-7
	Poisonous Plant Identification..	15-10
	Foot Care, Hydration, and Acclimatization	15-11
	Requesting Medical Evacuation	15-13

Contents

Appendix A RESOURCES A-1
Appendix B QUICK REFERENCE CARDS B-1
            GLOSSARY Glossary-1
            REFERENCES References-1
            INDEX Index-1

Figures

Figure 2-1. Terrain model 2-42
Figure 3-1. Sterile fire support overlay 3-7
Figure 3-2. Nonsterile fire support overlay 3-8
Figure 3-3. Example of a close air support request 3-14
Figure 4-1. Cobra head 4-9
Figure 4-2. Antenna base 4-10
Figure 4-3. Completed expedient 292-type antenna 4-12
Figure 4-4. Curvature of the Earth 4-13
Figure 5-1. Technique for lighting time fuze with a match 5-4
Figure 5-2. Improvised shape charge 5-6
Figure 5-3. Platter charge 5-7
Figure 5-4. Grapeshot charge 5-7
Figure 5-5. Various joining knots used in demolitions 5-8
Figure 5-6. British junction knot 5-9
Figure 5-7. Minimum safe distance for charges over 500 pounds 5-10

Contents

Figure 5-8. Formula for computing size of charge to breach concrete, masonry, and rock 5-13
Figure 5-9. Example of abatis 5-14
Figure 5-10. Formula for fallen tree obstacles or test shot 5-14
Figure 5-11. Timber-cutting ring charge 5-15
Figure 5-12. Timber-cutting charge (external).................................... 5-16
Figure 5-13. Formula for external timber-cutting charge 5-16
Figure 5-14. Timber-cutting charge (internal) 5-17
Figure 5-15. Formula for internal timber-cutting charge....................... 5-17
Figure 6-1. Formations... 6-1
Figure 6-2. Squad bounding overwatch 6-4
Figure 6-3. Platoon bounding overwatch.. 6-5
Figure 6-4. Linear danger area.. 6-10
Figure 6-5. Small open danger area 6-11
Figure 7-1. Area reconnaissance 7-6
Figure 7-2. Ambush formations 7-11
Figure 7-3. Hasty ambush.. 7-12
Figure 7-4. Deliberate ambush 7-15
Figure 7-5. Actions on the objective, raid 7-18
Figure 7-6. Occupation of the objective rally point 7-21
Figure 7-7. Patrol base... 7-23
Figure 8-1. 07-3-D9501. Assuming nearest covered position 8-2
Figure 8-2. 07-3-D9501. Control of the support element..................... 8-4

Figure 8-3. 07-3-D9505. Employing indirect fires to suppress enemy 8-11
Figure 8-4. 07-3-D9505. Moving element occupies overwatch and engages enemy 8-12
Figure 8-5. 07-3-D9505. Movement and fire technique 8-13
Figure 8-6. 06-3-D9502. React to ambush (near) (dismounted) 8-15
Figure 8-7. 07-3-D9502. Returning fire immediately 8-16
Figure 8-8. 07-3-D9502. Assaulting through enemy positions 8-17
Figure 8-9. 07-4-D9509. Clear a room. First two soldiers enter simultaneously 8-19
Figure 8-10. 07-4-D9509. Clear a room. Third soldier enters, clearing his sector 8-20
Figure 8-11. 07-4-D9509. Clear a room. Third soldier enters, dominating his sector 8-20
Figure 9-1. Examples of traditional (removable) protection used on rocks 9-6
Figure 9-2. Examples of fixed (permanent or semipermanent) protection used on rocks 9-7
Figure 9-3. Constructing a three-point, pre-equalized anchor using fixed artificial protection 9-8
Figure 9-4. Tensionless natural anchor 9-9
Figure 9-5. Rope terminology 9-10
Figure 9-6. Square knot 9-11
Figure 9-7. Round turn with two half hitches 9-11
Figure 9-8. Double figure-eight knot 9-12

Contents

Figure 9-9. End-of-the-rope clove hitch 9-13
Figure 9-10. Middle-of-the-rope clove hitch 9-14
Figure 9-11. Rappel seat 9-15
Figure 9-12. Rerouted figure-eight knot 9-16
Figure 9-13. Figure-eight slipknot 9-16
Figure 9-14. Munter hitch 9-17
Figure 9-15. Munter mule knot 9-18
Figure 9-16. Prusik knot 9-18
Figure 9-17. Bowline 9-19
Figure 9-18 Body belay 9-21
Figure 9-19. Mechanical belay devices 9-22
Figure 9-20. Air traffic controller 9-23
Figure 9-21. Transport-tightening system 9-26
Figure 9-22. Commando crawl method 9-27
Figure 9-23. Rappel seat (Tyrolean traverse) method 9-28
Figure 9-24. Z-Pulley system 9-30
Figure 9-25. Hasty rappel 9-31
Figure 9-26. Body rappel 9-32
Figure 9-27. Figure-eight descender 9-33
Figure 9-28. Close-up of carabiner wrap descender, and the seat hip rappel (shown with carabiner wrap seat hip descender) 9-33
Figure 9-29. Sling extension 9-34
Figure 9-30. Extended ATC rappel 9-35

Contents

Figure 10-1. Trajectory and maximum ordinate 10-3
Figure 10-2. Cone of fire and the beaten zone 10-4
Figure 10-3. Classes of fire, respect to ground 10-5
Figure 10-4A. Classes of fire, respect to target 10-7
Figure 10-4B. Classes of fire, respect to target 10-8
Figure 10-5. Classes of fire, respect to gun 10-10
Figure 11-1. 07-4-D9509. Clear a room, first two Soldiers enter 11-8
Figure 11-2. 07-4-D9509. Clear a room, third Soldier enters 11-8
Figure 11-3. 07-4-D9509. Clear a room, third Soldier enters dominating his sector 11-9
Procedures for Marking Buildings and Rooms................................ 11-9
Figure 12-1. Position of bridge team personnel 12-3
Figure 12-2. Transport-tightening system.. 12-5
Figure 12-3. Weapon rigging 12-10
Figure 12-4. Rucksack rigging.. 12-11
Figure 12-5. Equipment rigging 12-12
Figure 12-6. Crew positions, long count and short count................... 12-13
Figure 13-1. React to ambush (near) 13-6
Figure 13-2. React to ambush (far) 13-7
Figure 13-3. Mounted patrol forced to stop, method one 13-9
Figure 13-4. Mounted patrol forced to stop, method two 13-11
Figure 13-5. Break contact.. 13-13
Figure 13-6. Recovery/CASEVAC operations 13-15

Contents

Figure 14-1. Inverted "Y" 14-3
Figure 14-2. Heavy left or heavy right formation 14-5
Figure 14-3. Diamond formation 14-6
Figure 14-4. Vee formation 14-7
Figure 14-5. Echelon left or right formation 14-8
Figure 14-6. Trail formation 14-9
Figure 14-7. Staggered trail left or right formation 14-10
Figure 14-8. Large, one-sided pickup zone 14-11
Figure 14-9. UH-60L loading sequence 14-12
Figure 14-10. UH-60L unloading sequence 14-13
Figure 14-11. Tactical loading sequence 14-14
Figure 14-12. Tactical unloading sequence using door
    nearest cover, concealment 14-15
Figure 14-13. CH 47/CV-22 rear ramp off-load 14-16
Figure 15-1. Chin lift and jaw thrust methods 15-2
Figure 15-2. Nasal airway insertion 15-3
Figure 15-3. Mouth airway insertion 15-4
Figure 15-4. Percentage of body area 15-6
Figure 15-5. Poisonous plants 15-10
Figure A-1. React to indirect fire A-2
Figure A-2. React to direct fire contact A-3
Figure A-3. React to a near ambush A-4
Figure A-4. Break contact A-5
Figure A-5. Formations and order of movement A-6

Contents

Figure A-6. Linkup A-7
Figure A-7. Linear danger area A-8
Figure A-8. Large open danger area A-9
Figure A-9. Crossing a small open area A-10
Figure A-10. Squad attack A-11
Figure A-11. Crossing a small open area A-12
Figure A-12. Raid boards (middle) A-14
Figure A-13. Leader's reconnaissance A-19
Figure A-14. Diamond and vee formations A-21
Figure A-15. Clearing hallway junctions A-23
Figure A-16. Clearing a "T" intersection A-25
Figure A-17. Three-man flow clearing technique A-27
Figure B-1. IED/UXO card B-2
Figure B-2. Range card with final protective line B-3
Figure B-3. Standard Range Card B-4
Figure B-4. NATO 9-Line MEDEVAC Request card and MIST report B-5
Figure B-5A. DD Form 1380, Tactical Combat Casualty Care card (front) B-9
Figure B-6B. DD Form 1380, Tactical Combat Casualty Care card (back) B-10

Contents

Tables

Table 1-1. Leadership Requirements Model 1-1
Table 1-2. Tasks for the assumption of command 1-7
Table 2-1. Steps in the troop leading procedures 2-1
Table 2-2. Examples of specified and implied tasks 2-2
Table 2-3. Examples of restated missions 2-2
Table 2-4. Enemy 2-3
Table 2-5. Offensive considerations 2-3
Table 2-6. Example of a SALUTE report format 2-5
Table 2-7. Example of a warning order format 2-6
Table 2-8. Example of a warning order format (squad) 2-8
Table 2-9. Example of a squad OPORD format 2-13
Table 2-10. Example of an annotated FRAGORD format................. 2-18
Table 2-11. Example of an air movement annex format 2-21
Table 2-12. Example of a patrol base annex format 2-23
Table 2-13. Example of a waterborne insertion annex format 2-25
Table 2-14. Example of a covert gap crossing annex format 2-27
Table 2-15. Example of a truck annex format 2-29
Table 2-16. Intelligence coordination checklist 2-31
Table 2-17. Operations coordination checklist 2-31
Table 2-18 Fire support coordination checklist 2-32
Table 2-19. Coordination with forward unit checklist 2-33

Table 2-20. Adjacent coordination checklist 2-34
Table 2-21. Rehearsal area coordination checklist 2-34
Table 2-22. Army aviation coordination checklist 2-35
Table 2-23. Vehicular movement coordination checklist 2-38
Table 2-24. Actions by friendly forces and effects on enemy forces 2-39
Table 2-25. Purpose 2-39
Table 2-26. Elements of operations and subordinate tasks 2-39
Table 2-27. Tactical shaping operations and tasks 2-40
Table 3-1. Capabilities of mortars 3-2
Table 3-2. Capabilities of field artillery 3-3
Table 3-3 Risk estimate distances for unguided mortars and cannon artillery 3-5
Table 3-4. Contents of fire support overlay 3-6
Table 3-5. TTLODAC checklist 3-9
Table 3-6. Example of call-for-fire transmissions 3-12
Table 3-7. Fixed-wing close air support capabilities 3-15
Table 3-8. Unmanned aircraft systems close air support capabilities 3-17
Table 3-9. Army attack aviation call-for-fire format 3-18
Table 3-10. Rotary wing close air support capabilities 3-19
Table 4-1. Military radios 4-1
Table 4-2. Frequency ranges 4-4
Table 4-3. Quick reference table 4-11

Contents

Table 5-1. Characteristics of U.S. demolitions explosives 5-1
Table 5-2. Minimum safe distance for personnel in the open (bare charge) 5-10
Table 5-3. Breaching charges for reinforced concrete 5-11
Table 5-4. Conversion factors for materials other than reinforced concrete 5-12
Table 5-5. Material factor (K) for breaching charges 5-12
Table 5-6. Timber-cutting charge size 5-13
Table 6-1. Characteristics of movement techniques 6-2
Table 7-1. Reconnaissance methods 7-8
Table 9-1. Terrain classification table 9-2
Table 9-2. Sequence of climbing commands 9-24
Table 10-1. Specifications of machine guns 10-1
Table 10-2. Machine gun terms 10-2
Table 10-3. Classes of fire, respect to ground 10-4
Table 10-4. Classes of fire, respect to target 10-6
Table 10-5. Classes of fire, respect to gun 10-9
Table 13-1. Mounted tactical movement brief 13-3
Table 14-1. Air assault formations 14-4
Table 14-2. Specifications for the UH-60L Blackhawk 14-19
Table 14-3. Specifications for the CV-22 Osprey 14-20
Table 14-4. Specifications for the CH-47D Chinook 14-21
Table 15-1. ABCs of first aid 15-1
Table 15-2. Heat injuries 15-7

Table 15-3. Cold injuries 15-8
Table 15-4. Environmental injuries 15-9
Table 15-5. Hydration management and acclimatization 15-11
Table 15-6. Work, rest, and water consumption guidelines 15-12
Table A-1. Raid boards (left) A-13
Table A-2. Raid boards (middle) task organization A-15
Table A-3. Raid boards (right) SOP A-16
Table A-4. Ambush SOP (left) A-17
Table A-5. Ambush boards (middle) A-18
Table B-1. Explanation of the nine lines on a MEDEVAC
 Request card B-6

Preface

Training Circular (TC) 3-21.76 uses joint terms where applicable. Selected joint and Army terms and definitions appear in both the glossary and the text. Terms for which TC 3-21.76 is the proponent publication (the authority) are italicized in the text and are marked with an asterisk (*) in the glossary. Terms and definitions for which TC 3-21.76 is the proponent publication are boldfaced in the text. For other definitions shown in the text, the term is italicized and the number of the proponent publication follows the definition.

The principal audience for TC 3-21.76 are U.S. Army Rangers and combat arms units. Commanders and staffs of Army headquarters serving as joint task force or multinational headquarters should also refer to applicable joint or multinational doctrine concerning the range of military operations and joint or multinational forces. Trainers and educators throughout the Army will also use this publication.

Commanders, staffs, and subordinates ensure that their decisions and actions comply with applicable United States, international, and (in some cases) host-nation laws and regulations. Commanders at all levels ensure that their Soldiers operate according to the law of war and the rules of engagement (ROE). (See FM 27-10.)

This publication applies to the active Army, the Army National Guard (ARNG)/Army National Guard of the United States (ARNGUS), the United States Army Reserve (USAR), and the United States Marine Corp. Unless otherwise stated in this publication, masculine nouns and pronouns do not refer exclusively to men.

The proponent of this publication is United States Army Maneuver Center of Excellence (MCoE). The preparing agency is the Maneuver Center of Excellence, Fort Benning, Georgia. You may submit comments and recommended changes in any of several ways—U.S. mail, e-mail, fax, or telephone—as long as you use or follow the format of Department of the Army (DA) Form 2028, *(Recommended Changes to Publications and Blank Forms)*. Contact information is:

- E-Mail: usarmy.benning.tradoc.mbx.artb-s3-operations@mail.mil
- Phone: (706) 544-6448 (DSN 834)
- Fax: (706) 544- 6421 (DSN 834)

- U.S. Mail: Commander, Airborne and Ranger Training Brigade
  ATTN: ATSH-RB
  10850 Schneider Rd, Bldg. 5024
  Ft Benning, GA 31905

Ranger History

The history of the American Ranger is a long and colorful saga of courage, daring, and outstanding leadership. It is a story of men whose skills in the art of fighting have seldom been surpassed. Only the highlights of their numerous exploits are told here.

Rangers mainly performed defensive missions until, during King Phillip's War in 1675, Benjamin Church's Company of Independent Rangers (from Plymouth Colony) conducted successful raids on hostile Indians. In 1756, Major Robert Rogers, of New Hampshire, recruited nine companies of American colonists to fight for the British during the French and Indian War. Ranger techniques and methods of operation inherently characterized the American frontiersmen. Major Rogers was the first to capitalize on them and incorporate them into the fighting doctrine of a permanently organized fighting force.

The method of fighting used by the first Rangers was further developed during the Revolutionary War by Colonel Daniel Morgan, who organized a unit known as "Morgan's Riflemen." According to General Burgoyne, Morgan's men were "...the most famous corps of the Continental Army, all of them crack shots."

Francis Marion, the "Swamp Fox," organized another famous Revolutionary War Ranger element known as "Marion's Partisans." Marion's Partisans, numbering anywhere from a handful to several hundred, operated both with and independent of other elements of General Washington's Army. Operating out of the Carolina swamps, they disrupted British communications and prevented the organization of loyalists to support the British cause, substantially contributing to the American victory.

The American Civil War was again the occasion for the creation of special units such as Rangers. John S. Mosby, a master of the prompt and skillful use of cavalry, was one of the most outstanding of the Confederate Rangers. He believed that by resorting to aggressive action, he could compel his enemies to guard a hundred points. He would then attack one of the weakest points and be assured numerical superiority.

With America's entry into the Second World War, Rangers came forth to add to the pages of history. Major William O. Darby organized and activated the 1st Ranger battalion on June 19, 1942 at Carrickfergus, North Ireland. The members were all handpicked volunteers; 50 participated in the gallant Dieppe Raid on the northern coast of France with British and Canadian commandos. The 1st, 3rd, and 4th Ranger battalions participated with distinction in the North African, Sicilian, and Italian campaigns. Darby's Ranger Battalions spearheaded the Seventh Army landing at Gela and Licata during the Sicilian invasion, and played a key role in the subsequent campaign, which ended in the capture of Messina. They infiltrated German lines and mounted an attack against Cisterna, where they virtually annihilated an entire German parachute regiment during close in, night, bayonet, and hand-to-hand fighting.

The 2nd and 5th Ranger battalions participated in the D-day landings at Omaha Beach, Normandy. It was during the bitter fighting along the beach that the Rangers gained their official motto. As the situation became critical, the commander of the 29th Infantry Division stated that the entire force must clear the beach and advance inland. He then turned to Lieutenant Colonel Max Schneider, commander of the 5th Ranger Battalion, and said, "Rangers! Lead the way." The 5th Ranger battalion spearheaded the breakthrough. This enabled the Allies to drive inland, away from the invasion beaches.

The 6th Ranger battalion, operating in the Pacific, conducted Ranger-type missions behind enemy lines. These missions involved reconnaissance and hard hitting, long-range raids. These Rangers were the first American group to return to the Philippines, destroying key coastal installations prior to the invasion. A reinforced company from the 6th Ranger battalion formed the rescue force that liberated American and Allied prisoners of war (POWs) from the Japanese prison camp at Cabanatuan.

Another Ranger-type unit was the 5307th Composite Unit (Provisional), organized and trained as a long-range penetration unit for employment behind enemy lines in Japanese-occupied Burma. The unit commander was Brigadier General (later Major General) Frank D. Merrill. Its 2,997 officers and men became popularly known as "Merrill's Marauders."

Ranger History

The men of Merrill's Marauders were volunteers from the 5th, 154th, and 33rd Infantry regiments and from other Infantry regiments engaged in combat in the Southwest and South Pacific. These men responded to a call from Chief of Staff, General George C. Marshall, for volunteers for a hazardous mission. These volunteers were to have a high state of physical ruggedness and stamina, and were to come from jungle-trained and jungle-tested units.

Before joining the Northern Burma Campaign, Merrill's Marauders trained in India under British Major General Orde C. Wingate. From February to June 1943, they learned long-range penetration tactics and techniques like those developed and first employed by General Wingate. The operations of the Marauders were closely coordinated with those of the Chinese 22nd and 38th Divisions in a drive to recover northern Burma and clear the way for the construction of Led Road, which was to link the Indian railhead at Ledo with the old Burma Road to China. The Marauders marched and fought through jungle and over mountains from Hukwang Valley in Northwest Burma, to Myitkyina and the Irrawaddy River. In five major and 30 minor engagements, they met and defeated the veteran soldiers of the Japanese 18th Division. Operating in the rear of the main force of the Japanese, they prepared the way for the southward advance of the Chinese by disorganizing supply lines and communications. The climax of the Marauder's operations was the capture of Myitkyina airfield, the only all-weather strip in northern Burma. This was the final victory of "Merrill's Marauders," which disbanded in August 1944. Remaining personnel merged into the 475th Infantry Regiment, which fought its last battle on February 3 and 4, 1945, at Loi Kang Ridge, China. This Infantry regiment is the father of the 75th Ranger Regiment.

Soon after the Korean War started in June 1950, the 8th Army Ranger Company was formed with volunteers from American units in Japan. The company trained in Korea, and distinguished itself in combat during the drive to the Yalu River, performing task force and spearhead operations. During the massive Chinese intervention of November 1950, this small, vastly outnumbered unit withstood five enemy assaults on its position.

In September 1950, a Department of the Army (DA) message called for volunteers to train as Airborne Rangers. Five thousand regular Army paratroopers from the 82nd Airborne Division volunteered. Nine hundred were chosen to form the first eight Airborne Ranger companies. Nine more companies were formed from regular Army and National Guard Infantry division volunteers. These seventeen Airborne Ranger companies were activated and trained at Fort Benning, Georgia. Most received more training in the Colorado Mountains.

In 1950 and 1951, some 700 men of the 1st, 2nd, 3rd, 4th, 5th, and 8th Airborne Ranger companies fought to the front of every American Infantry division in Korea. Attacking by land, water, and air, these six Ranger companies raided, penetrated, and ambushed North Korean and Chinese forces. They were the first Rangers to make combat jumps. After the Chinese intervention, these Rangers were the first Americans to recross the 38th parallel. The 2nd Airborne Ranger Company was the only African-American Ranger unit in the history of the U.S. Army. The men of the six Ranger companies who fought in Korea paid the bloody price of freedom. One in nine of this gallant brotherhood died on the battlefields of Korea.

Other Airborne Ranger companies led the way while serving with Infantry divisions in the U.S., Germany, and Japan. These volunteers fought as members of line Infantry units in Korea. They volunteered for the Army, the Airborne, the Rangers, and for combat. The first men to earn and wear the coveted Ranger tab, these men are the original Airborne Rangers. One Ranger, Donn Porter, received the Medal of Honor posthumously. Fourteen Korean War Rangers rose to general officer. Dozens more became colonels, senior noncommissioned officers (NCOs), and civilian leaders.

In October 1951, Army Chief of Staff, General J. Lawton Collins, directed that Ranger training extend to all Army combat units. He directed the Commandant of the Infantry School to establish a Ranger Department. This new department would develop and conduct a Ranger course of instruction. His goal was to raise the standard of training in all combat units. The program built on lessons learned from World War II and the Korean conflict.

During the Vietnam Conflict, fourteen Ranger companies consisting of highly motivated volunteers served with distinction from the Mekong Delta to the demilitarized zone. Assigned to separate brigade, division, and field-force units, they conducted long-range reconnaissance and exploitation operations into enemy-held areas. They provided valuable combat intelligence. Initially designated as

Ranger History

long range reconnaissance patrol (LRRP), then long range patrol (LRP) companies, these units were later designated as C through P (there is no Juliet Company) Rangers, 75th Infantry.

After Vietnam, Army Chief of Staff, General Abrams, recognized the need for a highly trained and highly mobile reaction force. He activated the first battalion-sized Ranger units since World War II, the 1st and 2nd Battalions (Ranger), 75th Infantry. The 1st Battalion trained at Fort Benning, Georgia, and was activated February 8, 1974 at Fort Stewart, Georgia. The 2nd Battalion was activated on October 3, 1974. The 1st Battalion is now based at Hunter Army Airfield, Georgia. The 2nd Battalion is based at Fort Lewis, Washington.

General Abrams' farsighted decision and the combat effectiveness of the Ranger battalions were proven in the U.S. invasion of Grenada, "Operation Urgent Fury," October 1983. The mission was to protect American citizens and restore democracy. The Ranger battalions "led the way" with a daring, low-level airborne assault (from 500 feet) to seize the airfield at Point Salines. They continued operations for several days, eliminating pockets of resistance and rescuing American medical students. Due to this success, in 1984 the DA increased the strength of Ranger units to their highest level in 40 years. To do this, it activated another Ranger battalion and a Ranger Regimental Headquarters. After these units, the 3rd Battalion (Ranger), 75th Infantry, and headquarters company (Ranger), 75th Infantry, were activated, there were over 2,000 Soldiers assigned to Ranger units. On February 3, 1986, the 75th Infantry was renamed the 75th Ranger Regiment.

On December 20, 1989, the 75th Ranger Regiment was again called to show its effectiveness in combat. For the first time since reorganizing in 1984, the regimental headquarters and all three Ranger battalions deployed together. During "Operation Just Cause" in Panama, the 75th Ranger Regiment spearheaded the assault into Panama by conducting airborne assaults on the Torrijos/Tocumen airport and Rio Hato airfield. Their mission: to facilitate the restoration of democracy in Panama and to protect the lives of American citizens. Between December 20, 1989 and January 7, 1990, the regiment performed many follow-on missions in Panama.

Early in 1991, elements of the 75th Ranger Regiment deployed to Saudi Arabia in support of "Operation Desert Storm." In August 1993, elements of the 75th Ranger Regiment deployed to Somalia in support of "Operation Restore Hope," and returned in November 1993. In 1994, elements of the 75th Ranger Regiment deployed to Haiti in support of "Operation Uphold Democracy." In 2000 to 2001, elements of the 75th Ranger Regiment deployed to Kosovo in support of "Operation Joint Guardian."

Since September 11, 2001, the 75th Ranger Regiment has led the way in establishing democracy. In October 2001, elements of the 75th Ranger Regiment deployed to Afghanistan in support of "Operation Enduring Freedom." In March 2003, elements of the Regiment deployed in support of "Operation Iraqi Freedom." The 75th Ranger Regiment spearheaded the campaign against senior level members of Al Qaeda and the Taliban in both Theaters throughout the US deployment to Afghanistan and Iraq. Additionally, Ranger leaders throughout the entire force participated in more than 15 years of combat and advisor (or advise and assist) operations, serving as points of continuity among a transforming Army.

As in the past, Rangers continue to significantly contribute to the overall success of operations. The 75th Ranger regiment stands ready to execute its special operations mission in support of the United States' policies and objectives. Rangers throughout the force lead their formations, set the example for fellow Soldiers, and remain ready to defend the United States against its enemies. Rangers lead the way!

Ranger Creed

Ranger Creed

Recognizing that I volunteered as a Ranger, fully knowing the hazards of my chosen profession, I will always endeavor to uphold the prestige, honor, and high esprit de corps of the Rangers.

Acknowledging the fact that a Ranger is a more elite Soldier who arrives at the cutting edge of battle by land, sea, or air, I accept the fact that as a Ranger my country expects me to move further, faster, and fight harder than any other Soldier.

Never shall I fail my comrades I will always keep myself mentally alert, physically strong, and morally straight and I will shoulder more than my share of the task whatever it may be, one hundred percent and then some.

Gallantly will I show the world that I am a specially selected and well trained Soldier. My courtesy to superior officers, neatness of dress, and care of equipment shall set the example for others to follow.

Energetically will I meet the enemies of my country. I shall defeat them on the field of battle for I am better trained and will fight with all my might. Surrender is not a Ranger word. I will never leave a fallen comrade to fall into the hands of the enemy and under no circumstances will I ever embarrass my country.

Readily will I display the intestinal fortitude required to fight on to the Ranger objective and complete the mission, though I be the lone survivor.

Standing Orders, Rogers' Rangers

MAJOR ROBERT ROGERS, 1759

1. Don't forget nothing.
2. Have your musket clean as a whistle, hatchet scoured, sixty rounds powder and ball, and be ready to march at a minute's warning.
3. When you're on the march, act the way you would if you was sneaking up on a deer. See the enemy first.
4. Tell the truth about what you see and what you do. There is an army depending on us for correct information. You can lie all you please when you tell other folks about the Rangers, but don't never lie to a Ranger or officer.
5. Don't never take a chance you don't have to.
6. When we're on the march we march single file, far enough apart so one shot can't go through two men.
7. If we strike swamps, or soft ground, we spread out abreast, so it's hard to track us.
8. When we march, we keep moving till dark, so as to give the enemy the least possible chance at us.
9. When we camp, half the party stays awake while the other half sleeps.
10. If we take prisoners, we keep 'em separate 'til we have had time to examine them, so they can't cook up a story between 'em.
11. Don't ever march home the same way. Take a different route so you won't be ambushed.
12. No matter whether we travel in big parties or little ones, each party has to keep a scout 20 yards ahead, 20 yards on each flank, and 20 yards in the rear so the main body can't be surprised and wiped out.
13. Every night you'll be told where to meet if surrounded by a superior force.
14. Don't sit down to eat without posting sentries.
15. Don't sleep beyond dawn. Dawn's when the French and Indians attack.
16. Don't cross a river by a regular ford.
17. If somebody's trailing you, make a circle, come back onto your own tracks, and ambush the folks that aim to ambush you.
18. Don't stand up when the enemy's coming against you. Kneel down, lie down, hide behind a tree.
19. Let the enemy come till he's almost close enough to touch, then let him have it and jump out and finish him up with your hatchet.

Ranger History

Ranger Medal of Honor recipients

Name	Rank	Date	Unit
Millett, Lewis L., Sr.	Captain	Feb 7, 1951	Co. E, 2/27th Infantry
Porter, Donn F.*	Sergeant	Sep 7, 1952	Co. G, 2/14th Infantry
Mize, Ola L.	Sergeant	June 10-11, 1953	Co. K, 3/15th Infantry
Dolby, David C.	Staff Sergeant	May 21, 1966	Co. B, 1/8th (ABN) Cavalry
Foley, Robert F.	Captain	Nov 5, 1966	Co. A, 2/27th Infantry
Zabitosky, Fred M.	Staff Sergeant	Feb 19, 1968	5th Special Forces
Bucha, Paul W.	Captain	May 16-19, 1968	Co. D, 3/187 Infantry
Rabel, Laszlo*	Staff Sergeant	Nov 13 1968	74th Infantry (LRRP)
Howard, Robert L.	Sergeant First Class	Dec 30, 1968	5th Special Forces
Law, Robert D.*	Specialist 4	Feb 22, 1969	Co. I, 75th Infantry (Ranger)
Kerrey, J. Robert	Lieutenant	Mar 14, 1969	Seal Team 1
Doane, Stephen H.*	1st Lieutenant	Mar 25, 1969	Co. B, 1/5th Infantry
Pruden, Robert J.*	Staff Sergeant	Nov 22, 1969	Co. G, 75th Infantry (Ranger)
Littrell, Gary L.	Sergeant First Class	April 4-8, 1970	Advisory Team 21 (Ranger)
Lucas, Andre C.*	Lt Colonel	Jul 1-23, 1970	HHC 2/506 Infantry
Gordon, Gary I.*	Master Sergeant	Oct 3, 1993	Task Force Ranger
Shughart, Randall D.	Sergeant First Class	Oct 3, 1993	Task Force Ranger
Miller, Robert J.	Staff Sergeant	Jan 25, 2008	3rd Special Forces
Petry, Leroy A.	Sergeant First Class	May 26, 2008	2nd Bn, 75th Ranger Regiment
Swenson, William D.	Captain	Sep 8, 2009	Advisor, Task Force Phoenix
Groberg, Florent	Captain	Aug 8, 2012	4th BDE, 4th ID

LEGEND
Aug – August; BDE – brigade; Bn – battalion; Co – company; Dec – December; Feb – February; HHC – headquarters and headquarters company; ID – Infantry division; Jan – January; Jul – July; LRRP – long range reconnaissance patrol; Lt. – lieutenant; Mar – March; Nov – November; Oct – October; Sep – September
*Awarded posthumously

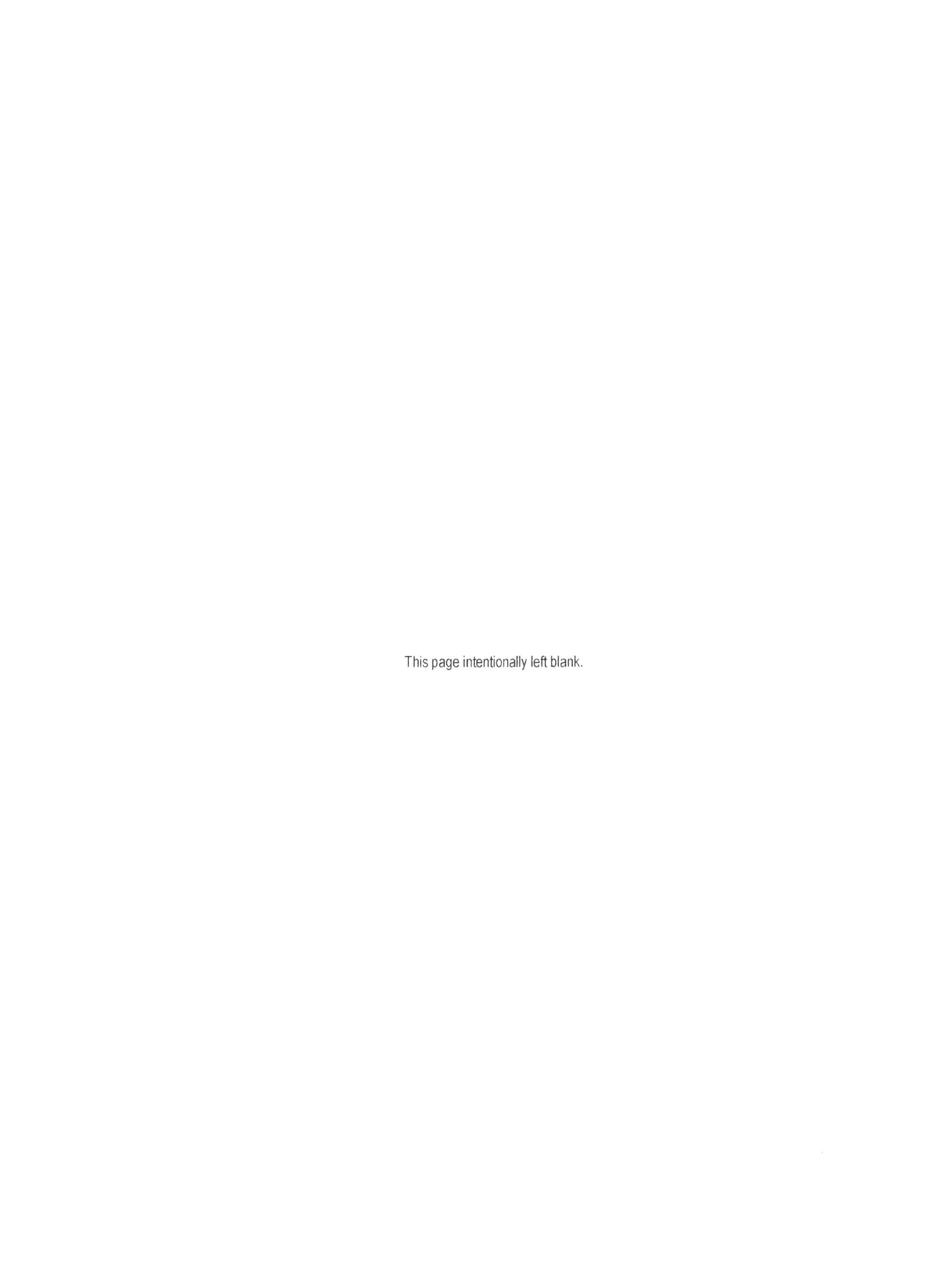

Chapter 1

Leadership

An Army leader is anyone who, by virtue of assumed role or assigned responsibility, inspires and influences people to accomplish organizational goals. Army leaders motivate people inside and outside the chain of command to pursue actions, focus thinking, and shape decisions for the greater good of the organization. Leadership is the process of influencing people by providing purpose, direction, and motivation to accomplish the mission and improve the organization.

PRINCIPLES

1-1. Leadership, the most essential element of combat power, gives purpose, direction, and motivation in combat. The leader balances and maximizes maneuver, firepower, and protection against the enemy. This chapter discusses how this is done by exploring the principles of leadership; the duties, responsibilities, and actions of an effective leader; and the leader's assumption of command.

1-2. The leadership requirements model establishes what leaders need to be, know, and do. A core set of requirements informs leaders about expectations. This is the process of influencing people by providing purpose, direction, and motivation to accomplish the mission and improve the organization. (See table 1-1 on pages 1-1 and 1-2.)

Table 1-1. Leadership Requirements Model

ATTRIBUTES		
Character Army values. Empathy. Warrior ethos/ service ethos. Discipline.	**Presence** Military and professional bearing. Fitness. Confidence. Resilience.	**Intellect** Mental agility. Sound judgment. Innovation. Interpersonal tact. Expertise.
Leads Leads others. Builds trust. Extends influence beyond the chain of command. Leads by example. Communicates.	**Develops** Creates a positive environment and fosters esprit de corps. Prepares self. Develops others. Stewards the profession.	**Achieves** Gets results.

Chapter 1

Table 1-1. Leadership requirements model (continued)

COMPETENCIES			
Oath to Constitution. Subordinate to law and civilian authority.	Combat power: unifier and multiplier.	Influence: commitment, compliance, and resistance.	Positive and harmful forms of leadership.
LEVELS OF LEADERSHIP	Direct – refine ability to apply competencies at a proficient level. Organizational – apply competencies to increasingly complex situations. Situational – actions adjusted to complex and uncertain environments.		
SPECIAL CONDITIONS OF LEADERSHIP	Formal – designated by rank or position (command is an example). Informal – take initiative and apply special expertise, when appropriate. Collective – synergistic effects achieved with multiple leaders aligned by purpose. Situational – actions adjusted to complex and uncertain environments.		
OUTCOMES			
Secured U.S. interests. Mission success. Sound decisions.	Expertly led organizations. Stewardship of resources. Stronger families		Fit units. Healthy climates. Engaged Soldiers and civilians.

1-3. To complete all assigned tasks, every Ranger in the patrol does their job. Each accomplishes specific duties and responsibilities and is part of the team. This includes the platoon leader (PL), platoon sergeant (PSG), squad leader (SL), weapons squad leader (WSL), team leader (TL), medic, radio operator, and forward observer (FO).

PLATOON LEADER

1-4. The PL is responsible for what the patrol does or fails to do. This includes tactical employment, training, administration, personnel management, and logistics. This is done by planning, making timely decisions, issuing orders, assigning tasks, and supervising patrol activities.

1-5. The platoon leader knows his Rangers and how to employ the patrol's weapons. The PL is responsible for positioning and employing all assigned or attached crew-served weapons and employment of supporting weapons. The PL also—

- Establishes the time schedule using backwards planning, considers time for execution, movement to the objective, and the planning and preparation phase of the operation.
- Takes the initiative to accomplish the mission in the absence of orders. Keeps higher headquarters (HQ) informed by using periodic situation reports (SITREPs).
- Plans with the help of the PSG, SLs, and other key personnel (team leaders, FOs, and attachment leaders).
- Stays abreast of the situation through coordination with adjacent patrols and higher HQ; supervises, issues fragmentary orders (FRAGORDs), and accomplishes the mission.
- Requests more support for the patrol from higher HQ, if needed to perform the mission.
- Directs and assists the PSG in planning and coordinating the patrol's sustainment effort and casualty evacuation (CASEVAC) plan.
- Receives on-hand status reports from the PSG and SLs, during planning.
- Reviews patrol requirements based on the tactical plan.
- Ensures that all-around security is maintained at all times.
- Supervises and spot-checks all assigned tasks, and corrects unsatisfactory actions.
- Is positioned to influence the most critical task for mission accomplishment, usually with the main effort, to ensure that the platoon achieves its decisive point during execution.

- Is responsible for positioning and employing all assigned and attached crew-served weapons.
- Commands through the SLs according to the intent of the commanders two levels higher.
- Conducts rehearsals.

PLATOON SERGEANT

1-6. The PSG is the senior noncommissioned officer (NCO) in the patrol and second in succession of command. The PSG helps and advises the PL, leads the patrol in the leader's absence, supervises the patrol's administration, logistics, and maintenance, and prepares and issues paragraph 4 of the operation order (OPORD). The PSG also—

- Organizes and controls the patrol command post (CP) according to the unit standard operating procedure (SOP); PLs guidance; and mission, enemy, terrain and weather, troops and support available-time available, and civil considerations (METT-TC) factors.
- Receives SLs requests for rations, water, and ammunition. Works with the company first sergeant (1SG) or executive officer (XO) to request resupply. Directs the routing of supplies and mail.
- Supervises and directs the patrol medic and patrol aid-litter teams in moving casualties to the rear.
- Maintains patrol status of personnel, weapons, and equipment; consolidates and forwards the patrol's casualty reports (DA Form 1156, *Casualty Feeder Card*) and receives and orients replacements.
- Monitors the morale, discipline, and health of patrol members.
- Supervises task-organized elements of patrol. This includes the—
  - Quartering parties.
  - Security forces during withdrawals.
  - Support elements during raids or attacks.
  - Security patrols during night attacks.
- Coordinates and supervises patrol resupply operations.
- Ensures that supplies are distributed according to the patrol leader's guidance and direction.
- Ensures that ammunition, supplies, and loads are properly and evenly distributed (a critical task during consolidation and reorganization).
- Ensures the CASEVAC plan is complete and executed properly.
- Ensures that the patrol adheres to the PLs time schedule.
- Assists the PL in supervising and spot-checking all assigned tasks, and corrects unsatisfactory actions.

Actions

1-7. During **movement and halts**, the PSG takes the actions necessary to facilitate movement; supervises rear security during movement; establishes, supervises, and maintains security during halts; performs additional tasks as required by the PL and assists in every way possible. The PSG also focuses on security and control of the patrol.

1-8. At **danger areas**, the PSG directs positioning of nearside security (usually conducted by the trail squad or team), and maintains accountability of personnel. During **actions on the objective**, the PSG—

- Assists with objective rally point (ORP) occupation.
- Supervises, establishes, and maintains security at the ORP.
- Supervises the final preparation of Soldiers, weapons, and equipment in the ORP according to the PLs guidance.
- Assists the patrol leader in control and security.
- Supervises the consolidation and reorganization of ammunition and equipment.
- Establishes marks, supervises the planned casualty collection point (CCP), and ensures that the personnel status including wounded in action (WIA) or killed in action (KIA) is accurately reported to higher HQ.
- Performs additional tasks assigned by the PL and reports status.

Chapter 1

1-9. During **actions in the patrol base (PB)**, the PSG assists in PB occupation, and establishing and adjusting the perimeter. The PSG enforces security in the PB, keeps movement and noise to a minimum, supervises and enforces camouflage, assigns sectors of fire, ensures designated personnel remain alert and equipment is maintained to high state of readiness.

1-10. The PSG requisitions supplies, water, and ammunition, and supervises the distribution. He supervises the priority of work and ensures its accomplishment, and performs additional tasks assigned by the PL. When creating a **security plan**, the PSG ensures—
- Crew-served weapons have interlocking sectors of fire.
- Claymore mines are emplaced to cover dead space.
- Range cards and sector sketches are complete. The security plan also includes—
  - Alert plan.
  - Evacuation plan.
  - Withdrawal plan.
  - Alternate PB.
  - Maintenance plan.
  - Hygiene plan.
  - Messing plan.
  - Water plan.
  - Rest plan.

SQUAD LEADER

1-11. The SL is responsible for what the squad does or fails to do. He is a tactical leader who leads by example. In addition to completing casualty feeder cards and reviewing the casualty reports completed by squad members, the SL—
- Directs the maintenance of the squad's weapons and equipment.
- Inspects the condition of Rangers' weapons, clothing, and equipment.
- Keeps the PL and PSG informed on status of the squad.
- Submits liquids, ammunition, casualties, and equipment (LACE) report to PSG.

Actions

1-12. During **actions throughout the mission**, the SL obtains status reports from team leaders and submits them to the PL and PSG, makes recommendations to the PL or PSG when problems are observed, and delegate's priority tasks to team leaders and supervises their accomplishment according to the SLs guidance.

1-13. The squad leader uses initiative in the absence of orders, follows the PLs plan, and makes recommendations. During **movement and halts**, the SL—
- Ensures heavy equipment is rotated among members and difficult duties are shared.
- Notifies the PL of the status of the squad.
- Maintains proper movement techniques while monitoring route, pace, and azimuth.
- Ensures the squad maintains security throughout the movement and at halts.
- Prevents breaks in contact.
- Ensures subordinate leaders are disseminating information, assigning sectors of fire, and checking personnel.

1-14. During **actions on the objective**, The SL ensures special equipment has been prepared for actions at the objective, and maintains positive control of the squad during the execution of the mission. The SL also positions key weapons systems during and after assault on the objective, obtains status reports from team leaders, ensures ammunition is redistributed, and reports the status to the PL. When performing **actions in the patrol base**, the SL—
- Ensures the PB is occupied according to the plan.
- Ensures that the sector of the PB is covered by interlocking fires, and makes final adjustments, if necessary.

Leadership

- Sends out listening posts (LPs) or observation posts (OPs) in front of the assigned sector. (METT-TC dependent).
- Ensures priorities of work are being accomplished, and reports accomplished priorities to the PL and PSG.
- Adheres to time schedule.
- Ensures personnel know alert and evacuation plans and locations of key leaders, OPs, and the alternate PB.

WEAPONS SQUAD LEADER

1-15. The weapons squad leader (WSL) is responsible for all that the weapons squad does or fails to do. The duties are the same as those of the SL. The WSL also controls the machine guns in support of the patrol's mission, and advises the PL on employment of the squad. The weapons squad leader also—

- Supervises machine gun teams to ensure they follow priorities of work.
- Inspects machine gun teams for correct range cards, fighting positions, and understanding of fire plan.
- Supervises maintenance of machine guns, ensuring that maintenance is performed correctly, deficiencies are corrected and reported, and that the performance of maintenance does not violate the security plan.
- Assists the PL in planning.
- Positions at halts and danger areas any machine guns not attached to squads, according to the patrol SOP.
- Rotates loads. Machine gunners normally get tired first.
- Submits LACE report to PSG.
- Designates sectors of fire, principal direction of fire, and secondary sectors of fire for all guns.
- Gives fire commands to achieve maximum effectiveness of firepower. The WSL—
  - Shifts fires.
  - Corrects windage or elevation to increase accuracy.
  - Alternates firing guns.
  - Controls rates of fire and fire distribution.
- Knows locations of assault and security elements, and prevents fratricide.
- Reports to the PL.

TEAM LEADER

1-16. The TL controls the movement of the fire team, and the rate and placement of fire. To do this, the TL leads from the front and uses the proper commands and signals. The TL maintains accountability of Rangers, weapons, and equipment, and ensures the Rangers maintain unit standards in all areas, and are knowledgeable of their tasks and the operation.

Chapter 1

1-17. The TL leads by example and has specific duties and responsibilities during **mission planning and execution**. The TL takes specific action when orders are issued. This includes the—
- Warning order (WARNORD):
    - Assists in control of the squad.
    - Monitors squad during issuance of the order.
- OPORD preparation:
    - Posts changes to schedule.
    - Posts and updates team duties on warning order board.
    - Submits ammunition and supply requests.
    - Picks up ammunition and supplies.
    - Distributes ammunition and special equipment.
    - Performs all tasks given in the SLs special instruction paragraph.
- OPORD issuance and rehearsal:
    - Monitors squad during issuance of the order.
    - Assists SL during rehearsals.
    - Takes actions necessary to facilitate movement.
    - Enforces rear security.
    - Establishes, supervises, and maintains security at all times.
    - Performs other tasks as the SL requires, and helps in every possible way, particularly in control and security.

Actions

1-18. During **action in the objective rally point**, the TL assists in the occupation of the ORP and helps supervise, establish and maintain security. The TL also—
- Supervises the final preparation of Rangers, weapons, and equipment in the ORP according to the SLs guidance.
- Assists in control of personnel departing and entering the ORP.
- Reorganizes perimeter after the reconnaissance party departs.
- Maintains communication with higher HQ.
- Upon return of the reconnaissance party, helps reorganize personnel, redistribute ammunition and equipment, and ensures that accountability of all personnel and equipment is maintained.
- Disseminates priority intelligence requirements (PIRs) to the team.
- Performs additional tasks assigned by the SL.

1-19. During **actions on the objective**, the TL ensures special equipment has been prepared and maintains positive control of the fire team during the execution of the mission. The TL also controls the movement and maneuver of individual fire team members, and assists in positioning key weapon systems during and after the assault on the objective. During actions in the PB, the TL—
- Inspects the perimeter to ensure team has interlocking sectors of fire and prepares the team sector sketch.
- Enforces the priority of work and ensures it is properly accomplished.
- Helps reorganize personnel, redistribute ammunition and equipment upon return of the reconnaissance party, and ensures accountability of all personnel and equipment is maintained.
- Disseminates the PIR to the team.
- Performs additional tasks assigned by the SL and assists in every way possible.

MEDIC

1-20. The medic assists the PSG in directing the aid and litter teams, and monitors the health and hygiene of the platoon. In addition to treating casualties, conducting triage, and assisting in CASEVACs under the control of the PSG, the medic—
- Aids the PL or PSG in field hygiene matters; personally checking the health and physical condition of platoon members.

Leadership

- Requests Class VIII—medical supplies, minimal amounts, through the PSG.
- Provides technical expertise to combat lifesavers and supervises them.
- Ensures casualty feeder reports are correct and attached to each evacuated casualty.
- Carries out other tasks assigned by the PL or PSG.

RADIO OPERATOR

1-21. The radio operator is responsible for establishing and maintaining communications with higher HQ and within the patrol. During planning, the radio operator—
- Enters the net at the specified time.
- Ensures that all frequencies, communications security fills, and net identifications (IDs) are preset in the squad and platoon radios.
- Informs the SL and PL of changes to call signs, frequencies, challenge and password, and number combination based on the appropriate time in the automated net control device (ANCD).
- Ensures the proper function of all radios, and troubleshoots and reports deficiencies to higher HQ.
- Weatherproofs all communications equipment.

1-22. In addition to serving as an en route recorder during all phases of the mission, the radio operator—
- Tracks time after the initiation of the assault.
- Records all enemy contact and reports it to higher in a size, activity, location, unit, time, and equipment (SALUTE) format.
- Reports all operation schedules (OPSKEDs) to higher HQ.

FORWARD OBSERVER

1-23. The FO works for the PL, serving as the eyes and ears of field artillery and mortars. The FO is mainly responsible for locating targets, and calling for and adjusting indirect fire support. The FO knows the terrain where the platoon is operating, the tactical situation, the mission, concept, and the unit's scheme of maneuver and priority of fires.

1-24. During planning, the FO selects targets to support the platoon's mission based on the company OPORD, platoon leader's guidance, and analysis of METT-TC factors, while preparing and using situation maps, overlays, and terrain sketches. During execution, the TL—
- Informs the fire support team HQ of platoon activities and of the fire support situation.
- Selects new targets to support the platoon's mission based on the company OPORD, the PLs guidance, and an analysis of METT-TC factors.
- Calls for and adjusts fire support.
- Operates as a team with the radio operator.
- Selects OPs.
- Maintains communications as prescribed by the fire support officer (FSO).
- Maintains the current eight-digit coordinate of his location at all times.

ASSUMPTION OF COMMAND

1-25. Any platoon or squad member might have to take command of his element in an emergency, so every Ranger must be prepared to do so. During an assumption of command, situation permitting, the Ranger assuming command accomplishes the tasks (not necessarily in this order), based on METT-TC shown in table 1-2.

Table 1-2. Tasks for the assumption of command

INFORMS	The unit's subordinate leaders of the command and notifies higher.

Chapter 1

CHECKS	Security.
CHECKS	Crew-served weapons.
PINPOINTS	Location.
COORDINATES AND CHECKS	Equipment.
CHECKS	Personnel status.
ISSUES	Fragmentary order (FRAGORD), if required.
REORGANIZES	As needed, maintaining unit integrity when possible.
MAINTAINS	Noise and light discipline.
CONTINUES	Patrol base activities, especially security, if assuming command in a patrol base.
RECONNOITERS	At the least, conducts map reconnaissance.
FINALIZES	Plan.

Chapter 2

Operations

This chapter provides techniques and procedures used by Infantry platoons and squads throughout the planning and execution phases of tactical operations. Specifically, it discusses the troop leading procedures, combat intelligence, combat orders, and planning techniques and tools needed to prepare a platoon for combat operations. These topics are time sensitive and apply to all combat operations. When time permits, leaders can plan and prepare in-depth. When less time is available, leaders rely on previously rehearsed actions, battle drills, and SOPs.

TROOP LEADING PROCEDURES

2-1. Table 2-1 shows the steps in the troop leading procedures (TLP). These steps are what a leader does to prepare the unit to accomplish a tactical mission.

2-2. The TLP starts when the leader is alerted for a mission, or receives a change or new mission. Steps 3 through 8 are done in any order, or at the same time.

Table 2-1. Steps in the troop leading procedures

1. Receive the mission.	5. Conduct reconnaissance.
2. Issue a warning order.	6. Complete the plan.
3. Make a tentative plan.	7. Issue the operations order.
4. Initiate movement.	8. Supervise and refine.

STEP 1. RECEIVE THE MISSION

2-3. The leader may receive the mission in an OPORD or a fragmentary order FRAGORD. The 1/3 - 2/3 rule only applies to the planning and preparation for an operation. Parallel planning occurs as the leader uses 1/3 of available planning and preparation time, and subordinates use the other 2/3. Emphasize conducting a hasty analysis with the primary focus on planning and preparation.

STEP 2. ISSUE A WARNING ORDER

2-4. The leader provides initial instructions in a WARNORD that contains enough information to begin preparation as soon as possible. The WARNORD mirrors the five-paragraph OPORD format and may include—
- Type of operation.
- General location of the operation.
- Initial operational timeline.
- Reconnaissance to initiate.
- Movement to initiate.
- Planning and preparation instructions (including planning timeline).
- Information requirements.
- Commander's critical information requirements (CCIR).

Chapter 2

STEP 3. MAKE A TENTATIVE PLAN

2-5. The leader develops an estimate of the situation to use as the basis for the tentative plan. This is the leader's mission analysis. METT-TC is used when developing the tentative plan.

    a. Conduct a detailed mission analysis:

- **Mission, intent, and concept** of higher commanders' concepts and intents two levels up. This information is found in the OPORD, paragraph 1b for two levels up, and in paras 2 and 3 for one higher up.
- **Unit tasks** are tasks that are clearly stated in the order (specified tasks) or tasks that become apparent as the OPORD is analyzed (implied tasks). (See table 2-2.)

Table 2-2. Examples of specified and implied tasks

SPECIFIED TASKS	IMPLIED TASKS
Retain Hill 545.	Provide security during movement.
Provide one squad to the 81-millimeter platoon to carry ammunition.	Conduct resupply operations.
Establish an observation post, vicinity GA124325 no later than 301500 Nov 16.	Coordinate with adjacent units.

- **Unit constraints.** The leader identifies any constraints placed on the unit. Constraints either prohibit or require an action. Leaders identify all constraints the OPORD places on their units' ability to execute their missions. The two types of constraints are proscriptive (required: mandates action) and prohibitive (not allowed: limits action).
- **Mission essential task(s).** After reviewing all the factors shown in previous paragraphs, the leader identifies the mission essential task(s). Failure to accomplish this task equals failure to accomplish the mission. The mission essential task should be in the maneuver paragraph.
- **Restated mission** clearly and concisely states the mission (purpose to be achieved) and the mission essential task(s) required to achieve it. It identifies WHO, WHAT (the task), WHEN (the critical time), WHERE (usually a grid coordinate), and WHY (the purpose the unit is to achieve). (See table 2-3.)

Table 2-3. Examples of restated missions

(*Who?*) 1st Platoon attacks. (*What?*) To seize. (*Where?*) Hill 482 vicinity NB 457371 Objective Blue. (*When?*) Not later than (NLT) 090500Z December 16 L 482 (*Why?*) In order to (IOT) enable the company's main effort to destroy enemy command bunker.
(*Who?*) 1st platoon, C company defends. (*What?*) To destroy from. (*Where?*) AB163456 to AB163486 to AB123486 to AB123456. (*When?*) NLT 181530Z October 16. (*Why?*) IOT prevent enemy forces from enveloping B company, 3-187th Infantry from the south.
LEGEND   IOT – in order to; NLT – not later than

    b. Analyze the situation and develop a course of action:

- **Suitable.** This accomplishes the mission and supports the commander's concept.
- **Acceptable.** The military advantage gained by executing the course of action must justify the cost in resources, especially casualties. This assessment is largely subjective.

Operations

c. With the restated mission from Step 1 to provide focus, the leader continues the estimate process using the remaining factors of METT-TC:
- What is known about the enemy? (See table 2-4.)
- How will terrain and weather affect the operation? Analyze terrain using observation and fields of fire, avenues of approach, key terrain, obstacles, and cover and concealment (OAKOC). (See table 2-5.)

Table 2-4. Enemy

COMPOSITION	This is an analysis of the forces and weapons that the enemy can bring to bear. Determine what weapons systems they have available, and what additional weapons and units are supporting him.
DISPOSITION	The enemy's disposition is how they are arrayed on the terrain, such as in defensive positions, in an assembly area, or moving in march formation.
STRENGTH	Percentage strength and number of passengers (PAXs).
RECENT ACTIVITIES	Identify recent and significant enemy activities that may indicate future intentions.
REINFORCEMENT CAPABILITIES	Determine positions for reserves and estimated time to counterattack or reinforce.
POSSIBLE COURSES OF ACTION	Determine the enemy's possible courses of action (COAs). Analyzing these COAs may ensure that the friendly unit is not surprised during execution.

- **Observation and fields of fire.** Determine locations that provide the best observation and fields of fire along the approaches, near the objective, or on key terrain. The analysis of fields of fire is mainly concerned with the ability to cover the terrain with direct fire.
- **Avenues of approach** are developed next and identified one level down. Aerial and subterranean avenues are also considered. Use table 2-5 for offensive considerations to avenues of approach.

Table 2-5. Offensive considerations

OFFENSIVE CONSIDERATIONS (FRIENDLY)	How can these avenues support my movement? What are the advantages and disadvantages of each? (Consider enemy, speed, cover, and concealment.) What are the likely enemy counterattack routes?
OFFENSIVE CONSIDERATIONS (ENEMY)	How can the enemy use these approaches? Which avenue is most dangerous? Least? (Prioritize each approach.) Which avenues would support a counterattack?

- **Key terrain** is any location or area that the seizure, retention, or control of that terrain affords a marked advantage to either combatant. Using the map and information already gathered, look for key terrain that dominates avenues of approach or the objective area. Next, look for decisive terrain that, if held or controlled, has an extraordinary impact on the mission.
- **Obstacles.** Identify the existing and reinforcing obstacles and hindering terrain that affects mobility.
- **Cover and concealment.** The analysis is often inseparable from the fields of fire and observation. Weapon positions need both to be effective and survivable. Infantry units are capable of improving poor cover and concealment by digging in and camouflaging their positions. When moving, the terrain is used to provide cover and concealment.

Chapter 2

STEP 4. INITIATE MOVEMENT; STEP 5. CONDUCT RECONNAISSANCE; AND STEP 6. COMPLETE THE PLAN

2-6. The unit may need to begin movement while the leader is still planning or forward reconnoitering. This step may occur anytime during the TLP. If time allows, the leader makes a personal reconnaissance.

2-7. When time does not allow, the leader makes a map reconnaissance. Sometimes, the leader relies on others (such as scouts) to conduct the reconnaissance. The leader completes the plan based on the reconnaissance and any changes in the situation.

STEP 7. ISSUE THE OPERATIONS ORDER

2-8. Platoon and SLs normally issue oral operation orders to aid subordinates in understanding the concept for the mission. If possible, leaders should issue the order with one or both of the following aids: within sight of the objective, on the defensive terrain, or on a terrain model or sketch.

2-9. Leaders may require subordinates to repeat all or part of the order, or demonstrate on the model or sketch, their understanding of the operation. They should also quiz their Rangers to ensure that all Rangers understand the mission.

STEP 8. SUPERVISE AND REFINE

2-10. The leader supervises the unit's preparation for combat by conducting rehearsals and inspections. Rehearsals include the practice of having SLs brief their planned actions in execution sequence to the PL.

2-11. The leader should conduct rehearsals on terrain that resembles the actual ground and in similar light conditions. Rehearsals are used to—
- Practice essential tasks (improve performance).
- Reveal weaknesses or problems in the plan.
- Coordinate the actions of subordinate elements.
- Improve Ranger understanding of the concept of the operation (foster confidence).

2-12. The platoon may begin rehearsals of battle drills and other SOP items before the receipt of the operation order. Once the order has been issued, it can rehearse mission-specific tasks. Some important tasks to rehearse include—
- Actions on the objective.
- Actions at the assault position.
- Breaching obstacles (mine and wire).
- Using special weapons or demolitions.
- Actions on unexpected enemy contact.

2-13. There are several different types of rehearsals: confirmation brief, reduced force, full force, and techniques. During a **confirmation brief rehearsal**, key leaders sequentially brief the actions required during an operation. Patrol leader rehearsals are conducted twice; right after a FRAGORD (confirmation brief) and again after subordinates develop their own plan. A **reduced force rehearsal** is conducted when time is a key constraint and security must be maintained. Key leaders normally attend. Mock ups, sand tables, and small-scale replicas are used. However, a full force rehearsal is the most effective type. It is first executed in daylight and open terrain, and then conducted in the same conditions as the operation. All Rangers participate and may use force-on-force.

2-14. The **techniques rehearsal** includes force-on-force, map (which has limited value and a limited number of attendees), radio (which cannot mass leaders but confirms communications), sand table or terrain model (this involves key leaders and includes all control measures), and the rehearsal of concept (ROC) drill (which is similar to the sand table or terrain model, but subordinates actually move themselves).

Operations

2-15. Squad leaders should conduct initial inspections shortly after receipt of the WARNORD. The platoon sergeant spot checks throughout the unit's preparation for combat. The platoon leader and platoon sergeant make a final inspection. Precombat checks and inspection include—
- Weapons and ammunition.
- Uniforms and equipment.
- Mission-essential equipment.
- Soldiers' understanding of the mission and individual responsibilities.
- Communications.
- Rations and water.
- Camouflage.
- Deficiencies noted during earlier inspections.

COMBAT INTELLIGENCE

2-16. Gathering information is one of the most important aspects of conducting a patrolling operation. All information is quickly, completely, and accurately reported. Use the SALUTE report format for reporting and recording information. (See table 2-6.)

Table 2-6. Example of a SALUTE report format

SIZE	Seven enemy personnel.
ACTIVITY	Traveling southwest.
LOCATION	GA123456.
UNIT/UNIFORM	Olive-drab uniforms with red six-point star on left shoulder.
TIME	210200JAN16.
EQUIPMENT	Carry one machine gun and one rocket launcher.

2-17. Try to include a field sketch with each report. Include only any aspects of military importance such as targets, objectives, obstacles, sector limits, or troop dispositions and locations (use symbols from ADRP 1-02). Use notes to explain the drawing, but they should not clutter the sketch. Leave off personnel, weapons, and equipment; these items go on the SALUTE report, not on this one.

2-18. The leader collects captured documents and turns them in with the reports, and marks each document with the time and place of capture. If prisoners are captured during a patrolling operation, they should be treated according to the Geneva Convention and handled by the 5-S rule. Immediately after returning from a mission, the unit is debriefed. The 5-S format includes—
- Search.
- Silence.
- Segregate.
- Safeguard.
- Speed to rear.

2-19. A WARNORD gives subordinate's advance notice of an upcoming operation. This gives them time to prepare. A warning order is brief but complete. Table 2-7 on pages 2-6 and 2-7, is an example WARNORD format. Table 2-8 on pages 2-8 through 2-12, is an example WARNORD for a squad.

Chapter 2

Note: A warning order only authorizes execution when it clearly says so.

Table 2-7. Example of a warning order format

WARNING ORDER _____ Roll call, pencil, pen, paper, Ranger Handbook (RHB), map, protractor, leader's monitor, hold all questions until the end. References: refer to higher HQ OPORD and identify map sheet for operation. Time zone used throughout the order: (optional) Task organization: optional, see para 1c.

1. **SITUATION.** Find this in higher HQ OPORD para 1a (1-3). Include the following information:
  a. **Area of interest:** outline the area of interest on the map.
     • Orient relative to each point on the compass (north, south, east, and west).
     • Box in the entire area of operation (AO) with grid lines.
  b. **Area of operations:** outline the area of operation on the map. Point out the objective and current location of your unit.
     • Trace your zone using boundaries.
     • Familiarize by identifying natural (terrain) and man-made features in the zone your unit is operating.
  c. **Enemy forces:** include significant changes in enemy composition, dispositions, and courses of action. Information not available for inclusion in the initial WARNORD can be included in subsequent warning orders (WHO, WHAT, WHERE).
  d. **Friendly forces:** optional, address only if essential to the WARNORD.
     • Give higher commander's mission (WHO, WHAT, WHEN, WHERE, WHY).
     • State higher commander's intention. (Higher HQ [go to map board] OPORD para 1b[2]), give task and purpose.
     • Point out friendly locations on the map board.
  e. **Attachments and detachments:** give initial task organization, only address major unit changes, and then go to the map board.

2. **MISSION.** State mission twice (who, what, when, where, why).

Table 2-7. Example of a warning order format (continued)

3. **EXECUTION.** Include the following information:

a. **Concept of operations:** provide as much information as available. The concept should describe the employment of maneuver elements. Give general direction, estimated distance, estimated time of travel, mode of travel, and major tasks to be conducted. Cover all movements, and specify points where the ground tactical plan starts and stops.

b. **Tasks to subordinate units:** provide specific tasks to subordinate units to aid in planning, preparing, and executing the mission. Planning guidance consists of tasks assigned to elements in the form of teams, special teams, and key individuals.

c. **Coordinating instructions:** include any information available at that time, if known. At least cover the following items:
- Uniform and equipment common to all.
- Consider the factors of METT-TC and tailor the load for each Ranger.
- Timeline. (State when, what, where, who and all specified times. Reverse plan. Use one-third to two-thirds rule).
- Give specific priorities in order of completion.
- Give information about coordination meetings.
- Rehearsals and inspections by priority.
- Earliest movement time.

4. **SUSTAINMENT.** Include any known logistics preparation for the operation.
  a. **Logistics:** include the following information:
  - **Maintenance:** include weapons and equipment direct exchange (DX) time and location.
  - **Transportation:** state method and mode of transportation for infiltration and exfiltration. Identify any coordination needed for external assets. Task subordinate leader (if needed) to generate load plan, number of lifts or serials, and bump plan.
  - **Supply:** only include classes of supply that require coordination or special instructions (such as rations, fuel, ammunition, or other items).

  b. **Army Health System support:** identify any medical equipment, support, or preventative medicine that needs to be coordinated.

5. **COMMAND AND SIGNAL.**
  a. **Command:** state the succession of command, if not covered in the unit's SOP.
  b. **Control:** include the following information:
  - **Command posts:** describe the employment of CPs, including the location of each CP and its time of opening and closing, as appropriate. Typically, at platoon level, the only reference to command posts is the company CP.
  - **Reports:** list reports not covered in the SOP.

  c. **Signal:** describe the concept of signal support, including current signal operating instructions (SOI) edition or refer to the higher OPORD. Give subordinates guidance on tasks to complete for preparation of the OPORD and the mission. Give time, place, and uniform for the OPORD. Give a time hack and ask for questions.

LEGEND
AO – area of operation; CP – command post; DX – direct exchange; HQ – headquarters; METT-TC - mission, enemy, terrain and weather, troops and support available-time available and civil considerations; OPORD – operation order; para – paragraph; RHB – Ranger Handbook; SOI - signal operating instructions; SOP - standard operating procedure; WARNORD – warning order

Chapter 2

Table 2-8. Example of a warning order format (squad)

WARNING ORDER (SQUAD) Roll call; Camp Darby special (CDS); 1:50,000 map; pen; paper; pencil; protractor; RHB; hold all questions; and TLs monitor task organization. **ATM-SEC/BRM-ASSLT/HQ SUPPORT**.

1. **SITUATION.** (BRIEF.)
    a. **Area of interest:** orient the map (N, S, E, and W). Our squad's area of interest is boxed in by the 86 gridline to the north, the 18 gridline to the east, the 77 gridline to the south, and the 13 gridline to the west.
    b. **Area of operation:** we will be operating in Zone C. Trace Zone C with boundaries. Familiarize Zone C with three natural and man-made features. Our objective (OBJ) is located here (point on map) at GA 152 796, and our current location is here (point on map) at GA 196 790.
    c. **Enemy:** use 3-Ws para 1c(1-3). Describe enemy recent locations and activities.
      • **WHO?** The Aragon Liberation Front. (ALF).
      • **WHAT?** Ambushed patrol.
      • **WHERE?** GA 156 804.
    d. **Friendly:** use 4-Ws para 2. Mission, intent, and concept one and two levels up. Task and purpose of adjacent patrols. Provide the big picture concept of the **higher HQ mission and intent.**
      • **Mission:**
         ▪ **WHO?** 1st platoon (PLT), B company (CO).
         ▪ **WHAT?** (Task.) Conduct area ambushes to destroy enemy forces.
         ▪ **WHERE?** On OBJ Black NLT 302300JAN2017.
         ▪ **WHY? (Purpose)** To prevent the enemy from maintaining control of OBJ Black.
      • **Intent:**
         ▪ Find, fix, and finish enemy forces in Zone C.
         ▪ Enemy personnel and equipment are destroyed.
    e. **Attachments and detachments.** MG TM 300530JAN2017

2. **MISSION.** Clear and concise, use 5-Ws, para 3, X2, task and purpose 1st squad (SQD), 1st PLT, B CO decisive operation (DO) conducts a point ambush to **destroy (TASK)** enemy personnel and equipment on OBJ Red (GA 152 793) NLT 302300JAN2017 in order to **prevent (PURPOSE)** the enemy from maintaining control of OBJ Red.

Table 2-8. Example of a warning order format (squad) (continued)

3. **EXECUTION.**

a. **Concept of operations:** (orient Rangers to sketch or terrain model). We are currently located at Camp Darby, GA 1962 7902. We will depart Camp Darby moving generally northwest for 6000 meters. We will be travelling by truck to our insertion point, GA 176 812, where we will dismount the trucks. The movement should take approximately 20 minutes. Our ground tactical plan will begin when we move generally southwest for 3000 meters. The movement should take three hours as we will be traveling by foot to our tentative ORP at Grid GA 154 795. Here will we finalize the preparing of men, weapons, and equipment (M, W, and E). We will then move generally southwest for 400 meters to our objective at Grid GA 152 793. It will take us 30 minutes to an hour to complete our movement by foot. From our objective we will travel generally northwest for 3000 meters, It will take four hours to move by foot to our linkup site, GA 152 819. Once complete with linkup, we will move generally southeast for 8000 meters. The movement should take approximately 30 minutes, as we will be travelling by truck back to Camp Darby. Once back at Camp Darby, we will debrief and prepare to conduct follow-on operations.

b. **Tasks to subordinate units:** nontactical and tactical instructions, METT-TC planning guidance, teams, special teams, and key individuals (control–movement–objective).

- **HQs:** second in the order of the march (OOM), M240 will provide supporting fires into the kill zone during actions on objective (AOO). Radiotelephone operator (RTO) is the recorder en route and during AOO. You will write para 5 of the SQD OPORD, ensure all radios are operational with proper frequencies loaded, also ensure we enter the net on time.
- **Alpha team (ATM):** first in the OOM, and is responsible for land navigation. The ATM is flank security for AOO, one-to-two Ranger enemy prisoner of war (EPW) team (TM), one-to-two Ranger aid and litter TM, one-to-two Ranger demolition (DEMO) TM, one-to-two Ranger ORP clearing TM, two two-man Ranger flank security TM for AOO, one-to-two Ranger linkup security TM, one squad automatic weapon (SAW) gunner to assault (ASSLT) element for AOO, one compass man, and one pace man. Alpha team leader (ATL) will be the security TM LDR for AOO. You are responsible for writing para 1 and linkup annex of SQD OPORD, draw all sketches, formation order of movement (FOOM), danger areas, battle drills, linkup, truck, AOO, terrain model, routes, and fire support overlay (sterile and nonsterile).
- **Bravo team (BTM):** third in the OOM and is ASSLT for AOO. One-to-two man EPW TM, one-to-two Ranger aid and litter TM, one-to-two Ranger DEMO TM, one-to-two Ranger surveillance and observation (S&O) TM, one grenadier (GREN) to security team for AOO, one compass man, and one pace man. The Bravo team leader (BTL) is the ASSLT TL for AOO. You are second in the chain of command and in charge at all times during my absence. You must write para 4 and the truck annex of the SQD OPORD, prepare supply, DX, and ammunition lists, draw and issue all items. Ensure that everyone does a test fire and that all equipment is tied down according to 4th Ranger training battalion (RTB) SOP. Update the squad status card and hand receipt.
- **TL:** updates the WARNORD board with all the correct information. As a task is accomplished, you will line it out. Post your change of command (COC), duty position (DP), and job description (JD) (special teams and key individuals). Come see me for further guidance at the conclusion of this warning order.

c. **Coordinating instructions:** depending on METT-TC, tailor the load according to the number of Rangers, not SOPs.

- Packing list is based on the Airborne and Ranger Training Brigade (ARTB) seasonal packing list.
- Write on note cards or paper and read off by item.

Table 2-8. Example of a warning order format (squad) (continued)

3. EXECUTION (continued)

WORN

PACKING LIST ON NOTE CARD

RUCK

PACKING LIST ON NOTE CARD

Time schedule

WHEN	WHAT	WHERE	WHO	REF
*0630	Warning order	Bay area	All	3D1
0700	Initial inspection	Bay area	All	
*0730	REQD AMMO, supply	CO TOC	BTL/RB	4B1
*0745	P/U AMMO, /supply	CO TOC	BTL/Detail	4B1
0800	Test fire	T/F area	All	
*0830	S-2, S-3, fires coord.	PLT bay	SL /RTO	3D1
0850	Adj. unit coord.	SQD bays	ATL, CM	3D1
*0900	Enter net	Bay area	RTO	
0930	Status card update	Bay area	SL/TL	RTO
0945	Terrain model complete	Bay area	SL/TL	
*1000	OPORD	Bay area	All	
1300	Rehearsal	Bay area	All	
1330	Final inspection	Bay area	All	
*1400	Truck linkup	Co TOC	All	3D1
1500	Depart Darby	Co TOC	All	
*1600	Insertion complete	TBD	All	3D1
1800	In ORP	TBD	All	
*2000	In position	GA 184774	All	
*2300	Mission complete	TBD	All	
*0200	Linkup complete	TBD	All	
0500	S-2 debrief	BN TOC	All	

* Specified times. Use the 1/3 - 2/3 rule and reverse planning.

Operations

Table 2-8. Example of a warning order format (squad) (continued)

4. **SUSTAINMENT.** Include the following information:
   a. **Logistics.**
   - **Maintenance:** weapons and equipment DX is at 0700 in the company CP.
   - **Transportation:** method of transportation for insertion will be truck, and truck for extraction. Bravo team leader will generate the load plan, bump plan, and number of chalks and lifts.
   - **Supply:**
     - CLI: each man will have two meals, ready to eat (MREs) and six quarts of water for the operation.
     - CLV: Bravo team leader will draw enough ammunition for each man to carry a basic load according to the squad SOP.

Example of a squad ammunition SOP

INDIVIDUAL WEAPONS (FOR EACH WEAPON SYSTEM)			SUPPLEMENTAL AMMUNITION (TOTAL)		
WPN	QTY	REMARKS	TYPE	QTY	REMARKS
M4	210	5.56-mm ball	Claymore	3	M18A1
M249	625	5.56-mm ball, link (4:1 mix)	Demo kit	2	C4, M81, M14
M240B	825	7.62-mm ball, link (4:1 mix)	M67 hand grenade	24	Two for each Soldier
M320	12	40-mm HEDP	5.56-mm tracer	126	42 for each leader
M320	3	40-mm shot	AT4	2	One for each team
M320	4	40-mm illumination	HC smoke	6	
M320	2	40-mm TP	Red smoke	2	
			Yellow smoke	4	

   b. Personnel services support: religious services will be held at 0800 in the chapel.
   c. Army Health System support: alpha team leader, coordinate for one additional combat lifesaver (CLS) bag.

5. **COMMAND AND SIGNAL.** Include the following information:
   a. **Command.**
   - **Location of commander or patrol leader:** the patrol leader will be located in the squad bay during phase one (mission preparation). Location of patrol leader for all other phases will be briefed in the OPORD.
   - **Succession of command:** state the succession of command if not covered in the unit's SOP. This is the SL, BTL, ATL, and RTO.
   b. **Control.**
   - **Command posts:** the platoon CP is located at GA 166 807, and the company CP is located at GA 196 790.
   - **Reports:** pertinent reports will be covered in the OPORD.
   c. **Signal:**
   - The battalion will be operating on 37.950 single channel/plain text (SC/PT), call sign Darby 741.
   - Our squad frequency is 77.000 SC/PT, call sign bravo-one-one.
   - All other signals such as frequencies, call signs, challenges, and passwords will be given during the OPORD.

Chapter 2

Table 2-8. Example of a warning order format (squad) (continued)

Additional Guidance: 1. Give subordinates additional guidance on tasks to complete for preparation of the OPORD and the mission. 2. Give time, place, and uniform of the OPORD. 3. Give a time hack and ask for questions. 4. If time permits, have your subordinate leader's confirmation brief you on the information that was put out to ensure understanding.
LEGEND 3-Ws – who, what, where; 4-Ws – who, what, where, why; 5-Ws – who, what, where, when, why; ADJ – adjacent; AMMO – ammunition; AOO – actions on objective; ATL – Alpha team leader; ATM – Alpha team; ASSLT – assault; BD – battle drill; BTL – Bravo team leader; BTM – Bravo team; CDS – Camp Darby special; CLI – Class I—food, rations, and water; CLS – combat lifesaver bag; CLV – Class V—ammunition; CM – compass man; CO – company; COC – chain of command; coord – coordination; CP - command post; DEMO – demolition; DX – direct exchange; DP – duty position; EPW – enemy prisoner of war; FOOM – formation order of movement; GA – Georgia; GREN – grenadier; HC – high concentration; HEDP – high explosive dual purpose; JD – job description; LDR – leader; METT-TC - mission, enemy, terrain and weather, troops and support-time available, and civil considerations ; mm – millimeter; MRE – meal, ready to eat; M, W, and E – men, weapons, and equipment; NLT – not later than; OBJ – objective; OOM – order of the march; OPORD – operation order; ORP – objective rally point; para – paragraph; PLT – platoon; QTY – quantity; P/U – pick up; REQD – required; RHB – Ranger Handbook; RTB – Ranger training battalion; RTO – radio-telephone operator; RUCK – rucksack; S-2 – intelligence and security officer; S-3 –operations officer; SAW – squad automatic weapon; SC/PT – single channel/plain text; SEC – security; SL – squad leader; S&O – surveillance and observation; SQD – squad; SOP – standard operating procedure; TBD – to be determined; T/F – test fire; TL – team leader; TM – team; TOC – tactical operations center; TP – training practice; WARNORD – warning order; WPN – weapon; X2 – times two.

OPERATION ORDER

2-20. An OPORD is a directive issued by a leader to subordinates in order to effect the coordinated execution of a specific operation. A five-paragraph format (see table 2-9 on pages 2-13 through 2-17) is used to organize the briefing, ensure completeness, and help subordinate leaders understand and follow the order. Use a terrain model or sketch, along with a map to explain the order.

2-21. The platoon or SL briefs the OPORD orally off notes that follow the five-paragraph format. Before the issuance of the OPORD, the leader ensures that pencil, pen, paper, Ranger Handbook (RHB), map, and protractor are in place. Leader's monitor subordinates, then calls roll and says, "PLEASE HOLD ALL QUESTIONS UNTIL THE END."

Table 2-9. Example of a squad OPORD format

OPERATION ORDER. (Plans and orders normally contain a code name and are numbered consecutively within a calendar year.)
References: the heading of the plan or order lists maps, charts, data, or other documents the unit needs to understand the plan or order. The user need not reference the SOP, but may refer to the SOP in the body of the plan or order. A map is referenced by map series number (and country or geographic area, if required), sheet number, and name, edition, and scale, if required. "Datum" refers to the mathematical model of the earth that applies to the coordinates on a particular map. It is used to determine coordinates. Different nations use different datum for printing coordinates on their maps. The datum is usually referenced in the marginal information of each map. Include the following information:
Time zone used throughout the order: if the operation takes place in one time zone, use that time zone throughout the order (including annexes and appendixes). If the operation spans several time zones, use Zulu time.
Task organization: describe the allocation of forces to support the commander's concept. Show task organization in one of two places: **just above paragraph 1,** or if the task organization is long or complex, **in an annex.** To organize—
- Go to the map.
- Apply the **orient, box, trace,** and **familiarize** technique only to the areas the unit is moving through. (Get this information from the platoon OPORD.)
- • Determine the effects of seasonal vegetation within the area of operations (AO).

1. SITUATION. Include the following information:
    a. **Area of interest:** describe the area of interest or areas outside of the area of operation that can influence your operation.
    b. **Area of operations:** describe the area of operations. Refer to the appropriate map and use overlays, as needed.
    - **Terrain:** using the OAKOC format, state how the terrain will affect friendly and enemy forces in the AO. Use the OAKOC from higher HQ OPORD. Refine it based on your analysis of the terrain in the AO.
    -) **Weather:** describe the aspects of weather that impact operations. Consider the five military aspects of weather to drive your analysis: visibility, winds, temperature/humidity, cloud cover, precipitation (V, W, T, C, P). State how the weather will affect both friendly and enemy forces in the AO.
    c. **Enemy forces:** the enemy situation in higher headquarters' OPORD (para 1c) forms the basis for this. Refine it by adding the detail your subordinates require.
    - State the enemy's **composition, disposition, and strength**.
    - Describe **recent activities** of the enemy.
    - Describe their known or suspected **locations and capabilities**.
    - Describe the enemy's most likely and most dangerous **course of action**.
        - Go to the map.
        - Point out on the map the location of recent known and suspected enemy activity.

Chapter 2

Table 2-9. Example of a squad OPORD format (continued)

1. **SITUATION** (continued).
   d. **Friendly forces:** get this information from paragraphs 1d, 2, and 3 of the higher headquarters' OPORD.
      (1) **Higher headquarters mission, intent, and concept.**
         - Higher headquarters two levels up.
            - Mission: state the *mission* of the higher unit (*two levels up*).
            - Intent: state intent two levels up.
         - Higher headquarters one level up.
            - Mission: state the *mission* of the higher unit (*one level up*).
            - Intent: state intent one level up.
      (2) **Mission of adjacent units:** state locations of units to the left, right, front, and rear. State those units' tasks and purposes and say how those units will influence yours, particularly adjacent unit patrols.
         - Show other unit's locations on map board.
         - Include statements about the influence each of the above patrols will have on your mission, if any.
         - Obtain this information from higher HQs OPORD. It gives each leader an idea of what other units are doing and where they are going. This information is in paragraph 3b(1), Execution, Concept of the Operation, Scheme of Movement and Maneuver.
         - Also, include any information obtained when the leader conducts adjacent unit coordination.
      e. **Attachments and detachments:** avoid repeating information already listed in task organization. However, when not in the task organization, list units that are attached or detached to the headquarters that issues the order. State when attachment or detachment will be in effect if that differs from when the OPORD is in effect, such as on order or on commitment of the reserve. Use the term "remains attached" when units will be or have been attached for some time.

2. **MISSION.** Who, what (task), when, where, why (purpose) from higher HQ maneuver paragraph.

3. **EXECUTION.**
   a. **Commander's intent:** state the intent, which is the clear, concise statement of what the force must do and the conditions the force must establish with respect to the friendly, enemy, terrain, and civil considerations that represent the commander's desired end state. This serves to allow subordinate and supporting commanders to achieve the commander's desired results without further orders, even when the operation does not unfold as planned.
   b. **Concept of operations:** write a clear, concise concept statement. Describe how the unit will accomplish its mission from start to finish. Base the number of subparagraphs, if any, on what the leader considers appropriate, the level of leadership, and the complexity of the operation. The following subparagraphs from ADP 5-0 show what might be required within the concept of the operation. Ensure that you state the purpose of the warfighting functions within the concept of the operation.

   WARFIGHTING FUNCTIONS

   > Fire support.
   > Movement and maneuver.
   > Protection.
   > Mission command.
   > Intelligence.
   > Sustainment.

Table 2-9. Example of a squad OPORD format (continued)

3. **EXECUTION** (continued).
 c. **Scheme of movement and maneuver:** describe the employment of maneuver units according to the concept of operations. Address subordinate units and attachments by name. State each one's mission as a task and purpose. Ensure the subordinate units' missions support that of the main effort. Focus on actions on the objective. Include a detailed plan and criteria for engagement and disengagement, an alternate plan in case of compromise or unplanned enemy force movement, and a withdrawal plan. The brief is to be sequential, going from start to finish, covering all aspects of the operation:
 - Brief from the start of the operation to mission completion.
 - Cover all primary and alternate routes, from insertion through AOO to linkup, and include extraction until the mission is complete.
 - Brief plan for crossing known danger areas.
 - Brief plan for reacting to enemy contact.
 - Brief any approved targets and CCPs as you brief the routes.

 d. **Scheme of fires:** state scheme of fires to support the overall concept and state who (which maneuver unit) has priority of fire. You can use the purpose, location, observer, trigger, communication method, and resources (PLOT-CR) format to plan fires. Refer to the target list worksheet and overlay here, if applicable. Discuss specific targets and point them out on the terrain model. (See Chapter 3, Fire Support.)

 e. **Casualty evacuation:** provide a detailed CASEVAC plan during each phase of the operation. Include CCP locations, tentative extraction points, and methods of extraction.

 f. **Tasks to subordinate units:** clearly state the missions or tasks for each subordinate unit that reports directly to the headquarters issuing the order. List the units in the task organization, including reserves. Use a separate subparagraph for each subordinate unit. State only the tasks needed for comprehension, clarity, and emphasis. Place tactical tasks that affect two or more units in coordinating instructions (subparagraph 3h). Platoon leaders may task their subordinate squads to provide any of the following special teams: reconnaissance and security, assault, support, aid and litter, EPW and search, clearing, and demolitions. You may also include detailed instructions for the platoon sergeant, RTO, compass-Soldier, and pace-Soldier.

 g. **Coordinating instructions:** this is always the last subparagraph under paragraph 3. List only the instructions that apply to two or more units, and which are seldom covered in unit SOPs. Refer the user to an annex for more complex instructions. The information listed below is required:
 - **Time schedule:** state time, place, uniform, priority of rehearsals, confirmation briefs, inspections, and movement.
 - **Commander's critical information requirements:** include PIR and friendly force information requirements (FFIRs).
   - **Priority intelligence requirements** includes all intelligence that the commander needs for planning and decision-making.
   - **Friendly force information requirements:** include what the commander needs to know about friendly forces available for the operation. It can include personnel status, ammunition status, and leadership capabilities.
 - **Essential elements of friendly information (EEFI):** these are critical aspects of friendly operations that, if known by the enemy, would compromise, lead to failure, or limit success of the operation.
 - **Risk-reduction control measures:** these are unique to the operation. They supplement the unit SOP and can include mission-oriented protective posture, operational exposure guidance, vehicle recognition signals, and fratricide prevention measures.
 - **Rules of engagement (ROE).**
 - **Environmental considerations.**
 - **Force protection.**

Chapter 2

Table 2-9. Example of a squad OPORD format (continued)

4. **SUSTAINMENT.** Describe the concept of sustainment to include logistics, personnel, and medical.
    a. **Logistics.** Include the following information:
    - **Sustainment overlay:** include current and proposed company trains locations, CCPs (include marking method), equipment collection points, helicopter landing zones (HLZs), ambulance exchange points (AXPs), and any friendly sustainment locations such as forward operating bases (FOBs) common operational pictures (COPs), or other methods.
    - **Maintenance:** include weapons and equipment, DX time, and location.
    - **Transportation:** state method and mode of transportation for insertion and extraction, load plan, number of lifts and serials, bump plan, recovery assets, and recovery plan.
    - **Supply:**
        - Class I—food, rations, and water.
        - Class III—petroleum, oils, and lubricants.
        - Class V—ammunition.
        - Class VII—major end items.
        - Class VIII—medical supplies, minimal amounts.
        - Class IX—repair parts.
        - Distribution methods.
    - **Field services:** include any services provided or required.
    b. **Personnel services support:** include the method of marking and handling EPWs.
    c. **Army Health System support:** include the following information:
    - **Medical mission command:** include location of medics. Identify medical leadership, personnel controlling medics, and method of marking patients.
    - **Medical treatment:** state how wounded or injured Soldiers will be treated (self-aid, buddy-aid, CLS bag, emergency medical technician [EMT], or other methods).
    - **Medical evacuation:** describe how dead or wounded, friendly and enemy personnel will be evacuated. Identify aid and litter teams. Include special equipment needed for evacuation.
    - **Preventive medicine:** identify any preventive medicine Soldiers may need for the mission (sun block, lip balm, insect repellant, in-country specific medicine, or other items).

Operations

Table 2-9. Example of a squad OPORD format (continued)

5. **COMMAND AND SIGNAL.** State where mission command facilities and key leaders are located during the operation.
   a. **Command.** Include the following information:
   - **Location of commander or patrol leader:** state where the commander intends to be during the operation, by phase if the operation is phased.
   - **Succession of command:** state the succession of command, if not covered in the unit SOP.

   b. **Control.** Include the following information:
   - **Command posts:** describe the employment of CPs, including the location of each CP and its time of opening and closing, as appropriate. Typically, at platoon level the only reference to command posts is the company CP.
   - **Reports:** list reports not covered in SOPs.

   c. **Signal:** describe the concept of signal support, including current SOI edition or refer to the higher HQ OPORD.
   - Identify the **SOI index** that is in effect.
   - Identify **methods of communication by priority.**
   - Describe **pyrotechnics and signals**, to include arm and hand signals (demonstrate).
   - Give **code words** such as OPSKEDs.
   - Give **challenge and password** (use behind friendly lines).
   - Give **number combination** (use forward of friendly lines).
   - Give **running password.**
   - Give **recognition signals** (near—far and day—night).

Actions after issuance of OPORD:
- Issue annexes.
- Highlight next hard time.
- Give time hack.
- Ask for questions.

LEGEND
AO – area of operations; AOO – actions on objective; AXP – ambulance exchange point; BENT – beginning evening nautical twilight; CASEVAC – casualty evacuation; CCP - casualty collection point; CLS – combat lifesaver; COP – common operational picture; CP – command post; DX – direct exchange; EEFI – essential elements of friendly information; EENT – ending evening nautical twilight; EMT – emergency medical technician; EPW – enemy prisoner of war; HLZ – helicopter landing zone; HQ – headquarters; OAKOC - observation and fields of fire, avenues of approach, key terrain, obstacles, and cover and concealment; OPORD – operation order; OPSKED – operation schedule; para – paragraph; PIR - priority intelligence requirement; PLOT-CR - purpose, location, observer, trigger, communication method, and resources; ROE – rules of engagement; RTO – radio-telephone operator; SOI - signal operating instructions; SOP – standard operating procedure; V, W, T, C, P - visibility, winds, temperature/humidity, cloud cover, precipitation

Chapter 2

FRAGMENTARY ORDER

2-22. A FRAGORD is an abbreviated form of an operation order, usually issued daily, which eliminates the need for restating portions of the OPORD. It is issued after an OPORD to change or modify that order or to execute a branch or sequel to that order. Table 2-10 on pages 2-18 through 2-20, shows an annotated FRAGORD format.

Table 2-10. Example of an annotated FRAGORD format

FRAGMENTARY ORDER_____
Time zone referenced throughout order:
Task organization:

1. **SITUATION.** *(Brief changes from base OPORD specific to this days' operation).* Include the following information:
   a. **Area of interest:** state any changes to the area of interest.
   b. **Area of operations:** state any changes to the area of operations. This includes the **terrain (note any changes that will affect operation in a new area of operations):** observation, fields of fire, cover and concealment, obstacles, key terrain, and avenues of approach.

Temp high	Sunrise	Moonrise
Temp low	Sunset	Moonset
Wind speed	*BMNT	Moon phase
Wind direction	**EENT	Percent illumination

This is the information the squad leader received from the platoon OPORD.
* Begin morning nautical twilight (BMNT).
** Ending evening nautical twilight (EENT).

   c. **Enemy:** include the following information:
   - Composition, disposition, and strength.
   - Capabilities.
   - Recent activities.
   - Most likely COA.
   d. **Friendly:** this includes:
   - Higher mission.
   - Adjacent patrols, task, or purpose.
   - Adjacent patrol objective and route (if known).

2. **MISSION.** Who, what (task), when, where, why (purpose) from higher HQ maneuver paragraph.

Table 2-10. Example of an annotated FRAGORD format (continued)

3. **EXECUTION.** Include the following information:
   a. **Commander's intent:** include any changes or state, "NO CHANGE."
   b. **Concept of operations:** include any changes or state, "NO CHANGE."
   c. **Scheme of movement and maneuver:** include any changes or state, "NO CHANGE."
   d. **Scheme of fires:** include any changes or state, "NO CHANGE."
   e. **Casualty evacuation:** include any changes or state, "NO CHANGE."
   f. **Tasks to subordinate units:** include any changes or state, "NO CHANGE."
   h. **Coordinating instructions:** include any changes or state, "NO CHANGE." Add the following information:
- Time schedule.
- Commander's critical information requirements.
  - Priority intelligence requirements.
  - Friendly force information requirements.
- Essential elements of friendly information.
- Risk-reduction control measures.
- Rules of engagement.
- Environmental considerations.
- Force protection.

4. **SUSTAINMENT.** Only cover changes from the base order. Use standard format and items that have not changed should be briefed, "NO CHANGE."
   a. **Logistics.** Include the following information:
- Sustainment overlay.
- Maintenance.
- Transportation.
- Supply:
  - Class I—food, rations, and water.
  - Class III—petroleum, oils, and lubricants.
  - Class V—ammunition.
  - Class VII—major end items.
  - Class VIII—medical supplies, minimal amounts.
  - Class IX—repair parts.
  - Distribution methods.
- Field services.

   b. **Personnel services support.** Include the following points:
- Method of marking and handling EPWs.
- Religious services.

   c. **Army Health System support.** Include the following points:
- Medical mission command.
- Medical treatment.
- Medical evacuation.
- Preventive medicine.

Chapter 2

Table 2-10. Example of an annotated FRAGORD format (continued)

5. **COMMAND AND SIGNAL.** Only brief changes to the base order. If there are changes, state where mission command facilities and key leaders are located during the operation.
   a. **Command.** Include the following information:
- **Location of commander or patrol leader:** state where the commander intends to be during the operation and by phase, if the operation is phased.
- **Succession of command:** state the succession of command, if not stated in the unit's SOP.

   b. **Control.** Include the following information:
- **Command posts:** describe the employment of CPs, including the location of each CP and its time of opening and closing, as appropriate. Typically, at platoon level the only reference to command posts is the company CP.
- **Reports:** list reports not covered in SOPs.

   c. **Signal:** describe the concept of signal support, including current SOI edition, or refer to higher HQs OPORD. Include the following information:
- Identify the **SOI index** that is in effect.
- Identify **methods of communication by priority.**
- Describe **pyrotechnics and signals**, to include arm and hand signals (demonstrate).
- Give **code words** such as OPSKEDs.
- Give **challenge and password** (use behind friendly lines).
- Give **number combination** (use forward of friendly lines).
- Give **running password.**
- Give **recognition signals** (near—far and day—night).

Field FRAGORD guidance:

1. The field FRAGORD should take no more than 40 minutes to issue, with 30 minutes for the target. The proposed planning guide is as follows:
- Paragraphs 1 and 2: 5 minutes.
- Paragraph 3: 20 to 30 minutes.
- Paragraphs 4 and 5: 5 minutes.

2. The FRAGORD should focus on actions on the objective. The PL may use subordinates to prepare para 1, 4, 5, and routes and fires for the FRAGORD. It is acceptable for subordinates to brief the portions of the FRAGORD they prepare.

3. Use of sketches and a terrain model are critical to allow rapid understanding of the operation and FRAGORD.

4. Rehearsals are critical as elements of the constrained planning model. When the FRAGORD is used with effective rehearsals, preparation time is reduced, allowing the PL more time for movement and reconnaissance.

5. Planning in a field environment reduces the amount of time leaders have for in-depth mission planning. The TLP give leaders a framework to plan missions and produce orders when time is short.

LEGEND
BENT – beginning evening nautical twilight; COA – course of action; CP – command post; EENT – ending evening nautical twilight; EPW – enemy prisoner of war; FRAGORD – fragmentary order; HQ – headquarters; OPORD – operation order; OPSKED – operation schedule; para – paragraph; PL – platoon leader; SOI - signal operating instructions; SOP – standard operating procedure; Temp – temperature; TLP – troop leading procedures

Operations

ANNEXES

2-23. Operation order annexes are issued after an OPORD only if more information is needed about truck movement, air assault, PBs, small boats, linkups, or stream crossings. Brevity is standard. Annexes are always issued after the operation order. Table 2-11 on pages 2-21 and 2-22, table 2-12 on pages 2-23 and 2-24, table 2-13 on pages 2-25 and 2-26, table 2-14 on pages 2-27 and 2-28, and table 2-15 on pages 2-29 and 2-30, are example formats for some types of annexes.

Table 2-11. Example of an air movement annex format

AIR MOVEMENT ANNEX.
1. **SITUATION.**
• **Enemy.** Include the following information:
    ▪ Enemy air capability.
    ▪ Enemy air defense artillery (ADA) capability.
    ▪ Include in weather: percent illumination, illumination angle, night vision device window, ceiling, and visibility.

2. **MISSION.** Who, what (task), when, where, why (purpose).

3. **EXECUTION.** Include the following information:
a. **Concept of operations.**
b. **Tasks to subordinate units.**
c. **Coordinating instructions.** Include the following information:
• **Pickup zone (PZ):**
    ▪ Name and number.
    ▪ Coordinates.
    ▪ Load time.
    ▪ Takeoff time.
    ▪ Markings.
    ▪ Control.
    ▪ Landing formation.
    ▪ Approach and departure direction.
    ▪ Alternate PZ name and number.
    ▪ Penetration points.
    ▪ Extraction points.
• **Landing zone (LZ):**
    ▪ Name and number.
    ▪ Coordinates.
    ▪ H-hour.
    ▪ Markings.
    ▪ Control.
    ▪ Landing formation and direction.
    ▪ Alternate LZ name and number.
    ▪ Deception plan.
    ▪ Extraction LZ.

Table 2-11. Example of an air movement annex format (continued)

3. EXECUTION (continued)
 • **Laager site:**
   ▪ Communications.
   ▪ Security force.
   ▪ Flight routes and alternates.
   ▪ Abort criteria.
   ▪ Downed aircraft/crew designated area of recovery (DAR).
   ▪ Special instructions.
   ▪ Cross-forward line of own troops (FLOT) considerations.
   ▪ Aircraft speed.
   ▪ Aircraft altitude.
   ▪ Aircraft crank time.
   ▪ Rehearsal schedule and plan.
   ▪ Actions on enemy contact (en route and on the ground).

4. SUSTAINMENT.
   a. **Logistics.** Include the following information:
   • Sustainment overlay: include forward area refuel and rearm points.
   • Maintenance: specific to aircraft.
   • Transportation.
   • Supply:
     ▪ Class I—food, rations, and water.
     ▪ Class III—petroleum, oils, and lubricants.
     ▪ Class V—ammunition.
     ▪ Class VII—major end items.
     ▪ Class VIII—medical supplies, minimal amounts.
     ▪ Class IX—repair parts.
     ▪ Distribution methods.

5. COMMAND AND SIGNAL.
   a. **Command.** Include the following information:
   • **Location of commander or patrol leader:** state where the commander intends to be during the operation, by phase if the operation is phased.
   • **Succession of command:** state the succession of command if not covered in the unit's SOP.
   b. **Control.** Include the following information: list reports not covered in SOPs.
   c. **Signal:** describe the concept of signal support, including current SOI edition, or refer to higher HQ OPORD.
   • Air and ground call signs and frequencies.
   • Air and ground emergency code.
   • Passwords and number combinations.
   • Fire net and quick-fire net.
   • Time zone.
   • Time check.
 	
 LEGEND
 ADA – air defense artillery; DAR – designated area of recovery; FLOT – forward line of own troops; HQ – headquarters; LZ – landing zone; NVD – night vision device; OPORD – operation order; PZ – pickup zone; SOI - signal operating instructions; SOP – standard operating procedure

Table 2-12. Example of a patrol base annex format

PATROL BASE ANNEX.
1. **SITUATION.** Include the following information:
   a. Enemy forces.
   b. Friendly forces.
   c. Attachments and detachments.

2. **MISSION.** Who, what (task), when, where, why (purpose).

3. **EXECUTION.** Include the following information:
   a. Concept of operations.
   b. Scheme of movement and maneuver.
   c. Scheme of fires.
   d. **Tasks to subordinate units.** Include the following information:
     (1) Teams:
        • Security.
        • Reconnaissance.
        • Surveillance.
        • Listening post/observation posts (LP/OPs).
     (2) Individuals.
   e. **Coordinating instructions.** Include the following information:
     (1) Occupation plan.
     (2) Operations plan:
        • Security plan.
        • Alert plan.
        • Priority of work.
        • Evacuation plan.
        • Alternate patrol base (used when primary base is unsuitable or compromised).

4. **SUSTAINMENT:** only brief specifics not covered in the base order.
   a. **Logistics.** Include the following information:
     • Sustainment overlay: include water, maintenance, hygiene, rations, and rest plans.
     • Maintenance.
     • Transportation.
     • Supply:
        ▪ Class I—food, rations, and water.
        ▪ Class III—petroleum, oils, and lubricants.
        ▪ Class V—ammunition.
        ▪ Class VII—major end items.
        ▪ Class VIII—medical supplies, minimal amounts.
        ▪ Class IX—repair parts.
        ▪ Distribution methods.
     • Field services.

Chapter 2

Table 2-12. Example of a patrol base annex format (continued)

4. **SUSTAINMENT:** only brief specifics not covered in the base order.
   a. **Logistics.** Include the following information:
      • Sustainment overlay: include water, maintenance, hygiene, rations, and rest plans.
      • Maintenance.
      • Transportation.
      • Supply:
          ▪ Class I—food, rations, and water.
          ▪ Class III—petroleum, oils, and lubricants.
          ▪ Class V—ammunition.
          ▪ Class VII—major end items.
          ▪ Class VIII—medical supplies, minimal amounts.
          ▪ Class IX—repair parts.
          ▪ Distribution methods.
          ▪ Field services.
   b. **Personnel services support.** Include the following information:
      • Method of marking and handling EPWs.
      • Religious services.
   c. **Army Health System support.** Include the following information:
      • Medical mission command.
      • Medical treatment.
      • Medical evacuation.
      • Preventive medicine.

5. **COMMAND AND SIGNAL.**
   a. **Command.** Include the following information:
      • **Location of commander or patrol leader:** state where the commander intends to be during the operation, by phase if the operation is phased.
      • **Succession of command:** state the succession of command, if not covered in the unit's SOP.
   b. **Control:** include the following information:
      • **Command posts:** describe the employment of CPs, including the location of each CP and its time of opening and closing, as appropriate. Typically, at platoon level the only reference to command posts is the company CP.
      • Reports. List reports not covered in SOPs.
   c. **Signal:** describe the concept of signal support, including current SOI edition or, refer to higher HQ OPORD.
      • Identify the **SOI index** that is in effect.
      • Identify **methods of communication by priority**.
      • Describe **pyrotechnics and signals**, to include arm and hand signals (demonstrate).
      • Give **code words** such as OPSKEDs.
      • Give **challenge and password** (use behind friendly lines).
      • Give **number combination** (use forward of friendly lines).
      • Give **running password.**
      • Give **recognition signals** (near—far and day—night).
 	
LEGEND
CP – command post; EPW – enemy prisoner of war; HQ – headquarters; LP/OP – listening post/observation post; OPORD– operation order; OPSKEP – operation schedule; SOI - signal operating instructions; SOP – standard operating procedure

Table 2-13. Example of a waterborne insertion annex format

WATERBORNE INSERTION ANNEX.
1. **SITUATION.** Include the following information:
   a. **Area of operations:**
   - Terrain:
     - River width.
     - River depth and water temperature.
     - Current.
     - Vegetation.
   - Weather:
     - Tide.
     - Surf.
     - Wind.
   b. **Enemy forces:** state any changes or additions to identification, location, activity, and strength.
   c. **Friendly forces (unit furnishing support).**
   d. **Attachments and detachments.**
   e. **Organization for movement.**

2. **MISSION.** Who, what (task), when, where, why (purpose).

3. **EXECUTION.** Include the following information:
   a. Concept of operations.
   b. Scheme of movement and maneuver.
   c. Scheme of fires.
   d. Tasks to subordinate units:
   - Security.
   - Tie-down teams:
     - Load equipment.
     - Secure equipment.
   - Designation of coxswains and boat commanders.
     - Selection of navigator(s) and observer(s).
   e. Coordinating instructions:
   - Formations and order of movement.
   - Route and alternate route.
   - Method of navigation.
   - Actions on enemy contact.
   - Rally points.
   - Embarkation plan.
   - Debarkation plan.
   - Rehearsals.
   - Time schedule.

Table 2-13. Example of a waterborne insertion annex format (continued)

4. **SUSTAINMENT:** only brief specifics not covered in base order.
   a. **Logistics.** Include the following information:
   - Sustainment overlay.
   - Maintenance.
   - Transportation. Include disposition of boats, paddles, and life jackets upon debarkation.
   - Supply:
     - Class I—food, rations, and water.
     - Class III—petroleum, oils, and lubricants.
     - Class V—ammunition.
     - Class VII—major end items.
     - Class VIII—medical supplies, minimal amounts.
     - Class IX—repair parts.
     - Distribution methods.
     - Distribution methods, including method of distribution of paddles and life jackets.

5. **COMMAND AND SIGNAL.**
   a. **Command.** Include the following information:
   - **Location of commander or patrol leader:** state where the commander intends to be during the operation, by phase if the operation is phased.
   - **Succession of command:** state the succession of command if not covered in the unit's SOP.

   b. **Control:** include the following information:
   - **Command posts:** describe the employment of CPs, including the location of each CP and its time of opening and closing, as appropriate. Typically, at platoon level the only reference to command posts IS the company CP.
   - **Reports:** list reports not covered in SOPs.

   c. **Signal:** describe the concept of signal support, including current SOI edition, or refer to higher HQ OPORD.
   - Identify the **SOI index** that is in effect.
   - Identify **methods of communication by priority.**
   - Describe **pyrotechnics and signals,** to include arm and hand signals (demonstrate).
   - Give **code words** such as OPSKEDs.
   - Give **challenge and password** (use behind friendly lines).
   - Give **number combination** (use forward of friendly lines).
   - Give **running password.**
   - Give **recognition signals** (near—far and day—night).

LEGEND
CP – command post; HQ – headquarters; OPORD – operation order; OPSKED – operation schedule; SOI - signal operating instructions; SOP – standard operating procedure

Table 2-14. Example of a covert gap crossing annex format

COVERT GAP CROSSING ANNEX.
1. **SITUATION.** Include the following information:
  a. Area of operations:
    • Terrain:
      ▪ River width.
      ▪ River depth and water temperature.
      ▪ Current.
      ▪ Vegetation.
      ▪ Obstacles.
    • Weather.
  b. Enemy forces (location, identification, and activity).
  c. Friendly forces.
  d. Attachments and detachments.

2. **MISSION.** Who, what (task), when, where, why (purpose).

3. **EXECUTION.** Include the following information:
  a. Concept of operations.
  b. Scheme of movement and maneuver.
  c. Scheme of fires.
  d. Tasks to subordinate units:
    • Elements.
    • Teams.
    • Individuals.
  e. Coordinating instructions:
    • Crossing procedure and techniques.
    • Security.
    • Order of crossing.
    • Actions on enemy contact.
    • Alternate plan.
    • Rallying points.
    • Rehearsal plan.
    • Time schedule.

Chapter 2

Table 2-14. Example of a covert gap crossing annex format (continued)

4. **SUSTAINMENT:** only brief specifics not covered in base order.
   a. **Logistics.** Include the following information:
     • Sustainment overlay.
     • Maintenance.
     • Transportation.
   • Supply:
     ▪ Class I—food, rations, and water.
     ▪ Class III—petroleum, oils, and lubricants.
     ▪ Class V—ammunition.
     ▪ Class VII—major end items.
     ▪ Class VIII—medical supplies, minimal amounts.
     ▪ Class IX—repair parts.
     ▪ Distribution methods.

5. **COMMAND AND SIGNAL.**
   a. **Command.** Include the following information:
     • **Location of commander or patrol leader:** state where the commander intends to be during the operation, by phase if the operation is phased.
     • **Succession of command:** state the succession of command if not covered in the unit's SOP.
   b. **Control:** include the following information:
     • **Command posts:** describe the employment of CPs, including the location of each CP and its time of opening and closing, as appropriate. Typically, at platoon level the only reference to command posts is the company CP.
     • **Reports:** list reports not covered in SOPs.
   c. **Signal:** describe the concept of signal support, including current SOI edition or refer to higher OPORD.
     • Identify the **SOI index** that is in effect.
     • Identify **methods of communication by priority.**
     • Describe **pyrotechnics and signals,** to include arm and hand signals (demonstrate).
     • Give **code words** such as OPSKEDs.
     • Give **challenge and password** (use behind friendly lines).
     • Give **number combination** (use forward of friendly lines).
     • Give **running password.**
     • Give **recognition signals** (near—far and day—night).

LEGEND CP – command post; OPORD – operation order; OPSKED – operation schedule; SOI - signal operating instructions; SOP – standard operating procedure

Table 2-15. Example of a truck annex format

TRUCK ANNEX. 1. **SITUATION.** Include the following information: • Enemy situation. • Friendly situation. • Attachments and detachments.
2. **MISSION.** Who, what (task), when, where, why (purpose).

3. **EXECUTION.** Include the following information:
• Concept of operations.
• Scheme of movement and maneuver.
• Scheme of fires.
• Tasks to subordinate units.
• Coordinating instructions:
  ▪ Times of departure and return.
  ▪ Loading plan and order of movement.
  ▪ Route (primary and alternate).
  ▪ Air guards.
  ▪ Actions on enemy contact (vehicle ambush) during movement, loading, and downloading.
  ▪ Actions at the de-trucking point.
  ▪ Rehearsals.
  ▪ Vehicle speed, separation, and recovery plan.
  ▪ Broken vehicle instructions.

4. **SUSTAINMENT:** only brief specifics not covered in base order.
  a. **Logistics.** Include the following information:
• Sustainment overlay.
• Maintenance.
• Transportation.
• Supply:
  ▪ Class I—food, rations, and water.
  ▪ Class III—petroleum, oils, and lubricants.
  ▪ Class V—ammunition.
  ▪ Class VII—major end items.
  ▪ Class VIII—medical supplies, minimal amounts.
  ▪ Class IX—repair parts.
  ▪ Distribution methods.
• Field services.

Table 2-15. Example of a truck annex format (continued)

b. **Personnel services support.** Include the following information:
      • Method of marking and handling EPWs.
      • Religious services.
   c. **Army Health System support.** Include the following information:
      • Medical mission command.
      • Medical treatment.
      • Medical evacuation.
      • Preventive medicine.

5. **COMMAND AND SIGNAL.**
   a. **Command.** Include the following information:
      • **Location of commander or patrol leader:** state where the commander intends to be during the operation, by phase if the operation is phased.
      • **Succession of command:** state the succession of command if not covered in the unit's SOP.
   b. **Control.** Include the following information:
      • **Command posts:** describe the employment of CPs, including the location of each CP and its time of opening and closing, as appropriate. Typically, at platoon level the only reference to command posts is the company CP.
      • **Reports:** list reports not covered in SOPs.
   c. **Signal:** describe the concept of signal support, including current SOI edition or refer to higher HQ OPORD. Include the following information:
      • Identify the **SOI index** that is in effect.
      • Identify **methods of communication by priority.**
      • Describe **pyrotechnics and signals**, to include arm and hand signals (demonstrate).
      • Give **code words** such as OPSKEDs.
      • Give **challenge and password** (use behind friendly lines).
      • Give **number combination** (use forward of friendly lines).
      • Give **running password.**
      • Give **recognition signals** (near—far and day—night).

LEGEND CP – command post; EPW – enemy prisoner of war; HQ – headquarters; OPORD – operation order; OPSKED – operation schedule; SOI - signal operating instructions; SOP – standard operating procedure

Operations

COORDINATION CHECKLISTS

2-24. The checklists shown in table 2-16, table 2-17 on pages 2-31 and 2-32, and table 2-18 on page 2-32, along with table 2-19 on page 2-33, tables 2-20 and 2-21 on page 2-34, table 2-22 on pages 2-35 through 2-37, and table 2-23 on page 2-38, include items that a platoon or squad leader checks when planning for a combat operation. In some cases, the platoon or squad leader coordinates directly with the appropriate staff section. In most cases, the company commander or PL provides this information. The platoon or squad leader can carry copies of these checklists to keep from overlooking anything that may be vital to the mission.

Table 2-16. Intelligence coordination checklist

INTELLIGENCE COORDINATION CHECKLIST.
The unit one level higher constantly updates intelligence. This ensures that the platoon leader's plan reflects the most recent enemy activity. Include the following information:
- Identification of enemy unit.
- Weather and light data.
- Terrain update:
  - Aerial photos.
  - Trails and obstacles not on map.
- Known or suspected enemy locations.
- Weapons.
- Probable course of action.
- Most dangerous course of action.
- Recent enemy activities.
- Reaction time of reaction forces.
- Civilians on the battlefield.
- Update to the commander's critical information requirement (CCIR).

Table 2-17. Operations coordination checklist

OPERATIONS COORDINATION CHECKLIST.
The platoon or squad leader coordinates with the company commander or platoon leader to confirm the mission and operational plan, receive last minute changes, and update subordinates in person or issue a FRAGORD. Include the following information:
- Mission confirmation brief.
- Identification of friendly units.
- Changes in the friendly situation.
- Route selection, LZ, PZ, and drop zone (DZ) selection.
- Commander's critical information requirement (CCIR):
  - Changes to PIR.
  - Changes to friendly force information requirement (FFIR).
  - Changes to essential elements of friendly information (EEFIs).
  - Changes to ROE.
- Linkup procedures:
  - Contingencies.
  - Quick reaction force (QRF).
  - QRF frequency.

Chapter 2

Table 2-17. Operations coordination checklist (continued)

- Transportation and movement plan.
- Resupply (with S-4).
- Signal plan.
- Departure and reentry of forward units.
- Special equipment requirements.
- Adjacent units in the area of operations.
- Rehearsal areas.
- Method of insertion and extraction.

LEGEND
DZ – drop zone; EEFI - essential element of friendly information; FFIR - friendly force information requirement; FRAGORD – fragmentary order; LZ – landing zone; PIR - priority intelligence requirement; PZ – pickup zone; ROE – rules of engagement; QRF – quick reaction force; S-4 – logistics officer

Table 2-18. Fire support coordination

checklist   FIRE SUPPORT COORDINATION CHECKLIST.
The platoon or squad leader coordinates the following with the FO at squad level and FSO at platoon level. Include the following information:
- Mission confirmation brief.
- Identification of supporting unit.
- Mission and objective.
- Route to and from the objective (include alternate routes).
- Time of departure and expected time of return.
- Unit target list (from fire plan).
- Type of available support (artillery, mortar, naval gunfire, and aerial support, including Army, Navy, and Air Force) and their locations.
- Ammunition available (including different fuzes).
- Priority of fires.
- Control measures:
    - Checkpoints.
    - Boundaries.
    - Phase lines.
    - Fire support coordination measures.
    - Priority targets (target list).
    - Restrictive fire area (RFA).
    - Restrictive fire line (RFL).
    - No-fire area (NFA).
    - Precoordinated authentication.
- Communication (include primary and alternate means, emergency signals, and code words).

LEGEND
FO – forward observer; FSO – fire support officer; NFA – no-fire area; RFA – restrictive fire area; RFL – restrictive fire line

Operations

Table 2-19. Coordination with forward unit checklist

COORDINATION WITH FORWARD UNIT CHECKLIST.
A platoon or squad that requires foot movement through a friendly forward unit coordinates with that unit's commander for a safe and orderly passage. If no time and place has been designated for coordination with the forward unit, the platoon or squad leader should set a time and place to coordinate with the S-3. He talks with someone at the forward unit who has the authority to commit the forward unit to assist the platoon or squad during departure. Coordination is a two-way exchange of information. . Include the following information:
  a. Identification (yourself and your unit).
  b. Size of the platoon or squad.
  c. General area of operations.
  d. Known or suspected enemy positions or obstacles.
  e. Possible enemy ambush sites.
  f. Latest enemy activity.
  g. Detailed information on friendly positions, such as crew-served weapons and final protective fire (FPF).
  h. Fire and barrier plan:
  - Support the unit can furnish. How long and what can they do?
    - Fire support.
    - Litter teams.
    - Navigational signals and aids.
    - Guides.
    - Communications.
    - Reaction units.
    - Other.
  - Call signs and frequencies.
  - Pyrotechnic plan.
  - Challenge and password, running password, number combination.
  - Emergency signals and code words.
  - If the unit is relieved, pass the information to the relieving unit.
  - Recognition signals.

LEGEND
FPF – final protective fire; S-3 – operations officer

Table 2-20. Adjacent coordination checklist

ADJACENT UNIT COORDINATION CHECKLIST.
Immediately after the operation order or mission briefing, the platoon or squad leader should check with other platoon and squad leaders who will be operating in the same areas. If the leader is unaware of any other units operating is the area, he should check with the S-3 during the operations coordination. The S-3 can help arrange this coordination, if necessary. The platoon and squad leaders should exchange the following information with other units operating in the same area:
- Identification of the unit.
- Mission and size of unit.
- Planned times, and points of departure and reentry.
- Route(s).
- Fire support and control measures.
- Frequencies and call signs.
- Challenge and password, running password, number combination.
- Pyrotechnic plan.
- Any information that the unit may have about the enemy.
- Recognition signals.

LEGEND
S-3 – battalion or brigade operations staff officer

Table 2-21. Rehearsal area coordination checklist

REHEARSAL AREA COORDINATION CHECKLIST.
The assistant patrol leader coordinates the use of the rehearsal area to facilitate the unit's safe, efficient, and effective use of the area before its mission: .Include the following information:
- Identification of the unit.
- Mission.
- Terrain similar to objective site.
- Security of the area.
- Availability of aggressors.
- Use of blanks, pyrotechnics, and ammunition.
- Mock-ups available.
- Time the area is available (preferably, when light conditions approximate light conditions of patrol).
- Transportation.
- Coordination with other units using the area.

Operations

Table 2-22. Army aviation coordination checklist

ARMY AVIATION COORDINATION CHECKLIST.
The patrol leader coordinates this with the company commander or S-3 Air to facilitate the time and detailed and effective use of aviation assets as they apply to the tactical mission:
1. **SITUATION.** Include the following information:
   a. **Enemy:**
   - Air capability.
   - Air defense artillery (ADA) capability.
   - Include in weather the percent of illumination, illumination angle, night vision device (NVD) window, ceiling, and visibility.

   b. **Friendly:**
   - Unit(s) supporting operation, and axis of movement/corridor/routes.
   - ADA status.

2. **MISSION.** Who, what (task), when, where, why (purpose).

3. **EXECUTION.**
   a. **Concept of the operation.** Overview of what requesting unit wants to accomplish with the air assault/ air movement. Include the following information:
   b. **Tasks to combat units:**
   - Infantry.
   - Attack aviation.

   c. **Tasks to combat support units:**
   - Artillery.
   - Aviation (lift).

   d. **Coordinating instructions:**
   (1) Pickup zone:
   - Direction of landing.
   - Time of landing and flight direction.
   - Locations of PZ and alternate PZ.
   - Loading procedures.
   - Marking of PZ (panel, smoke, signal mirror [SM], lights).
   - Flight route planned (start point [SP], air control point [ACP], release point [RP]).
   - Formations (PZ, en route, LZ).
   - Code words:
     - PZ secure (before landing), PZ clear (lead bird and last bird).
     - Alternate PZ (at PZ, en route, LZ), names of PZ and alternative PZ.
   - Tactical (TAC) air and artillery.
   - Number of PAXs for each helicopter and for entire lift.
   - Equipment carried by individuals.
   - Marking of key leaders.
   - Abort criteria (PZ, en route, LZ).

Table 2-22. Army aviation coordination checklist (continued)

3. **EXECUTION.** (Continued.)
   (2) Landing zone:
      • Direction of landing.
      • False insertion plans.
      • Time of landing (LZ time).
      • Locations of LZ and alternate LZ.
      • Marking of LZ (panel, smoke, SM, lights).
      • Formation of landing.
      • Code words, LZ name, alternate LZ name.
      • TAC air and artillery preparation, fire support coordination:
          ▪ Secure LZ?
          ▪ Do not secure LZ?

4. **SUSTAINMENT.** Only brief specifics not covered in base order to include number of aircraft for each lift and number of lifts, whether the aircraft will refuel/rearm during mission, special equipment carried by personnel, aircraft configuration, and bump plan. Include the following information:
  a. **Logistics:**
      • Sustainment overlay.
      • Maintenance.
      • Transportation.
      • Supply:
          ▪ Class I—food, rations, and water.
          ▪ Class III—petroleum, oils, and lubricants.
          ▪ Class V—ammunition.
          ▪ Class VII—major end items.
          ▪ Class VIII—medical supplies, minimal amounts.
          ▪ Class IX—repair parts.
          ▪ Distribution methods.
      • Field services.
  b. **Personnel services support:**
      • Method of marking and handling EPWs.
      • Religious services.
  c. **Army Health System support:**
      • Medical mission command.
      • Medical treatment.
      • Medical evacuation.
      • Preventive medicine.

Table 2-22. Army aviation coordination checklist (continued)

5. COMMAND AND SIGNAL.
   a. **Command.** Include the following information:
      - **Location of commander or patrol leader:** state where the commander intends to be during the operation, by phase if the operation is phased. Also, include locations of air mission commander, ground tactical commander, and air assault task force commander.
      - **Succession of command:** state the succession of command, if not covered in the unit's SOP.
   b. **Control:** include the following information:
      - **Command posts:** describe the employment of CPs, including the location of each CP and its time of opening and closing, as appropriate. Usually at platoon level, the only reference to command posts is the company CP.
      - **Reports:** list reports not covered in SOPs.
   c. **Signal:** describe the concept of signal support, including current SOI edition or refer to higher OPORD:
      - Identify the **SOI index** that is in effect.
      - Identify **methods of communication by priority.**
      - Describe **pyrotechnics and signals**; including arm and hand signals (demonstrate).
      - Give **code words** such as OPSKEDs.
      - Give **challenge and password** (use behind friendly lines).
      - Give **number combination** (use forward of friendly lines).
      - Give **running password.**
      - Give **recognition signals** (near—far and day—night).

LEGEND
ACP – air control point; ADA – air defense artillery; CP – command post; LZ – landing zone; NVD – night vision device; OPORD – operation order; OPSKED – operation schedule; PAX – passenger; PZ – pickup zone; RP – release point; S-3 – battalion or brigade operations staff officer; SM – signal mirror; SOI - signal operating instructions; SOP – standard operating procedure; SP – start point; TAC – tactical

Chapter 2

Table 2-23. Vehicular movement coordination checklist

VEHICULAR MOVEMENT COORDINATION CHECKLIST.
The platoon sergeant or first sergeant coordinates this with the supporting unit to facilitate the effective, detailed, and efficient use of vehicular support and assets. Include the following information:
- Identification of the unit.
- Supporting unit identification.
- Number and type of vehicles, and tactical preparation.
- Entrucking point.
- Departure time.
- Preparation of vehicles for movement:
    - Driver responsibilities.
    - Platoon and squad responsibilities.
    - Special supplies and equipment required.
- Availability of vehicles for preparation, rehearsals, and inspection (times and locations).
- Routes:
    - Primary.
    - Alternate.
    - Checkpoints.
- De-trucking points:
    - Primary.
    - Alternate.
- Order of march.
- Speed.
- Communications (frequencies, call signs, and codes).
- Emergency procedures and signals.

ACTIONS, PURPOSE, OPERATIONS, AND TASKS

2-25. Table 2-24 lays out the actions taken by friendly forces and the effect those actions have on the enemy. The desired or intended result of the tactical operation stated in terms related to the enemy or the desired situation is the purpose. Purpose is the WHY of the mission statement and often follows the words "in order to." It is the most important component of the mission statement. (See table 2-25.)

2-26. An operation is a military action or the carrying out of a military action to gain the objectives of any battle or campaign. These is accomplished by using tasks. A task is a specific, clearly defined, decisive, and measurable activity or action accomplished by a Ranger or organization that contributes to the achievement of encompassing missions or other requirements. (See table 2-26 on pages 2-39 and 2-40.) There are also shaping tasks that are vital to the security of Soldiers and success of missions. (See table 2-27 on page 2-40.)

Operations

Table 2-24. Actions by friendly forces and effects on enemy forces

ACTIONS BY FRIENDLY FORCES	EFFECT ON ENEMY FORCES
Attack by fire	Block
Breach	Canalize
Bypass	Contain
Clear	Defeat
Control	Destroy
Counterreconnaissance	Disrupt
Disengagement	Fix
Exfiltrate	Isolate
Follow and assume	Neutralize
Follow and support	Suppress
Occupy Retain Secure Seize Support by fire	Turn

Table 2-25. Purpose

Allow Cause Create Deceive Deny	Divert Enable Envelop Influence Open	Prevent Protect Support Surprise

Table 2-26. Elements of operations and subordinate tasks

ELEMENTS OF DECISIVE ACTION AND SUBORDINATE TASKS	
Offensive Tasks	Defensive Tasks

Movement to contact:
   Search and attack
   Cordon and search
	Area defense

Attack (also known as special purpose attacks):
   Ambush
   Counterattack
   Demonstration
   Feint
   Raid
   Spoiling attack
	Mobile defense

Table 2-26. Elements of operations and subordinate tasks (continued)

ELEMENTS OF DECISIVE OPERATIONS AND SUBORDINATE TASKS	
Offensive Tasks	Defensive Tasks

Exploitation
Pursuit
	Retrograde operations:
	  Delay
	  Withdrawal
	  Retirement

Forms of offensive maneuver:
  Envelopment
  Flank attack
  Frontal attack
  Infiltration
  Penetration
  Turning movement
	Forms of defense:
	  Defense of a linear obstacle
	  Perimeter defense
	  Reverse slope defense

Table 2-27. Tactical shaping operations and tasks

TACTICAL SHAPING TASKS
Passage of lines

Reconnaissance operations:
  Zone
  Area
  Route
  Reconnaissance in force

Relief in place

Security operations:
  Screen
  Guard
  Cover
  Area (includes route and convoy)
  Local

Troop movements:
  Administrative movement
  Approach march
  Road march

Operations

TERRAIN MODEL

2-27. During the planning process, the terrain model (see figure 2-1) offers an effective way to visually communicate the patrol routes and detail actions on the objective. At a minimum, the model is used to display routes to the objective and to highlight prominent terrain features the patrol encounters during movement. A second terrain model of the objective area is prepared. It should be large enough and detailed enough to brief the patrol's actions on the objective.

a. **Checklist.** Make sure the following items are included in the terrain models:
- North seeking arrow.
- Scale.
- Grid lines.
- Objective location.
- Exaggerated terrain relief and water obstacles.
- Friendly patrol locations.
- Targets (indirect fires, including grid and type of round).
- Routes, primary and alternate.
- Planned release points (RPs): ORP, linkup release point (LURP), RP.
- Danger areas (roads, trails, open areas).
- Legend.
- Blowup of objective area.

b. **Construction.** Some field expedient techniques that can be used to construct terrain models are:
- Use a 3" x 5" card from a meal, ready to eat (MRE) box, or piece of paper to label the objective or key sites.
- Use string from the guts of 550 cord, or use colored tape to make grid lines. Identify the grids with numbers written on small pieces of paper.
- Replicate trees and vegetation using moss, green or brown spray paint, pine needles, crushed leaves, or cut grass.
- Use blue chalk; blue spray paint, blue yarn, tin foil, or MRE creamer to designate bodies of water.
- Make North-seeking arrows from sharpened twigs, pencils, or colored yarn.
- Use red yarn, 5.56-mm rounds, toy Rangers, or poker chips to designate enemy positions.
- Construct friendly positions such as security elements, support by fire (SBF), and assault elements using 5.56-mm rounds, toy Rangers, poker chips, small MRE packets of sugar and coffee, or preprinted acetate cards.
- Use small pieces of cardboard or paper to identify target reference points (TRPs) and indirect fire targets. Show the grids for each point.
- Construct breach, SBF, and assault positions using the same methods, again using colored yarn or string for easy identification.
- Construct bunkers and buildings using MRE boxes, tongue depressors, or sticks.
- Construct perimeter wire from a spiral notebook.
- Construct key phase lines with colored string or yarn.
- Use colored tape or yarn to replicate trench lines by digging a furrow and coloring it with colored chalk or spray paint.

Note: Clearly identify in a legend all symbols used on the terrain model.

Chapter 2

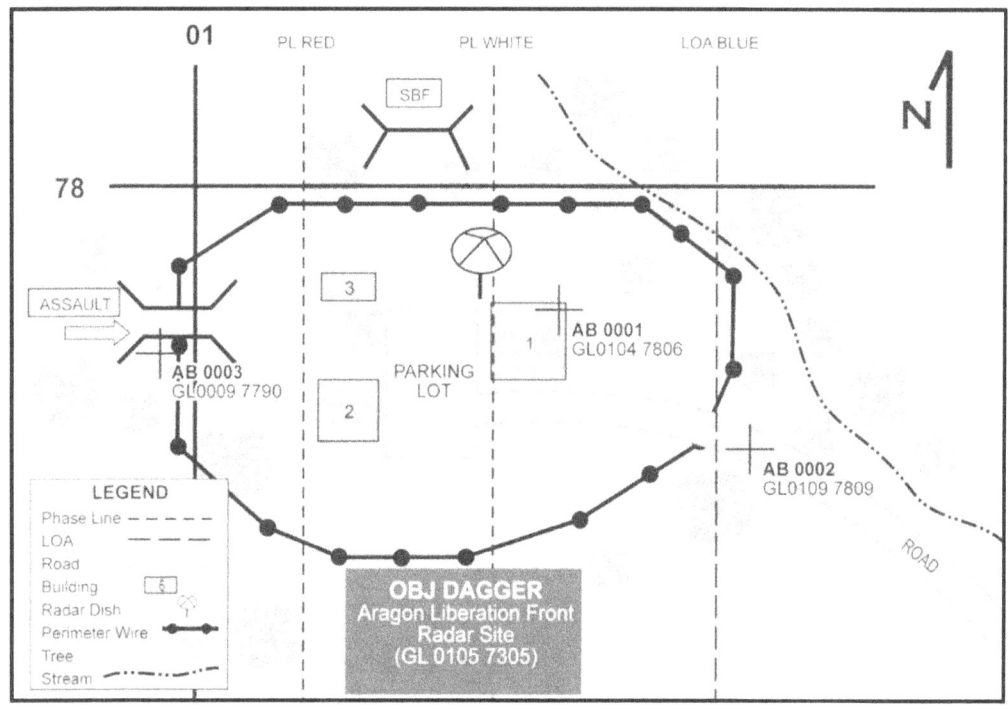

Figure 2-1. Terrain model

LEGEND
AB – target reference number; GL – map grid zone designator; LOA – line of advance; OBJ – objective; PL – phase line; SBF – support by fire

Chapter 3

Fire Support

Indirect fire support can greatly increase the combat effectiveness and survivability of any Infantry unit. The ability to plan for and effectively use this asset is a task that every Ranger and small unit leader should master. Fire support assets can help a unit by suppressing, fixing, destroying, or neutralizing the enemy. Leaders should consider employing indirect fire support throughout every offensive and defensive operation. This chapter discusses plans, tasks, capabilities, risk estimate distances (REDs), target overlays, close air support (CAS) elements and sequence of calls for fire, and an example of call-for-fire transmissions. (See Appendix A of this publication for more information.)

BASIC TASKS AND TARGETING

3-1. The effectiveness of the fire support system depends on successful performance of its four basic tasks. This includes support forces in contact, support the battle plan, synchronize the fire support system, and sustain the fire support system. Targeting objectives are the overall effects the leader hopes to achieve with fire support assets. **Decide** is the first functional step in the targeting process. A decision defines the overall focus and sets priorities for collecting intelligence and planning the attack. The leader addresses targeting priorities for each phase or critical event of an operation. At all echelons, one or more alternative course of action is analyzed. Each is based on—

- Mission analysis.
- Current and projected battle situations.
- Anticipated opportunities.

3-2. **Detect** is the second critical function. The intelligence officer (G-2 or S-2) directs the effort to detect the high payoff targets (HPTs) identified in the decide step. To identify the exact who, what, when, and how of target acquisition, the intelligence officer works closely with the—

- Analysis and control element.
- Field artillery intelligence officer.
- Targeting officer and FSO, or both

INTERDICTION

3-3. Interdiction is an action to divert, disrupt, delay, intercept, board, detain, or destroy the enemy's military surface capabilities (such as vessels, vehicles, aircraft, people, and cargo) before they can be used effectively against friendly forces, or to otherwise achieve friendly objectives. Table 3-1 on page 3-2, and table 3-2 on page 3-3, outline the capabilities of mortars and field artillery. Interdiction—

- **Limits:** reduces enemy options. For example, direct air interdiction and fire support to limit enemy avenue(s) of approach and fire support.
- **Disrupts:** stops effective interaction between the enemy and their support systems, reduces enemy efficiency, and increases vulnerability.
- **Delays:** disrupts, diverts, or destroys enemy capabilities or targets. Changes when the enemy reaches a point on the battlefield, or changes their ability to project combat power from it.
- **Diverts:** creates a distraction that forces the enemy to tie up critical resources. For example, attack targets that cause the enemy to move capabilities or assets from one area or activity to another.

Chapter 3

- **Destroys:** ruins the structure or condition of a vital enemy target. Destruction can be defined as an objective by stating a number or percentage of an enemy asset or target that the weapon system(s) can realistically achieve. For example, artillery normally says that destruction comprises a 30 percent reduction in capability or structural integrity; maneuver combat forces normally use 70 percent.
- **Damages:** this can be a subjective or objective assessment of battle damage, or it can describe the damage to the objective as light, moderate, or severe.

Table 3-1. Capabilities of mortars

WEAPON	MUNITION AVAILABLE	MAX RANGE (meters)	MIN RATE (rounds per minute)	BURST RADIUS (meters)	SUSTAINED RATE (rounds per minute)
60 mm	HE, WP, Illum	3500 m (HE)	70 m (HE)	30 for four minutes.	30 m
81 mm	HE, WP, Illum	5600 m (HE)	70 m (HE)	25 for two minutes.	38 m
120 mm	HE, Smoke, Illum	7200 m (HE)	180 m (HE)	15 for one minute.	60 m
LEGEND HE – high explosive; Illum – illumination; m – meter; mm – millimeter; WP – white phosphorous					

Fire Support

Table 3-2. Capabilities of field artillery

ARTILLERY	AMMUNITION		RANGE (METERS)			RATES OF FIRE (ROUNDS PER MINUTE)	
	Projectile	Fuze	Maximum	DPICM	RAP	Sustained	Maximum
105-mm M119-series	HE, HC, WP, Illum APICM	PD, VT, MT, ET, MTSQ, delay	11,500 with charge 7. 14,000 with charge 8.	12,000	19,5000	Three for 30 minutes	Every three minutes
155-mm M109A5			18,000 or 22,000 with M795 HE, M825 smoke	18,000 or 28,200 with M864	30,100	2	Every three minutes
155-mm M109 A5/A6	HE, HC, WP, Illum APICM, DPICM, M825, smoke, SCATMINE		18,000 or 21,7000 with M795 HE, M825 smoke, M982 Excalibur: Block Ia-1 – 24 km Block Ia-2 – 37+km Block Ib – 40+km	17,900 or 28,100 with M864 base bleed	30,000	Zones 3-7: one round per minute Zone 8: one round per minute until limited by tube temperature sensor	Every three minutes
155-mm M777-series			22,200 with M201A1 modular charge 8S or 22,500 with M232 modular charge Zone 5: 24,500 with M982 Excalibur block 1-1a	N/A		Two according to thermal warming device	Every two minutes

NOTE: Excalibur is not authorized for the M109A5.
See ATP 3-09.32, Fire Multi-Service Tactics, Techniques, and Procedures for the Joint Application of Firepower, Appendix I for a detailed discussion of "danger close."

LEGEND
APICM – antipersonnel improved conventional munitions; DPICM – dual purpose improved conventional munitions; ET – electronic time; HC – hexachloroethane smoke; HE – high explosive; Illum – illumination; MT mechanical time; MTSQ – mechanical time superquick; N/A – not available; PD – point detonating; RAP – rocket assisted projectile; SCATMINE – scatterable mine; VT – variable time; WP – white phosphorus

Chapter 3

> **DANGER CLOSE**
>
> When the target is within 600 meters of any friendly troops (for mortars and field artillery), announce, DANGER CLOSE in the *method of engagement* portion of the call-for-fire.
>
> When adjusting five-inch or smaller naval guns on targets within 750 meters, announce, DANGER CLOSE. For larger naval guns announce, DANGER CLOSE, for targets within 1000 meters. Failure to adhere to this guidance can result in fratricide.
>
> Avoid making corrections using the bracketing method of adjustment. Doing so can cause serious injury or death. Use only the creeping method of adjustment during danger close missions. Make corrections of no more than 100 meters by creeping the rounds to the target.

RISK ESTIMATE DISTANCE

3-4. Risk estimate distance applies to combat only. Minimum safe distances (MSDs) apply to training. RED takes into account the bursting radius of particular munitions and the characteristics of the delivery system. It associates this combination with a percentage representing the likelihood of becoming a casualty. It is the percentage of risk.

3-5. RED is defined as the minimum distance friendly troops can approach the effects of friendly fires without suffering appreciable casualties of 0.1 percent point of impact or higher. Table 3-3 on page 3-5 gives the REDs for mortars and canon artillery, and includes the dual purpose improved conventional munitions. (Refer to ATP 3-09.32, Fire Multi-Service Tactics, Techniques, and Procedures for the Joint Application of Firepower, for more information.)

> **WARNING**
>
> Use RED formulas only in combat to determine acceptable risk levels, and to identify the risk to Rangers at various distances from the targets.
>
> When training, use minimum safe distances (MSDs).

Fire Support

Table 3-3 Risk estimate distances for unguided mortars and cannon artillery

UNGUIDED MORTAR RISK ESTIMATE DISTANCES					
System	Description	Danger Close	Range	0.1% Pi	
				Standing	Prone
M224	60mm mortar	600m	1/3	115m / 378 ft	115m / 378 ft
			2/3	125m / 410 ft	120m / 394 ft
			max	145m / 476 ft	145m / 146 ft
M29 M29A1	81mm mortar	600m	1/3	170m / 558 ft	160m / 525 ft
			2/3	195m / 640 ft	190m / 624 ft
			max	195m 640 ft	185m / 607 ft
M120 M327	120mm mortar	600m	1/3	280m / 919 ft	260m / 853 ft
			2/3	395m / 1296 ft	365m / 1198 ft
			max	430m / 1411 ft	410m / 1345 ft
UNGUIDED CANON RISK ESTIMATE DISTANCES					
System	Description	Danger Close	Range	0.1% Pi	
				Standing	Standing
M119 M119A2	105mm Howitzer HE (M1 Comp B/M760)	600m	1/3	29m / 952 ft	270m / 886 ft
			2/3	300m / 984 ft	285m / 935 ft
			max	455m / 1493 ft	430m / 1411 ft
	105mm Howitzer HERA (M913 HERA / M297 HERA)	600m	1/3	250m / 820 ft	230m / 886 ft
			2/3	410m / 1345 ft	285m / 935 ft
			max	650m / 2132 ft	430m / 1411 ft
M109A6 M777A2	155mm Howitzer HE (M107 Comp B/M795)	600m	1/3	300m / 984 ft	285m / 935 ft
			2/3	460m / 1509 ft	440m / 1444 ft
			max	695m / 2280 ft	665m / 2182 ft
	155mm Howitzer DPICM (M483A1)	600m	1/3	270m / 886 ft	260m / 853 ft
			2/3	325m / 1066 ft	310m / 1017 ft
			Max	510m / 1673 ft	490m / 1608 ft
	155mm Howitzer DPICM (M864)	600m	1/3	325m / 1066 ft	305m / 1001 ft
			2/3	500m / 1640 ft	485m / 1591 ft
			Max	825m / 2706 ft	775m / 2542 ft
	155mm Howitzer RAP (M945A1 RAP)	600m	1/3	360m / 2076 ft	360m / 1181 ft
			2/3	530m / 1739 ft	520m / 1706 ft
			max	1045m / 3428 ft	965m / 3166 ft

LEGEND
% - percent; DPICM – dual purpose improved conventional munitions; ft – feet; HE - high explosive; m - meter; max - maximum; mm - millimeter; RAP - rocket assisted projectile

Chapter 3

3-6. Casualty criterion is the five-minute assault criterion for a prone Ranger in winter clothing and helmet. Physical incapacitation means that a Ranger is physically unable to function in an assault within a five-minute period after an attack. A point of impact value of less than 0.1 percent can be interpreted as being less than or equal to one chance in one thousand.

3-7. Using echelonment of fires within the specified RED for a delivery system requires the unit to assume some risks. The maneuver commander determines by delivery system how close to the forces the fires are allowed to fall. Although the decision is made at this risk level, the commander relies heavily on the FSOs expertise.

TARGET OVERLAYS

3-8. Table 3-4 shows the contents of fire support overlay. The sterile fire support overlay is depicted in figure 3-1 on page 3-6. The nonsterile fire support overlay is shown in figure 3-2 on page 3-7. This includes—
- Index marks to position overlay on map
- Target symbols.

Table 3-4. Contents of fire support overlay

Unit and official capacity of person making overlay:	Routes, primary and alternate.
Date the overlay was prepared:	Phase lines and checkpoints used by the patrol.
Map sheet number:	Spares.
Effective period of overlay (day, time, group [DTG]):	Index marks to position overlay on map.
Priority target.	Objective.
Objective rally point (ORP) location.	Target symbols.
Call signs and frequencies (primary and alternate):	Description, location, and remarks column, complete.

Figure 3-1. Sterile fire support overlay

Chapter 3

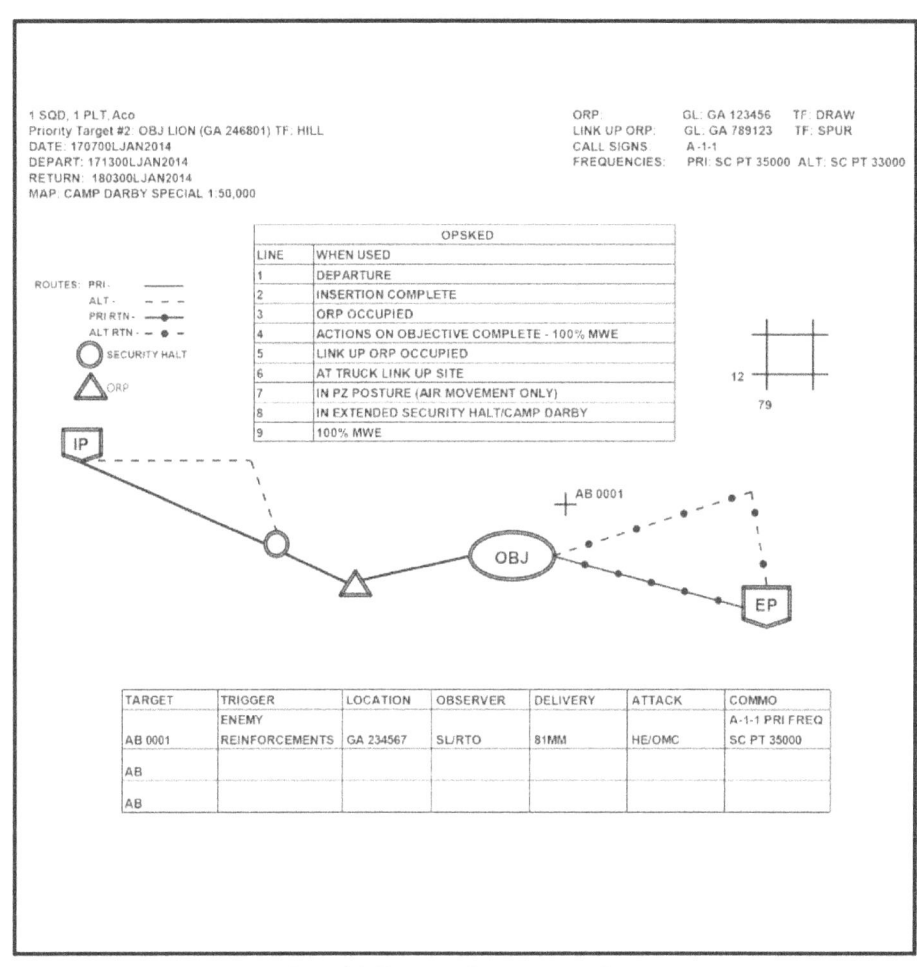

Figure 3-2. Nonsterile fire support overlay

LEGEND
ALT – alternative; COMMO - communication; EP – extraction point; HE/OMC – high explosive/on my command; IP – insertion point; MWE – men, weapons, and equipment; OBJ – objective; OPSKED – operation schedule; ORP – objective rally point; PLT – platoon; PZ – pickup zone; PRI – primary; RTN – return; RTO – radiotelephone operator; SC PT – single channel plain text; SL – squad leader; SQD – squad; TF – terrain feature.

NOTE: AB 0001 is a target reference number and GA GL is a Georgia grid zone designator. These are map coordinates.

Fire Support

3-9. **Target, trigger, location, observer, delivery system, attack guidance, and communications network (TTLODAC) checklist:** using one of these (see table 3-5) helps ensure the leader's fire support plan is complete. It is used to identify all aspects of individual targets before coordination and the OPORD.

Table 3-5. TTLODAC checklist

TARGET	NUMBER OR TYPE OF TARGET
Trigger	When to fire the target
Location	Minimum of six-digit grid
Observer	Primary and alternate
Delivery system	Mortar, artillery, air
Attack guidance	Ammo, special instructions
Communications net	Company TAC, Arty CLF
LEGEND Ammo – ammunition; Arty CLF – artillery coordinated fire line; net – network; TAC – tactical air controller	

CALL-FOR-FIRE

3-10. Definite steps are taken when making a call-for-fire, starting with call signs and the WARNORD. These steps are precise in order to preserve the safety of Rangers while accurately hitting the target.

3-11. An example of call-for-fire transmissions is shown in table 3-6 on page 3-11. Asterisks (*) below indicate required elements for a basic call-for-fire mission. These steps are:
   a. **Observer's identification. Call signs.***
   b. **Warning order:***
   - Type of mission:
     - Adjust fire.
     - Fire for effect.
     - Suppress.
     - Immediate suppression/immediate smoke.
   - Size of element to fire for effect. When observer does not specify size element to fire, battalion fire direction center (FDC) decides.
   c. **Method of target location:***
   - Polar plot.
   - Shift from a known point.
   - Grid.
   d. **Location of target:** *
   - **Grid coordinates.** Six digit, or if greater accuracy is required, eight digit.
   - **Shift from a known point.** Send observer target direction:
     - Mils (nearest 10).
     - Degrees.
     - Cardinal direction.
     - Send lateral shift, right/left, nearest 10 meters (m).

Chapter 3

- Send range shift, add/drop, nearest 100 m.
- Send vertical shift, up/down, nearest 5 m; use only if it exceeds 35 m.
  - **Polar plot:**
    - Send direction to nearest 10 mils.
    - Send distance to nearest 100 m.
    - Send vertical shift to nearest 5 m.

e. **Description of target:** *
  - Type.
  - Activity.
  - Number.
  - Degree of protection.
  - Size and shape (length, width, or radius).

f. **Method of engagement:**
  - **Type of adjustment.** When observer does not request a specific type of fire control adjustment, issue area fire:
    - Precision fire = point target.
    - Area fire = moving target.
  - **Danger close.** This condition exists when friendly troops are within—
    - 600 m for mortars.
    - 600 m for artillery.
    - 750 m for naval guns, five inches or smaller.
  - **Mark.** Used to orient observer or to indicate targets.
  - **Trajectory:**
    - Low angle (standard).
    - High angle (mortar fire, or if requested).
  - **Ammunition.** Use high explosive (HE) quick, unless specified by the observer:
    - Projectile (HE, illumination, improved conventional munitions [ICM], SMOKE).
    - Fuze (quick, timed, and other options).
    - Volume of fire (observer may request the number of rounds to be fired).
  - **Distribution:**
    - 100 m sheaf (standard).
    - Converged sheaf (used for small, hard targets).
    - Special sheaf (any length, width, and attitude).
    - Open sheaf (separate bursts).
    - Parallel sheaf (linear target).

g. **Method of fire and control:**
  - **Method of fire.** Specific guns and a specific interval between rounds. Normally adjust fire: one gun is used with a five-second interval between rounds.
    - **Method of control:**
      - AT MY COMMAND, FIRE. Remains in effect until observer orders, CANCEL AT MY COMMAND.
      - CANNOT OBSERVE. Observer cannot see the target.
      - TIME ON TARGET. Observer tells FDC when he wants the rounds to impact.
      - CONTINUOUS ILLUMINATION. If this was not already calculated by the FDC, the observer indicates interval between rounds in seconds.
      - COORDINATED ILLUMINATION. Observer tells FDC to set interval between ILLUM and HE shells.
      - CEASE LOADING.
      - CHECK FIRING. Halt immediately.

Fire Support

- CONTINUOUS FIRE. Load and fire as fast as possible.
- REPEAT. Fire another round(s), with or without adjustments.

h. **Correction of errors.** When the FDC has made an error when reading backfire support data, the observer announces, CORRECTION and transmits the correct data in its entirety.

i. **Message to observer:**
- Battery(ies) to fire for effect.
- Adjustment of battery.
- Changes to the initial call-for-fire.
- Number of rounds (per tube) to be fired for effect.
- Target numbers.
- Additional information:
    - Time of flight. Moving target mission.
    - Probable error in range. 38 m or greater (normal mission).
    - Angle "T" 500 mils or greater.

j. **Authentication.** Challenge and reply.

Chapter 3

Table 3-6. Example of call-for-fire transmissions

GRID MISSION	
Observer	Firing Unit
F24, THIS IS J42, ADJUST FIRE, OVER.	J42, THIS IS F24, ADJUST FIRE, OUT.
	GRID WM180513, DIRECTION 0530, OUT.
INFANTRY PLATOON DUG IN, OVER.	INFANTRY PLATOON DUG IN, OUT.
SHOT OUT.	SHOT OVER.
SPLASH OUT.	SPLASH, OVER.
END OF MISSION, 15 CASUALTIES, PLATOON DISPERSED, OVER.	END OF MISSION, 15 CASUALTIES, PLATOON DISPERSED, OUT.
SHIFT FROM KNOWN POINT	
Observer	Firing Unit
J42, THIS IS F24, ADJUST FIRE, SHIFT AB1001, OVER.	F24, THIS IS J42, ADJUST FIRE, SHIFT AB1001, OUT.
DIRECTION 2420, RIGHT 400, ADD 400, OVER.	DIRECTION 2420, RIGHT 400, ADD 400, OUT.
FIVE T-72 TANKS AT POL SITE, OVER.	FIVE T-72 TANKS AT POL SITE AUTHENTICATE JULIET NOVEMBER, OVER.
I AUTHENTICATE TANGO, OVER.	
SHOT OUT.	SHOT OVER.
END OF MISSION, TWO TANKS DESTROYED, THREE IN WOODLINE, OVER.	END OF MISSION, 15 CASUALTIES, PLATOON DISPERSED, OUT.
POLAR	
Observer	Firing Unit
J42, THIS IS F24, ADJUST FIRE, POLAR, OVER.	F24, THIS IS J42, ADJUST FIRE, POLAR, OUT.
DIRECTION 2300, DISTANCE 4000, OVER.	DIRECTION 2300, DISANCE 4000, OUT.
INFANTRY PLATOON DUG IN, OVER.	INFANTRY PLATOON DUG IN, OUT.
SHOT OUT.	SHOT OVER.
SPLASH OUT.	SPLASH OVER.
END OF MISSION, 15 CASUALTIES, PLATOON DISPERSED, OVER.	END OF MISSION, 15 CASUALTIES, PLATOON DISPERSED, OUT.

Fire Support

CLOSE AIR SUPPORT

3-12. There are two types of CAS requests: planned and immediate. Planned requests are processed by the Army chain to corps for approval. Immediate requests are initiated at any level and processed by the battalion operations officer (S-3), FSO, and air liaison officer.

3-13. Figure 3-3 is an example of calling for CAS. Close air support capabilities are detailed in table 3-7 on pages 3-14 and 3-15, and table 3-8 on page 3-16. Table 3-7 details fixed-wing CAS capabilities, and table 3-8 details Unmanned Aircraft System (UAS) CAS capabilities. The following is the format for requesting immediate CAS:
- Observer identification.
- Warning order (request close air).
- Target description. At a minimum, this includes type and number of targets, activity or movement, and point or area targets.
- Target location (grid) should include elevation.
- Desired time on target (TOT).
- Desired effects on target.
- Final control.
- Remarks; report—
  - Friendly locations.
  - Wind direction and hazards.
  - Threats such as air defense artillery (ADA) or small arms.

Chapter 3

```
1. INITIAL POINT (IP): _____NP459854 (or) X-Ray_____
2. HEADING (IP TO TARGET: _____069_____ MAGNETIC
                         (OFFST: LEFT/RIGHT)
3. DISTANCE (IP TO TARGET): _____9.8_____ (NAUTICAL MILES)
4. TARGET ELEVATION: _____1140_____ (FEET ABOVE MEAN SEA LEVEL)
5. TARGET DESCRIPTION: _____5 tanks attacking west_____
6. TARGET LOCATION: _____NP675920_____
                    (UTM, LAT/LONG, VISUAL REFERENCES, AND OTHER DATA)
7. TYPE OF MARK: _Laser_  CODE: _372_
8. LOCATION OF FRIENDLIES: _____1000m SW of target_____
9. EGRESS: _____NW to avoid artillery suppression_____
   (REMARKS) _____
   _____
   TIME ON TARGET (TOT): _____
   TIME TO TARGET (TTT) STANDBY:_____PLUS: _____
                                (MINUTES)        (SECONDS)
OMIT DATA NOT REQUIRED.
LINE NUMBERS ARE NOT TRANSMITTED.
ALL UNITS OF MEASURE ARE STANDARD.
SPECIFY IF OTHER UNITS OF MEASURE ARE BEING USED.
```

LEGEND
IP – initial point; LAT – latitude/longitude; NW – northwest; SW – southwest; TOT – time on target; TTT – time to target; UTM – Universal Transverse Mercator

Figure 3-3. Example of a close air support request

Fire Support

Table 3-7. Fixed-wing close air support capabilities

\multicolumn{7}{c}{FIXED-WING CAPABILITIES AND COMMUNICATION EQUIPMENT}
Aircraft	Ordnance	Sensors and Markings	Datalink	Frequency Band	Frequency Hopping	COMSEC
F-15E Strike Eagle	JDAM, LJDAM, LGB, CBU/WCMD, AGM-130/158, JSOW, 20-mm cannon	Link 16	FLIR, SAR, Sniper, LITENING, Terrain Following Radar	VHF AM/FM, UHF, SATCOM	HQ II, SINCGARS	VINSON
AV-8B Harrier II	LGB, AGM-65-E, GP bombs, CBU, JDAM, LJDAM, 2.75" rockets, 5" Zuni	LITENING, CCD TV, FLIR, SAR, LUU-2/19	None	UHF, VHF-AM/FM	HQ II, SINCGARS	VINSON
A-10C THUNDERBOLT	LGB, AGM-65, GP bombs, CBU/WCMD, JDAM, 2.75" rockets, 30-mm cannon	LITENING, Sniper Pave Penny, Quickdraw, IZLID, LUU-2/19	SADL, VMF	UHFx2, VHF-AM/FM, SATCOM	HQ II, SINCGARS	VINSON
EC-130 Compass Call	Electronic attack	Link 16	SIGINT	3xUHF, 1xVHF, 2xHF, 2xSATCOM, (AITG terminal also configurable to additional UHF, VHF, FM, SINCGARS, HQ, DAMA), IRC, NCCT chat, STE phone	HQ II	

Chapter 3

Table 3-7. Fixed-wing close air support capabilities (continued)

\multicolumn{7}{c}{FIXED-WING CAPABILITIES AND COMMUNICATION EQUIPMENT}						
Aircraft	Ordnance	Sensors and Marking	Datalink	Frequency Band	Frequency Hopping	COMSEC
AC-130J Ghostrider	105-mm Howitzer, 40-mm cannon	IDS, PLS, LLL TV, beacon tracking radar, IZLID, ATI, HIBEAM, LTD	Link 16	UHFx2 SATCOM, HFx2, VHF-AM/FMx3	HQ II, No, No, SINCGARS	VINSON ANDVT
B-1B Lancer	JDAM (with pattern capability), GP bombs, CBU/WCMD, JASSM	DCI/JRE	SAR, Sniper, GMTI/T, PSS-SOF	UHF, VHF SATCOM Voice, HF	HQ II, SINCGARS	VINSON

LEGEND
AGM – air-to-ground missile; AITG – airborne integrated terminal group; ALLTV - all light level television; AM – amplitude modulation; ANDVT – advanced narrowband digital voice terminal; ANW2 – adaptive networking wideband waveform; AT – advanced targeting; ATI – ambient temperature illuminator; ATFLIR – advanced targeting forward looking infrared; CBU – cluster bomb unit; CCD – charge-coupled device; COMSEC – communications security; DAMA – demand assigned multiple access; DCI – digital communications improvement; DMS – dual mode seeker-equipped; EGBU – enhanced guided bomb unit; EO – electro-optical; FLIR – forward looking infrared; FM – frequency modulation; GMTI – ground moving target indicator; GMTI/I – ground moving target indicator and tracking; GP – general purpose; HPW – high performance waveform; HARM – high-speed antiradiation missile; HF – high frequency; HIBEAM – high beam; HMS – helmet mounted sight; HTS – HARM Targeting System; IDS – infrared detection set; IR – infrared; IRC – internet relay chat; IZLID – infrared zoom laser illuminator designator; JASSM – joint air to surface standoff missile; JDAM – joint direct attack munition; JHMCS – joint helmet mounted cueing system; JRE – joint range extension; JSOW – joint standoff weapon; LIA – laser illuminator assembly; LJDAM – laser JDAM; LGB – laser guided bomb; LLLTV – low light level television; LUU – illuminating unit; LTD – laser target designator; mm – millimeter; MTV – mobile tactical video; NCCT – network centric collaborative targeting; PLS – personnel locator system; PSS-SOF – Precision Strike Unit – Special Operations Forces; SADL – situation awareness data link; SAR – synthetic aperture radar; SATCOM – satellite communications; SLAM (ER) – standoff land attack missile (expanded response); SOPGM – standoff precision guided munitions; SIGINT – signals intelligence; SINCGARS – single-channel ground-air radio system; SAR – synthetic aperture radar; STE – secure transmission equipment; TV – television; UHF – ultra high frequency; VHF – very high frequency; VMF – variable message format; VULOS – VHF/UHF/line of sight; WCMD – wind corrected munition dispenser.
NOTE: HQ – HAVEQUICK is a frequency-hopping system that protects military UHF radio traffic; LITENING is a multisensor targeting and surveillance system; VINSON is an encrypted UHF/VHF communication system

Fire Support

Table 3-8. Unmanned aircraft systems close air support capabilities

UAS CAPABILITIES AND COMMUNICATION EQUIPMENT						
Aircraft	Ordnance	Sensors and Marking	DATALINK	Frequency Band	Frequency Hopping	COMSEC
Hunter	None	MOSP (EO, IR, ELRF, LD), CRP	C-Band (LOS)	UHF/VHF		
MQ-1B Predator	Hellfire	MTS, LTM, LTD	Link 16, PACWIND	UHF, VHF-AM/FM, SATCOM, VDL	HQ II, SINCGARS	
MQ-9 Reaper	Hellfire, GBU-12, GBU-38	MTS, SAR, GMTI, LTM, LTD	Link 16	UHF, VHF-AM/FM, SATCOM, VDL	HQ II, SINCGARS	VINSON
MQ-1C Gray Eagle	Hellfire	CSP (EO, IR, ELRF/LD), CRP, SAR/GMTI	TCDL (Ku band), SATCOM	VHF/UHF, SATCOM	HQ II	
RQ-7B Shadow	None	IR, EO, IRLP, LRF/D, LTD		VHF FM, UHF2, UHF2 SATCOM2 (CRP is VHF FM only)	HQ II, SINCGARS	VINSON
RQ-11B Raven	None	IR, EO, IR Pointer	None	None	None	None

LEGEND
AM – amplitude modulation; C-Band (LOS) – communications data link, line of sight; COMSEC – communications security; CRP – communications relay payload; CSP – common sensor payload; ELRF – eyesafe laser range finder; EO- electro-optical; FM – frequency modulation; GBU – guided bomb unit; GMTI - ground moving target indicator; IR – infrared; LD – laser designator; LOS – line of sight; LTD – laser target designator; LTM – laser target marker; MOSP – multi-optronic stabilized payload; MTS – multispectral targeting; SAR – synthetic aperture radar; SATCOM – satellite communications; SINCGARS – single-channel ground-air radio system; TCDL – tactical common data link; UAS - Unmanned Aircraft System; UHF – ultra high frequency; VDL – video downlink; VHF – very high frequency

NOTE: HQ – HAVEQUICK is a frequency-hopping system that protects military UHF radio traffic; MQ and RQ are the designation for UASs; PACWIND is an LOS full-motion video signal; VINSON is an encrypted UHF/VHF communication system

Chapter 3

ARMY ATTACK AVIATION

3-14. Army attack aviation is defined as a hasty or deliberate attack in support of units engaged in close combat. During an attack, armed helicopters engage enemy units with direct fire that impacts nearby friendly forces. Targets may range from a few hundred meters to a few thousand meters. Army attack aviation is coordinated and directed by a team, platoon, or company-level ground unit Soldiers using standardized Army attack aviation procedures in unit SOPs. (See table 3-9.)

3-15. During the planning process the team, platoon, or company-level leadership is responsible for ensuring enough time is allowed to conduct rehearsals between the ground unit and the aviation unit in order to ensure all participants in the mission have situational awareness of the plan, routes, capabilities, and limitations of what each unit can provide and support. Table 3-10 on page 3-18 details rotary wing CAS capabilities.

Table 3-9. Army attack aviation call-for-fire format

ARMY ATTACK AVIATION CALL-FOR-FIRE FORMAT
1. Observer and warning order: "*LONGBOW 6*, THIS IS *OBSERVER 2*, FIRE MISSION, OVER."   (aircraft call sign)   (observer call sign)
2. Friendly location and mark: "MY POSITION *NP359654*, MARKED BY *STROBE*."   (TRP, grid, other)   (strobe, beacon, IR strobe, other)
3. Target location: TARGET LOCATION *NP459854*"   (bearing [magnetic] and range [meters], TRP, grid, other)
4. Target description and mark: "*3 TANKS*, MARKED BY *INFRARED POINTER*"   (target description)   (infrared pointer, tracer, other)
5. Remarks: "*NONE*."   (threats, danger close clearance, restriction, at my command, other)
NOTES: 1. Clearance. If airspace has been cleared between the employing aircraft and the target, transmission of this brief *is* clearance to fire unless "DANGER CLOSE" or "AT MY COMMAND" is stated."   2. Danger close. For danger close fire, the observer or commander must accept responsibility for increased risk. State "CLEARED DANGER CLOSE" in line 5 and pass the initials of the on-scene ground commander. This clearance may be preplanned.   3. At my command. For positive control of the aircraft, state "AT MY COMMAND" on line 5. The aircraft will call "READY TO FIRE," when ready.
LEGEND   IR – infrared; TRP – target reference point

Table 3-10. Rotary wing close air support capabilities

ROTARY WING CAPABILITIES AND COMMUNICATION EQUIPMENT						
Aircraft	Ordnance	Sensors and Marking	Datalink	Frequency Band	Frequency Hopping	COMSEC
AH-6	7.62 MG, .50-caliber MG, 2.75" rockets	FLIR, IR Pointer		VHF-AM/FM2, UHF-AM/FM, SATCOM	HQ II, SINCGARS	VINSON, ANDVT
AH-64D/E	Hellfire (laser or RF), 2.75" rockets, 30-mm cannon	FLIR (LTD3), MMW, radar, DTV, IZLID		UHF-AM, VHF-FMX2, VHF-AM, SATCOM, VMF/BFT	HQ II, SINCGARS	VINSON

LEGEND
AH – attack helicopter; AM – amplitude modulation; ANDVT - advanced narrowband digital voice terminal; COMSEC – communications security; DTV – day television; FLIR – forward looking infrared; FM – frequency modulation; IR – infrared; IZLID – infrared zoom laser illuminator designator; LTD – laser target designator; MG – machine gun; mm - millimeter; MMW – millimeter wave; RF – radio frequency; SATCOM – satellite communications; SINCGARS - single-channel ground-air radio system; UHF – ultra high frequency; VHF – very high frequency

NOTE: HQ – HAVEQUICK is a frequency-hopping system that protects military UHF radio traffic; VINSON is an encrypted UHF/VHF communication system

This page intentionally left blank.

Chapter 4

Communications

The basic requirement of combat communications is to provide rapid, reliable, and secure interchange of information. Communications are vital to mission success. This chapter helps the Ranger squad and platoon maintain effective communications and correct any radio antenna problems. It also discusses military radio communications equipment and automated ANCDs.

EQUIPMENT

4-1. This section discusses military radio communications equipment and automated net control devices (ANCDs). Each military radio has a receiver and transmitter. Rangers use several different types of radios. (See table 4-1 on pages 4-1 through 4-3.) Radios vary from high frequency (HF), very high frequency (VHF), ultra high frequency (UHF), and tactical satellite (TACSAT).

4-2. Knowing what each radio can do is crucial in planning and requesting the most reliable and effective communications equipment for a particular mission. Military operations use four primary frequency ranges. (See table 4-2 on page 4-4.)

Table 4-1. Military radios

CHARACTERISTICS	AN/PRC-152	AN/PRC-148	AN/PRC-119F
Description	Multiband, hand-held receiver and transmitter	Multiband interteam or intrateam radio	Multiband, multimission manpack
Frequency Range			
HF			
VHF low	Yes	Yes	Yes (up to 5W)
VHF high	Yes	Yes	
UHF	Yes	Yes	
TACSAT	Yes	Yes (up to 5W)	
Power output	Up to 10W TACSAT	Up to 5W	Up to 10W
Battery Requirements	Rechargeable lithium-ion (included with the radio)	Rechargeable lithium-ion (included with the radio)	Any one of these: BB-390 BB-2590 BB-590 BB-5590
Scanning	10 user-programmed nets (TACSAT or LOS frequencies)	10 user-programmed nets (TACSAT or LOS frequencies)	Four channels in FM mode

Chapter 4

Table 4-1. Military radios (continued)

CHARACTERISTICS	AN/PRC-152	AN/PRC-148	AN/PRC-119F
Data Transmission			
LOS AM/FM	Yes	Yes	Yes
Commercial	Yes	Yes	
Optional internal	Yes		Yes
TACSAT		Yes	
DAGR	Yes	Yes	
TACSAT		Yes	
PLGR	Yes	Yes	
NMEA-183	Yes	Yes	
Internal	Yes	Yes	
Optional Internal			
Dimensions and Weight	2.9 x 9.6 x 2.5 inches with battery attached. 2.6 lbs. with internal GPS and battery.	2.7 x 7.8 x 1.5 inches. 2.2 lbs. with battery.	3.4 x 5.3 x 10.2 inches. 7.7 lbs. without battery.
Disadvantages	Lower power output than AN/PRC-117F(c).	Lower power output than AN/PRC-117F(c) and AN/PRC-152.	Lower power output than AN/PRC-117F(c). Limited frequency range. Cannot communicate with USAF aircraft.
Encryption Types:			
ANDVT	Yes	Yes	
Vinson	Yes	Yes	Yes
KG-84	Yes		
Fascinator	Yes	Yes	
Method:			
SINCGARS	Yes	Yes	Yes
HAVE QUICK I	Yes	Yes	
Serial Tone			
ECCM	Yes	Yes	
Frequency Hop	Yes	Yes	Yes

Communications

Table 4-1. Military radios (continued)

CHARACTERISTICS	AN/PRC-152	AN/PRC-148	AN/PRC-119F
Fill Devices			
KYK-13	Yes	Yes	Yes
KOI-18	Yes	Yes	Yes
KYX-15	Yes	Yes	Yes
SKL	Yes	Yes	Yes
ANCD, AN/CYZ-10	Yes	Yes	Yes
SKL, AN/PYQ-10	Yes	Yes	Yes
Immersion Depth	Twenty meters	Two to twenty meters	One meter
SA Reporting Capability	Yes	Yes with remote control unit (RCU)	If equipped with optional internal GPS
Special Features	Ship-to-shore, ground-to-ground, air-to-ground capable.	Ship-to-shore, ground-to-ground, air-to-ground capable.	Does not apply.
Terrain Restrictions	LOS: open to slightly rolling terrain. TACSAT: any terrain.	LOS: open to slightly rolling terrain. TACSAT: any terrain.	LOS: open to slightly rolling terrain.

LEGEND
AM - amplitude modulation; ANCD - automated net control device; ANDVT - advanced narrowband digital voice terminal; DAGR - defense advanced GPS receiver; ECCM - electronic counter-countermeasures; FM - frequency modulation; GPS - Global Positioning System; HF - high frequency; lbs. - pounds; LOS - line of sight; PLGR - precision lightweight GPS receiver; RCU - remote control unit; SA – situational awareness; SINCGARS - Single Channel Ground and Airborne Radio System; SKL - simple key loader; TACSAT - tactical satellite; UHF - ultra high frequency; USAF - United States Air Force; VHF - very high frequency
NOTE: VINSON is an encrypted UHF/VHF communication system

Chapter 4

Table 4-2. Frequency ranges

HIGH FREQUENCY	VERY HIGH FREQUENCY, LOW	VERY HIGH FREQUENCY	ULTRA HIGH FREQUENCY
1.6 to 29.999 MHz	30.000 to 89.999 MHz	90.000 to 224.999 MHz	225.000 to 512.000 MHZ
Long range LOS[1]	LOS[2]	LOS[3]	LOS

[1] Long range LOS; capable of round-the-world communication due to longer physical wavelengths, which cause HF transmissions to bounce off the terrain and reflect off Earth's ionosphere, unlike VHF and UHF transmissions that are absorbed into the ionosphere. This keeps the frequency essentially trapped between the ground and Earth's atmosphere. For this volatile capability to offer effective communications, several factors must be ideal.
[2] LOS frequencies are limited to direct LOS for maximum effectiveness. Curvature of the earth, mountainous terrain, and dense vegetation degrade LOS radio maximum range capabilities.
[3] Modern military communications rely on UHF SATCOM or dedicated TACSAT communication for round-the-world real-time secure voice and data communication.

LEGEND
HF - high frequency; LOS - line of sight; MHz - megahertz; SATCOM - satellite communication; TACSAT - tactical satellite; UHF - ultra high frequency; VHF - very high frequency

MAN-PACK RADIO ASSEMBLY (AN/PRC-119F)

4-3. To assemble a man-pack radio, first check and install a battery. Once that is done, go on to position the antenna, set up the handset, set presets and frequencies, and scan. This is achieved by following these steps:
    a. Inspect the battery box for dirt or damage.
    b. Stand radio on its side with the battery cover facing up.
    c. Check battery life condition (rechargeable BB 390 batteries).
    d. Place battery in box.
    e. Close and latch the battery cover.
    f. Return radio to upright position.
    g. If a used battery is installed, enter the battery life condition into the radio.
- Set function (FCTN) to load (LD).
- Press battery (BAT), and then clear (CLR).
- Enter number recorded on side of battery.
- Press store (STO).
- Set FCTN to squelch on (SQ ON).

Communications

h. Inspect and position the antenna.
- Inspect whip antenna connector on antenna and on radio for damage.
- Screw whip antenna into base.
- Hand-tighten.
- Carefully mate antenna base with radio transmitter antenna (RT ANT) connector.
- Hand tighten.
- Position antenna as needed by bending gooseneck.

Note: Keep the antenna straight, if possible. If the antenna is bent to a horizontal position, the radio may have to be turned before receiving and transmitting messages.

i. Set up the handset.
- Inspect the handset for damage.
- Push handset on audio/data (AUD/DATA) and twist clockwise to lock in place.

j. Pack.
- Place radio transmitter (RT) in field pack with the antenna on the left shoulder.
- Fold top flap of field over RT and secure flap to field pack using straps and buckles.

k. Set presets.
- Channel (CHAN): 1.
- MODE: single channel (SC).
- Radio frequency power (RF PWR): high (HI).
- Volume (VOL): mid-range.
- Dimensions (DIM): full clockwise.
- FCTN: LD.
- DATA RATE: OFF.

l. Single-channel loading frequencies.
- Obtain Ranger signal operating instructions (SOI).
- Set FCTN: LD.
- Set mode: SC.
- Set CHAN: manual cue (MAN, CUE), or set channel (one to six) where the frequency is stored.
- Press frequency (FREQ): display shows 000001, or frequency RT is currently turned on.
- Press clear (CLR): display shows five lines.
- Enter the number of the new frequency. If a mistake is made with a number, press CLR.
- Press STO: display blinks.
- Set FCTN: SQ ON.

m. Clearing of frequencies.
- Set MODE: SC.
- Set CHAN: MAN, CUE, or set channel (one to six) where the frequency is to be stored.
- Press FREQ.
- Press CLR.
- Press LOAD, STO.
- Set FCTN: SQ ON.

Chapter 4

n. Scanning of multiple frequencies:
- Load all desired frequencies using "single channel loading frequencies" instructions.
- Set CHAN: CUE.
- Set SC: frequency hop (FH).
- Set FCTN: SQ ON.
- Press STO: (display says SCAN).
- Press the number eight: more than one frequency can now be scanned.

Basic Troubleshooting

4-4. Basic troubleshooting skills are needed to correct the simple communications problems that occur during a mission. Being able to troubleshoot quickly can make the difference between successful accomplishment of the mission and mission failure. This includes:

  a. **Check radio settings:**
  - Radio frequency: load proper frequency.
  - Power output: set to HIGH power.
  - Time if using frequency hop: reset time.
  - Cryptographic (crypto) fill if using cipher text: reload cryptographic fill from the automated net control device (ANCD).
  - Control knob: ensure radio is in ON position.

  b. **Check radio assembly and battery:**
  - Check antenna fitting: attach long whip or field expedient antenna.
  - Check hand-mike fitting: ensure contacts are clean and fitting is properly secured to radio.
  - Check battery: install fresh battery.

  c. With line of sight (LOS) radios, moving to higher ground may be necessary in order to make radio contact, especially in densely vegetated or uneven terrain.

ANTENNAS

4-5. This section discusses repair techniques, construction and adjustment, field expedient antennas, antenna length and orientation, and improvement of marginal communications. Antennas are sometimes broken or damaged, causing communications degradation or failure. If a spare antenna is available, replace the bad one.

4-6. If there is no spare, the squad or platoon might have to construct an emergency antenna. The following information suggests some ways to repair antennas and antenna supports, and construction and adjustment of emergency antennas.

RADIO TRANSMITTER

Serious injury or death can result from contact with the radiating antenna of a medium-power or high-power transmitter.

TURN OFF the transmitter while adjusting the antenna.

Communications

WHIP AND WIRE ANTENNAS

4-7. When a whip antenna breaks in two, connect the broken part to the part attached to the base by joining the sections. To restore the antenna to its original length, add a piece of wire that is nearly the same length as the missing part of the whip. Lash the pole support securely to both sections of the antenna. Before connecting the two antenna sections to the pole support, thoroughly clean them to ensure good contact. If possible, solder the connections.

4-8. Emergency repair of a wire antenna may involve the repair or replacement of the wire used as the antenna or transmission line; or the repair or replacement of the assembly used to support the antenna. When one or more wires of an antenna are broken, the antenna can be repaired by reconnecting the broken wires. To do this, lower the antenna to the ground, clean the ends of the wires, and twist the wires together. Whenever possible, solder the connection.

4-9. If the antenna receives damage beyond repair, construct a new one. Make sure that the length of the substitute antenna wires are the same length as those of the original. Antenna supports may also require repair or replacement. Anything can be used as a substitute for the damaged support, provided it is insulated and strong enough.

4-10. If the radiating element is not properly insulated, then field antennas can short to the ground and no longer work. Many common items make good field expedient insulators. The best are plastic or glass. Plastic spoons, buttons, bottle necks, and plastic bags make good insulators. Although wood and rope are less effective insulators, they are better than nothing. The radiating element (the antenna wire) should only touch this supporting (nonconductive) insulator and the antenna terminal. It should remain physically separated from everything else.

CONSTRUCTION AND ADJUSTMENT

4-11. There are specific methods to construct and adjust antennas. The best wire for antennas is copper or aluminum. However, in an emergency, use any wire that can be found. The exact length of most antennas is critical. Make sure that the emergency antenna is the same length as the original antenna.

4-12. Antennas can usually survive heavy windstorms if supported by a tree trunk or strong branch. To keep the antenna tight and from breaking or stretching when the trees sway, attach a spring or old inner tube to one end of the antenna. Another technique is to pass a rope through a pulley or eyehook. Attach the rope to the end of the antenna, and heavily weight the rope to keep the antenna tight. To ensure the rope or wire guidelines do not interfere with the operation of the antenna, cut the wire into several short lengths and connect the pieces with insulators.

4-13. An improvised antenna may change the performance of a radio set. A distant station may be used to test the antenna. If the signal received from this station is strong, the antenna is operating satisfactorily. If the signal is weak, adjust the height and length of the antenna and the transmission line to receive the strongest signal at a given setting on the volume control of the receiver. This is the best method of tuning an antenna when transmission is dangerous or forbidden.

4-14. In some radio sets, use the transmitter to adjust the antenna. First, set the controls of the transmitter to normal. Then, tune the system by adjusting the antenna height, the antenna length, and the transmission line length to obtain the best transmission output.

Expedient 292-Type Antenna

4-15. Developed for use in the jungle, when properly used, these antennas can improve communications. Their weight and bulk render them impractical for most squad or platoon operations. However, the unit can carry just the masthead and antenna sections and mount them onto wood poles or trees. An expedient version can be constructed using any insulated wire and other available material. For example, almost any plastic, glass, or rubber items can serve as insulators. If these are unavailable, dry wood can work.

Chapter 4

4-16. At the radio set, remove about one inch of insulation from each end of the wire. Connect the ends to the positive side of the cobra head connector. Be sure the connections are tight or secure. (See figure 4-1, and figure 4-2 on page 4-10.) Set up the correct frequency (see table 4-3 on page 4-11), turn on the set, and proceed with communications.

4-17. Use the planning considerations discussed in the next paragraph to determine the length of the elements (one radiating wire and three ground plane wires) for the desired frequency. (See figure 4-3 on page 4-12 for an expedient version set up.) Cut these elements (A) from Claymore mine wire, or similar wire. The heavier the gauge, the better, but insulated copper core wire works best. Cut spacing sticks (B) the same length as the ground plane wires. Place the sticks in a triangle and tie their ends together with wire, tape, or rope. Attach an insulator (C) to each corner and one end of each ground-plane wire to each insulator. Bring the loose ends of the ground-plane wires together, attach them to an insulator (C), and tie securely. Strip about three inches of insulation from each wire and twist them together.

4-18. Tie one end of the radiating element wire to the other side of insulator, and the other end to another insulator (B). Strip about three inches of insulation from the radiating element (C). Cut enough wire to reach from the proposed location of the antenna to the radio set. Keep this line as short as possible because excess length reduces the efficiency of the system. Tie a knot at each end to identify it as the "hot" lead. Remove insulation from the "hot" wire and tie it to the radiating element wire at insulator (C). Remove insulation from the other wire and attach it to the bare ground-plane element wires at insulator (C). Tape all connections and do not allow the radiating element wire to touch the ground plane wires.

4-19. Attach a rope to the insulator on the free end of the radiating element and toss the rope over the branches of a tree. Pull the antenna as high as possible, keeping the lead-in routed down through the triangle. Secure the rope to hold the antenna in place.

Communications

Figure 4-1. Cobra head

Chapter 4

Figure 4-2. Antenna base

Table 4-3. Quick reference table

OPERATING FREQUENCY IN MEGAHERTZ (MHz)	ELEMENT LENGTH (RADIATING ELEMENTS AND GROUND-PLANE ELEMENTS)
30	2.38 meters (7 feet, 10 inches)
32	2.23 meters (7 feet, 4 inches)
34	2.1 meters (6 feet, 11 inches)
36	1.98 meters (6 feet, 6 inches)
38	1.87 meters (6 feet, 2 inches)
40	1.78 meters (5 feet, 10 inches)
43	1.66 meters (5 feet, 5 inches)
46	1.55 meters (5 feet, 1 inch)
49	1.46 meters (4 feet, 9 inches)
52	1.37 meters (4 feet, 6 inches)
55	1.3 meters (4 feet, 3 inches)
58	1.23 meters (4 feet, 0 inches)
61	1.17 meters (3 feet, 10 inches)
64	1.12 meters (3 feet, 8 inches)
68	1.05 meters (3 feet, 5 inches)
72	0.99 meters (3 feet, 3 inches)
76	0.94 meters (3 feet, 1 inch)

Chapter 4

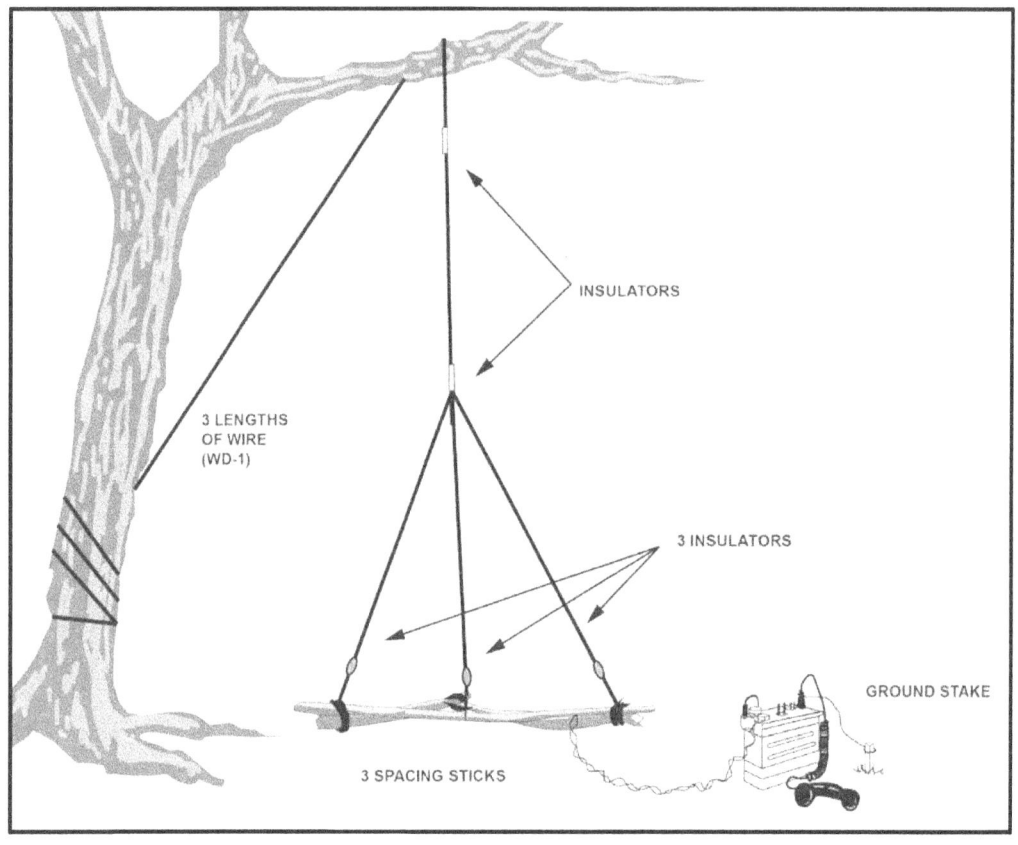

Figure 4-3. Completed expedient 292-type antenna

LEGEND
WD-1 is a type of military wire.

Communications

ANTENNA LENGTH PLANNING CONSIDERATIONS

4-20. The length of an antenna is considered in the construction of field expedients. At a minimum, a quarter of the frequency wavelength should be used as the length of the field expedient antenna. Another important factor in LOS communications is the height of the antenna in relation to the receiving station. The higher the antenna, the greater the range the radio transmission has.

4-21. Terrain and curvature of the earth affect LOS communication by absorbing VHF and UHF communications into the earth's surface. This is overcome by increasing antenna height, power output, and radio frequency. Since radio frequencies are predesignated and power output is limited to the capabilities of the radio set, antenna length and height are the two variables that can be manipulated to increase radio communication range. Using the following formulas, it is possible to plan for the use of field-expedient antennas, determine the best location to gain and maintain communication, and plan for communication windows, as necessary.

4-22. To calculate the physical length of an antenna in feet, use the following equation. It gives the antenna length in feet for a one-quarter wavelength of the frequency. To determine the antenna length in feet for a full wavelength antenna, multiply the antenna length by four.

X = 234/ freq
(X = the length of the antenna in feet; freq = the radio frequency used)
   EXAMPLE:
       234/ 38.950 = 6.01 feet (quarter-wavelength antenna)
       6.01 feet x 2 = 12.02 feet (half-wavelength antenna)
       6.01 feet x 4 = 24.04 feet (full-wavelength antenna)
LEGEND
freq - frequency

4-23. Curvature of the earth allows a person standing 5 feet, 7 inches tall looking across a flat surface to see objects approximately 4.7 kilometers (km) in the distance. (See figure 4-4.) Anything beyond this is below the horizon (dead space). To overcome this, the person moves to a higher elevation in order to see beyond 4.7 kilometers.

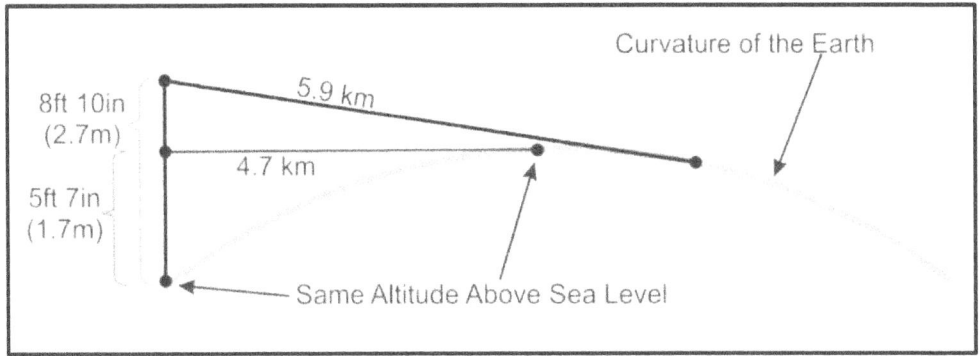

Figure 4-4. Curvature of the Earth

LEGEND
ft – feet; in – inch; km – kilometer; m – meter

Chapter 4

4-24. LOS communication is subject to this same principle. Use the following formula to calculate the required antenna height for a given distance (keep in mind that when on low ground such as a valley, draw, or depression, the height of the antenna is greater; when on high ground the antenna height may be shorter). Use the following formula to compute height of antenna to compensate for curvature of the Earth:

> X = 234/frequency (freq)
> (X = the length of the antenna in feet; freq = the radio frequency used)
> Distance in km (dkm) from receiving station = square root of √ (12.7 x Am) where Am is the antenna height in meters.
>
> *EXAMPLE:*
>
> Known height: dkm = √(12.7 x Am), dkm = √(12.7 x 1.7m) or
> Unknown height: Am = 0.07874 x (dkm)^2, Am = 0.07874 x (4.7km)^2
>
> LEGEND
> Am – antenna height in meters; dkm – distance in km; freq – frequency; km - kilometer

Chapter 5

Demolitions

This chapter introduces Rangers to the characteristics of explosives (low and high), initiation systems, modernized demolition initiator (MDI) components, detonation systems, safety considerations, expedient explosives, breaching charges, and timber-cutting charges. (Refer to TM 3-34.82 for more information.)

5-1. Low explosives have a detonating velocity up to 1300 feet per second, which produces a pushing or shoving effect. High explosives have a detonating velocity of 3280 to 27,888 feet per second, which produces a shattering effect. (See table 5-1 on pages 5-1 and 5-2.)

Table 5-1. Characteristics of U.S. demolitions explosives

NAME	APPLICATIONS	DETONATION VELOCITY MIN/SEC	FT/SEC	REFACTOR*	FUME TOXICITY	WATER RESISTANCE
Ammonium nitrate	Cratering charge	2700	8800	0.42	Dangerous	Poor
PETN	Det cord blasting caps demolition charges	8300	27,200	1.66	Slightly dangerous	Excellent
RDX	Blasting caps composition explosive	8350	27,400	1.60	Dangerous	Excellent
Trinitrotoluene (TNT)	Demolition charge composition explosive	6900	22,600	1.00	Dangerous	Excellent
Tetryl	Booster charge composition explosive	7100	23,300	1.25	Dangerous	Excellent
Nitroglycerin	Commercial dynamite	7700	25,200	1.50	Dangerous	Good
Black powder	Time fuze	400	1300	0.55	Dangerous	Poor
Amatol 80/20	Bursting charge	4900	16,000	1.17	Dangerous	Poor
Composition A3	Booster charge Bursting charge	8100	26,500	---	Dangerous	Good
Composition B	Bursting charge	7800	25,600	1.35	Dangerous	Excellent
Composition C4 (M112)	Cutting and breaching charges	8040	26,400	1.34	Slightly dangerous	Excellent
Composition H6	Cratering charge	7190	23,600	1.33	Dangerous	Excellent
Tetrytol 75/25	Demolition charge	7000	23,000	1.20	Dangerous	Excellent

Chapter 5

Table 5-1. Characteristics of U.S. demolitions explosives (continued)

NAME	APPLICATIONS	DETONATION VELOCITY		REFACTOR*	FUME TOXICITY	WATER RESISTANCE
		MIN/SEC	FT/SEC			
Pentolite 50/50	Booster and bursting charges	7450	24,400	---	Dangerous	Excellent
M1 Dynamite	Demolition charge	6100	20,000	0.92	Dangerous	Fair
Det cord	Priming demolition charge	6100 to 7300	20,000 to 24,000	---	Slightly dangerous	Excellent
Sheet explosive M118 and M186	Cutting charge	7300	24,000	1.14	Dangerous	Excellent
Bangalore torpedo M1A2	Demolition charge	7800	25,600	1.17	Dangerous	Excellent
Shaped charges M2A3, M2A4, and M3A1	Cutting charge	7800	25,600	1.17	Dangerous	Excellent
LEGEND Det cord – detonating cord; Ft/Sec – feet per second; Min/Sec – minutes to seconds; PETN – pentaerytritol tetranitrate; RDX - hexahydro-trinitro-triazine (explosive/propellant); TNT – trinitrotoluene. *TNT = 1.00 relative effectiveness						

INITIATING (PRIMING) SYSTEMS

5-2. The best way to prime demolition systems is with MDIs. These are blasting caps attached to various lengths of time fuze or shock tube. They can be used with a fuze igniter and detonating cord to create many firing systems. In the absence of MDI, field expedient methods may be used.

5-3. A shock tube is a thin, plastic tube of extruded polymer with a layer of special explosive material on the interior surface. Explosive material propagates a detonation wave that moves along the shock tube to a factory-crimped and sealed blasting cap. Detonation is normally contained within the plastic tubing. However, burns may occur if the shock tube is held. Advantages of a shock tube:
- It is extremely reliable.
- Offers instant electric initiation, and it also prevents radio transmitters, static electricity, and such from accidentally causing an initiation.
- It may be extended using leftover sections from previous operations.

Demolitions

5-4. Five types of MDI blasting caps are available to replace the M6 electric and M7 nonelectric blasting cap. Three are high-strength, and two are low-strength. High-strength blasting caps can prime all standard military explosives (including detonating cord) and initiate the shock tube for other MDI blasting caps. The five blasting caps are:

a. **M11:**
- Factory-crimped to 30 feet of shock tube.
- A movable "J" hook is attached for quick and easy attachment to detonation cord.
- A red flag is attached one meter from the blasting cap and a yellow flag two meters from the blasting cap.

b. **M14:**
- Factory-crimped to 7.5 feet of time fuze.
- May be initiated using a fuze igniter or match.
- Burn time for total length is about five minutes.
- Yellow bands indicate calibrated one-minute time intervals.

c. **M15:**
- Two blasting caps factory-crimped to 70 feet of shock tube.
- Factory-crimped to 7.5 feet of time fuze.
- Each blasting cap has delay elements to allow for staged detonations.
- Low-strength blasting caps. Used as a relay device to transmit a shock tube detonation impulse from an initiator to a high-strength blasting cap.

d. **M12** is factory-crimped to 500 feet of shock tube on a cardboard spool.

e. **M13** is factory-crimped to 1000 feet of shock tube.

5-5. If fuze igniter is unavailable, light the time (blasting) fuze with a match. Split the fuze at the end (see figure 5-1 on page 5-4) and place the head of an unlit match in the powder train. Light the inserted match head with a flaming match, or rub the abrasive on the matchbox against it. This may take several attempts in windy conditions.

5-6. The M81 fuse igniter is used to ignite the time-blasting fuze. It is also used to initiate the shock tube of MDI blasting caps.

Note: High altitudes and colder temperatures increase burn time.

Chapter 5

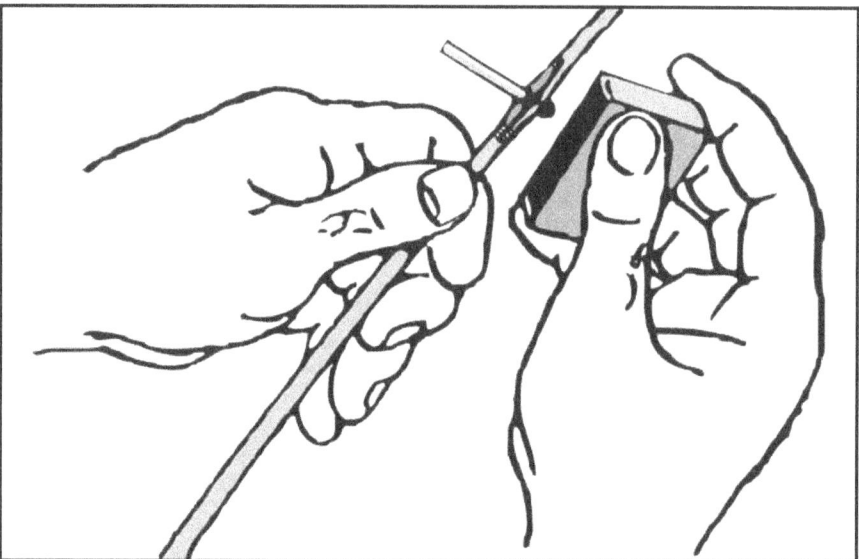

Figure 5-1. Technique for lighting time fuze with a match

DETONATION (FIRING) SYSTEMS

5-7. The two types of firing systems are MDI alone, or MDI plus detonating cord. An MDI-alone firing system is one in which the initiation set, transmission, and branch lines are constructed using MDI components, and the explosive charges are primed with MDI blasting caps. Construct the charge in the following manner:
- 1. Emplace and secure explosive charge, such as C4, trinitrotoluene (TNT), or cratering charge, on target.
- 2. Place a sandbag or other easily identifiable marker over the M11, M14, or M15 blasting cap.
- 3. Connect to an M12 or M13 transmission line, if desired.
- 4. Connect blasting cap with shock tube to an M14 cap with time fuze. Cut time blasting fuze to desired delay time.
- 5. Prime the explosive charge by inserting the blasting cap into the charge.
- 6. Visually inspect firing system for possible misfire indicators such as cracks, bulges, or corrosion.
- 7. Return to the firing point and secure a fuze igniter to the cut end of the time fuze.
- 8. Remove the safety cotter pin from the igniter's body.
- 9. Actuate the charge by grasping the igniter body with one hand while sharply pulling the pull ring.

5-8. Construct the charge using the above steps for MDI stand-alone system. Incorporate detonating cord branch lines into the system using the "J" hooks of the M11 shock tube. Taping the ends of the detonation cord reduces the effect of moisture on the system.

Demolitions

SAFETY

5-9. MDI is not recommended for below ground use, except in quarry operations with water-gel or slurry explosives. Use detonating cord when it is necessary to bury primed charges. Do not use M1 dynamite with the M15 blasting cap. The M15 delay blasting cap should be used only with water-gel or slurry explosives.

5-10. Do not handle misfires downrange until the required 30-minute waiting period for both primary and secondary initiation systems has elapsed, and other safety precautions have been accomplished.

5-11. Never yank or pull hard on the shock tube. This may actuate the blasting cap. Do not dispose of used shock tubes by burning because of potentially toxic fumes given off from the burning plastic.

5-12. Always use protective equipment when handling demolitions. Minimum protection consists of leather gloves, ballistic eye protection, and a helmet.

EXPEDIENT EXPLOSIVES

5-13. There are three types of expedient explosives: improvised shaped charge, platter charge, and the grapeshot charge. All three are created using common items.

5-14. An **improvised shaped charge** (see figure 5-2 on page 5-6) concentrates the energy of the explosion released into a small area, making a tubular or linear fracture in the target. The versatility and simplicity of these charges make them effective against targets, especially those made of concrete or those with armor plating.

5-15. Bowls, funnels, cone-shaped glasses, (champagne glasses with stem removed) are used as cones. Champagne or cognac bottles are excellent materials to use. Charge characteristics are:
- **Cavity liners** are made of copper, tin, or zinc. If none is available, cut a cavity out of the plastic explosive.
- **Cavity angle** works with 30-to-60 degree angles. The cavity angle in most high-explosive antitank (HEAT) ammunition is 42-to-45 degrees.
- **Explosive height (in container)** is twice the height of the cone measured from the base of the cone to the top of the explosive.
- **Standoff** is normally one and a half times the cone's diameter.
- **Detonation point** is the exact top center of the charge. Cover the blasting cap with a small amount of C4, if any part of the blasting cap is exposed.

5-16. Remove the narrow neck of a bottle or the stem of a glass by wrapping it with a piece of soft, absorbent twine, or by soaking the string in gasoline and lighting it. Place two bands of adhesive tape, one on each side of the twine, to hold the twine firmly in place. The bottle or stem is turned continuously with the neck up, to heat the glass uniformly.

5-17. A narrow band of plastic explosive placed around the neck and burned gives the same result. After the twine or plastic has burned, submerge the neck of the bottle in water and tap it against some object to break it off. Tape the sharp edge of the bottle to prevent cutting hands while tamping the explosive in place. Do not immerse the bottle in water before the plastic has been completely burned or it could detonate.

Chapter 5

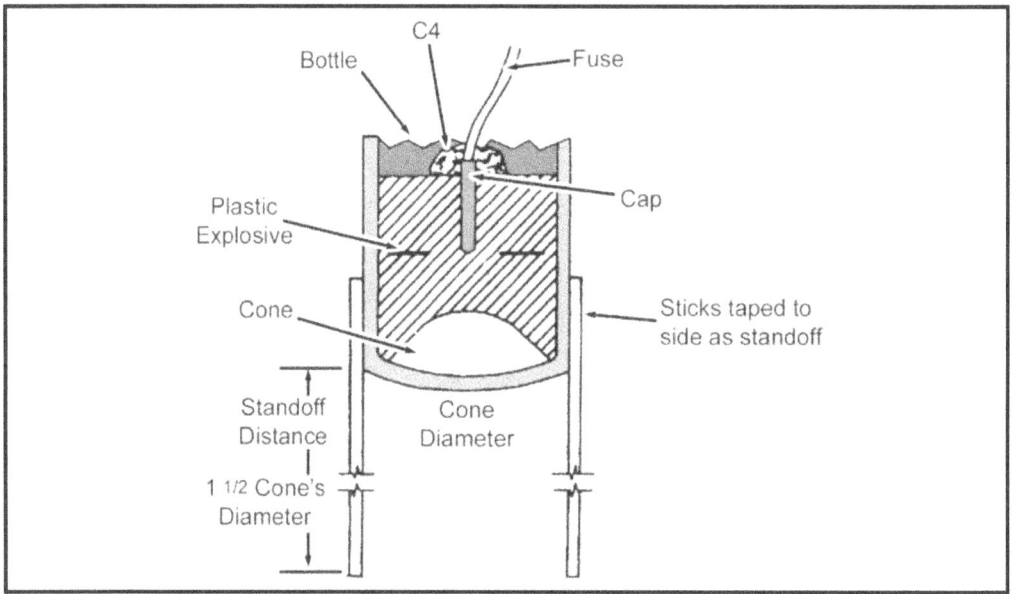

Figure 5-2. Improvised shape charge

5-18. The **platter charge** device (see figure 5-3) turns a metal plate into a powerful blunt-nosed projectile. The plate should be steel, preferably round (but square can work), and weigh from two-to-six pounds. The weight of the explosive should equal the weight of the platter.

5-19. Uniformly pack the explosive behind the platter. You only need a container if the explosives fail to remain firmly against the platter. Tape to anchor the explosives, if needed. Prime the charge at the exact, rear center of the charge. If any part of the blasting cap is exposed, cover it with a small quantity of C4.

5-20. Aim charge at the direct center of the target, ensuring that the charge is on the opposite side of the platter from the target. Effective range is 35 yards for a small target. With practice, a Ranger might hit a 55-gallon drum at 25 yards 90 percent of the time. A gutted fuze igniter can serve as an expedient aiming device.

Demolitions

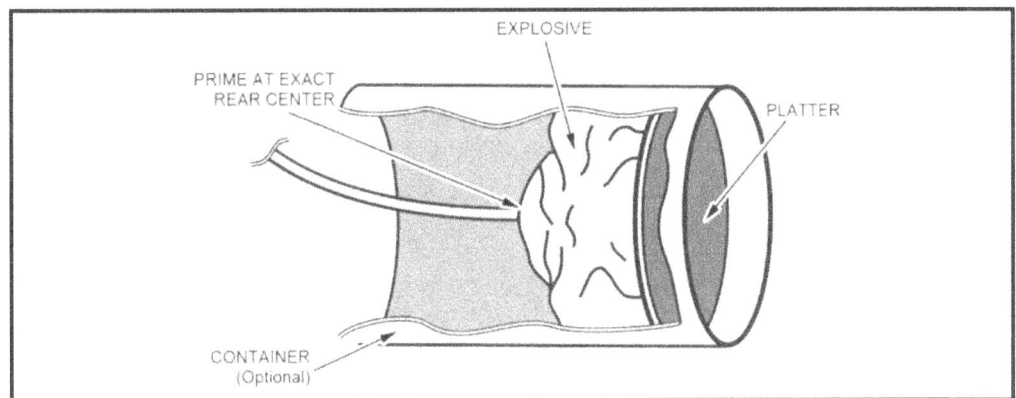

Figure 5-3. Platter charge

5-21. To use this antipersonnel fragmentation mine, or **grapeshot charge** (see figure 5-4), create a hole in the center, bottom of the container, for the blasting cap. Place explosives evenly on the bottom of the container. Remove all voids and air pockets by pressing the C4 into place using a nonsparking instrument.

5-22. Place buffer material directly over the top of the explosives. Place projectiles over the top of the buffer materials, then cover to prevent spilling from movement. Aim at the target from approximately 100 feet. Use a small amount of C4 on any exposed portion of the blasting cap.

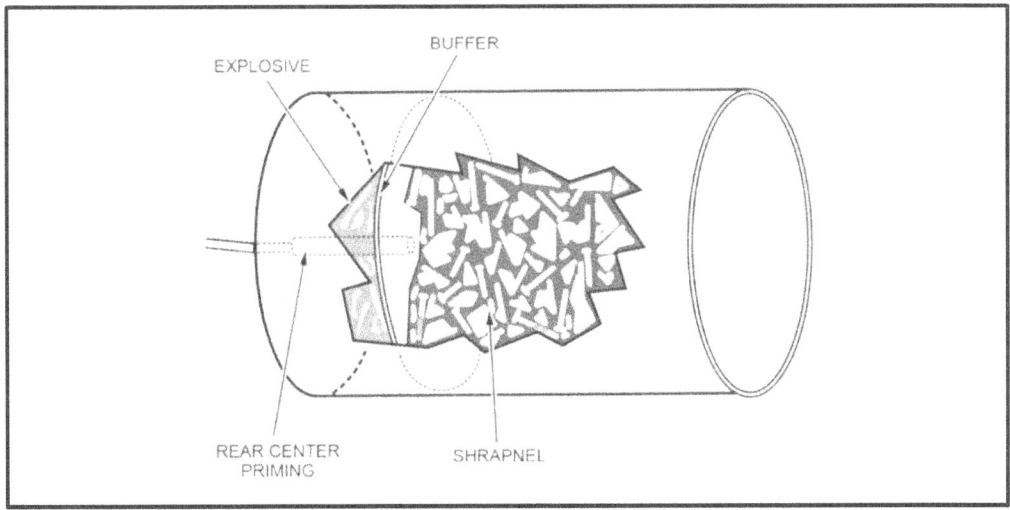

Figure 5-4. Grapeshot charge

Chapter 5

DEMOLITION KNOTS AND MINIMUM SAFE DISTANCES

5-23. Several knots are used in demolitions. Figure 5-5, and figure 5-6 on page 5-9, show a few simple knots that can join demolitions to detonation cord.

5-24. Rangers must remain especially aware of their situations when using **demolitions**. Table 5-2 on page 5-10 depicts minimum safe distances for employing up to 500 pounds. For charges over 500 pounds of demolitions, see figure 5-7 on page 5-10.

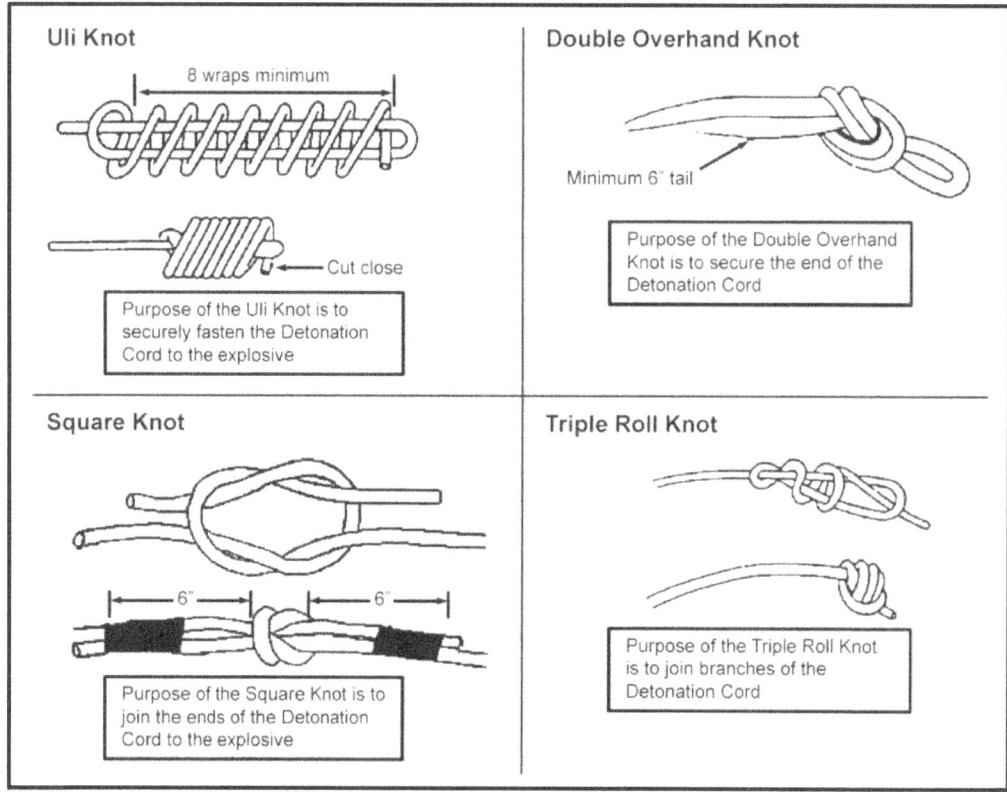

Figure 5-5. Various joining knots used in demolitions

LEGEND
6" – 6 inches

Demolitions

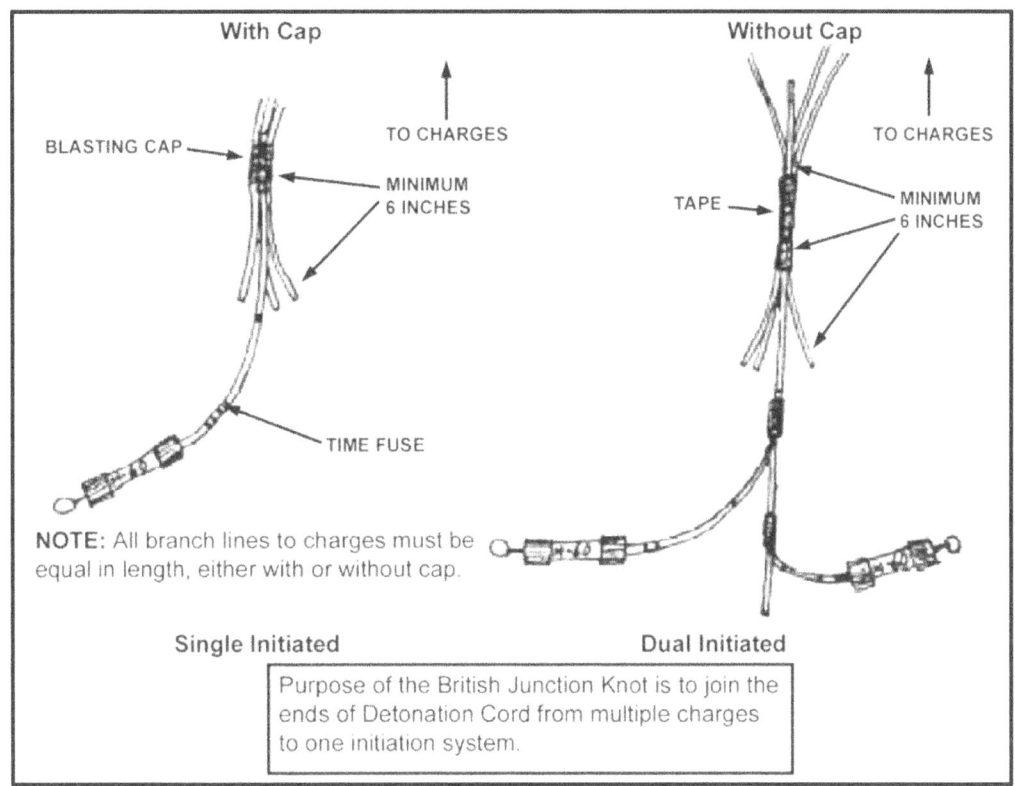

Figure 5-6. British junction knot

Chapter 5

Table 5-2. Minimum safe distance for personnel in the open (bare charge)

EXPLOSIVE WEIGHT (POUNDS)	SAFE DISTANCE FEET	SAFE DISTANCE METERS	EXPLOSIVE WEIGHT (POUNDS)	SAFE DISTANCE FEET	SAFE DISTANCE METERS
27 or less	985	300	175	1838	560
30	10021	311	200	1920	585
35	1073	327	225	1999	609
40	1123	342	250	2067	630
45	1168	356	275	2136	651
50	1211	369	300	2199	670
60	1287	392	325	2258	688
70	1355	413	350	2313	705
80	1415	431	375	2369	722
90	1474	449	400	2418	737
100	1526	465	425	2461	750
125	1641	500	500	2625	800
150	1752	534	—	—	—

Minimum Safe Distance for Charges Over 500 Pounds

Safe distance (meters) = 100 $\sqrt[3]{\text{pounds of explosive}}$

Safe distance (feet) = 300 $\sqrt[3]{\text{pounds of explosive}}$

Figure 5-7. Minimum safe distance for charges over 500 pounds

CHARGES ON TARGETS

For charges on targets, the minimum radius of danger is 1000 meters. The minimum safe distance when in a missile-proof shelter from the point of detonation is 100 meters.

Demolitions

CHARGES

5-25. Breaching charges are used to break down barriers. There are several types of breaching charges discussed in the remainder of this chapter. Among them are breaching charges for reinforced concrete, materials other than reinforced concrete, and timber-cutting charges. For additional information on breaching charges, refer to TM 3-34.82.

5-26. In table 5-3, the left column represents the thickness of a reinforced concrete wall. The remaining seven columns show the number of packages of C4 required to breach the wall using the charge placements shown in the drawings above the columns. Use table 5-3, and tables 5-4 and 5-5 on page 5-12, for breaching charges.

5-27. Use the formula in figure 5-8 on page 5-13 to calculate the charges (and table 3-5 on page 3-8 for more information). Multiply the number of packages of C4 from table 5-3 by the conversion factor from table 5-4 for materials other than reinforced concrete. Figure 5-8 on page 5-13 is the formula to compute the size of a charge needed for concrete, masonry, and rock.

Table 5-3. Breaching charges for reinforced concrete

REINFORCED CONCRETE THICKNESS (FEET)	PLACEMENT METHODS						
	C = 1.0	C = 1.0	C = 1.0	C = 1.8	C = 2.0	C = 2.0	C = 3.6
	Packages of M112 (C-4)						
2.0	1	5	5	9	10	10	17
2.5	2	9	9	17	18	18	33
3.0	2	13	13	24	26	26	47
3.5	4	21	21	37	41	41	74
4.0	5	31	31	56	62	62	111
4.5	7	44	44	79	88	88	157
5.0	9	48	48	85	95	95	170
5.5	12	63	63	113	126	126	226
6.0	13	82	82	147	163	163	293
6.5	17	104	104	18	207	207	372
7.0	21	111	111	200	222	222	399
7.5	26	137	137	245	273	273	490
8.0	31	166	166	298	331	331	595
LEGEND C – tampering factor; C-4 – composition C-4 explosive							

Chapter 5

Table 5-4. Conversion factors for materials other than reinforced concrete

MATERIAL	CONVERSION FACTOR
Earth	0.1
Ordinary masonry Hard pan Shale Ordinary concrete Rock Good timber Earth construction	0.5
Dense concrete First-class masonry	0.7

Table 5-5. Material factor (K) for breaching charges

MATERIAL	BREACHING RADIUS (R)	MATERIAL FACTOR (K)
Earth	All values	0.07
Poor masonry, shale, hardpan, good timber, and earthen construction	Less than 1.5 meters (m) (5 feet [ft]) 1.5 m (5 ft) or more	0.32 0.29
Good masonry, concrete block, and rock	0.3 m (1 ft) or less Over 0.3 m (1 ft) to less than 0.9 m (3 ft) 0.9 m (3 ft) to less than 1.5 m (5 ft) 1.5 M (5 ft) to less than 2.1 m (7 ft) 2.1 m (7 ft) or more	0.88 0.48 0.40 0.32 0.27
Dense concrete and first-class masonry	0.3 m (1 ft) or less Over 0.3 m (1 ft) to less than 0.9 m (3 ft) 0.9 m (3 ft) to less than 1.5 m (5 ft) 1.5 M (5 ft) to less than 2.1 m (7 ft) 2.1 m (7 ft) or more	1.14 0.62 0.52 0.41 0.35
Reinforced concrete (factor does not consider cutting steel)	0.3 m (1 ft) or less Over 0.3 m (1 ft) to less than 0.9 m (3 ft) 0.9 m (3 ft) to less than 1.5 m (5 ft) 1.5 M (5 ft) to less than 2.1 m (7 ft) 2.1 m (7 ft) or more	1.76 0.96 0.80 0.63 0.54

LEGEND
ft – feet or foot; K – material factor; m – meter; R - radius

Demolitions

$$P = R^3 KC$$

Where—

P = TNT required (in pounds).

R = Breaching radius (in feet).

K = Material factor, which reflects the strength, hardness, and mass of the material to be demolished.

C = Tampering factor, which depends on the location and tamping of the charge.

Figure 5-8. Formula for computing size of charge to breach concrete, masonry, and rock

5-28. Table 5-6 lists the sizes of C4 for the different types of timber-cutting charges. Figure 5-9 is a depiction of abatis. An abatis is an obstacle formed by felled trees that impedes the enemy's ability to advance on an avenue of approach. It is made by cutting trees that remain attached to their stumps. Figures 5-10 through 5-15 on pages 5-15 through 5-17, display the types of formulas and charges to use with each situation.

Table 5-6. Timber-cutting charge size

CHARGE TYPE	PACKAGES OF C4 REQUIRED (1.25-POUND PACKAGES) BY TIMBER DIAMETER (INCHES)											
	6	8	10	12	15	18	21	24	27	30	33	36
Internal	1	1	1	1	1	1	2	2	2	3	3	4
External	1	1	2	3	4	5	7	9	11	14	17	20
Abatis								7	9	11	14	16
NOTE: Packages required are rounded UP the next whole package.												

Chapter 5

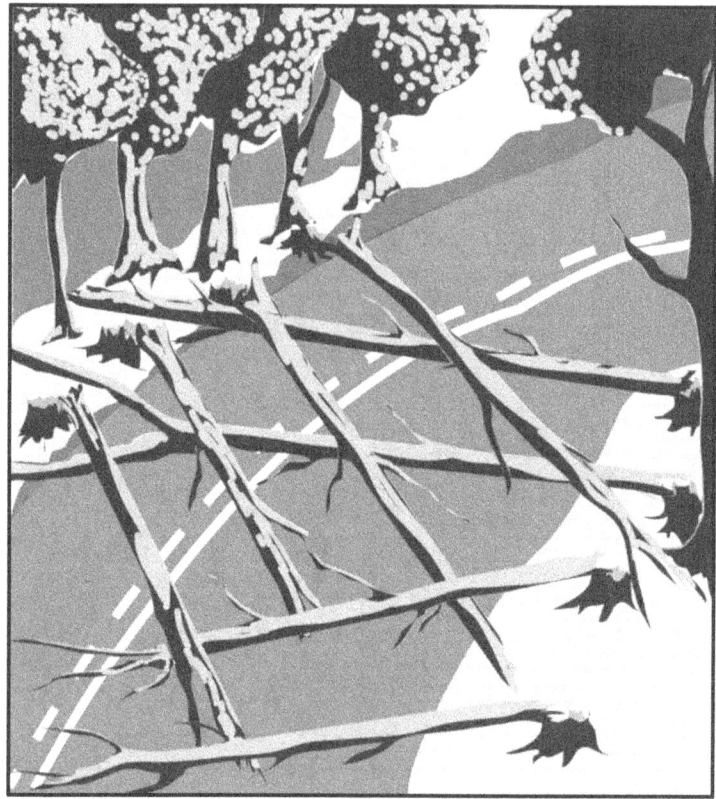

Figure 5-9. Example of abatis

$$P = D^2/50 = P = 0.02D^2$$

Where—

P = TNT required per tree (in pounds).

D = Diameter or least dimension of dimensioned timber (in inches).

Figure 5-10. Formula for fallen tree obstacles or test shot

Demolitions

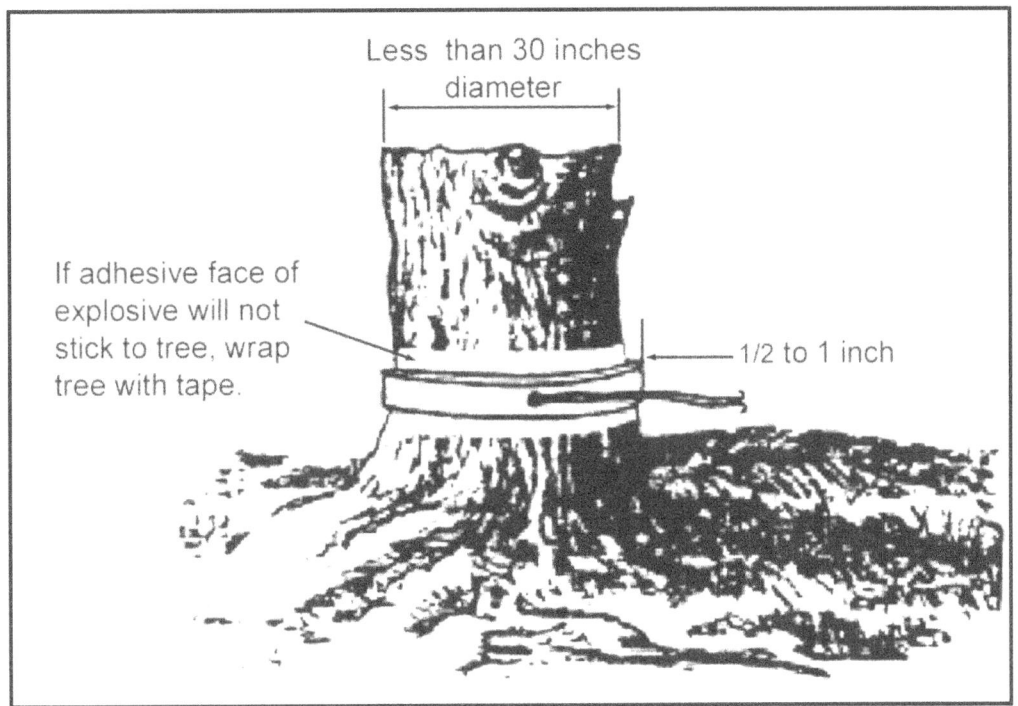

Figure 5-11. Timber-cutting ring charge

Chapter 5

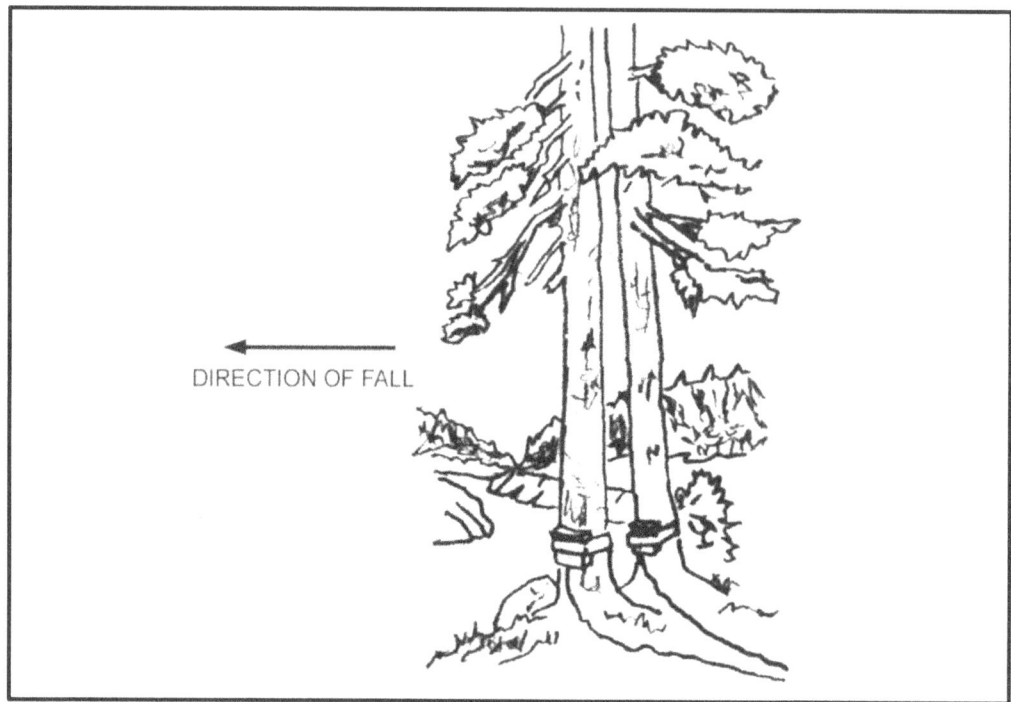

Figure 5-12. Timber-cutting charge (external)

$$P = D^2/40 \text{ or } P = 0.025D^2$$

Where—

P = TNT required per target (in pounds).
D = Diameter or least dimension of dimensioned timber (in inches).

Figure 5-13. Formula for external timber-cutting charge

Demolitions

Figure 5-14. Timber-cutting charge (internal)

$$P = D^2/250 \text{ or } P = 0.004 D^2$$

Where—

P = TNT required per target (in pounds).

D = Diameter or least dimension of dimensional timber (in inches)

Figure 5-15. Formula for internal timber-cutting charge

This page intentionally left blank.

Chapter 6

Movement

To survive on the battlefield, stealth, dispersion, and security is enforced in all tactical movements. The leader is skilled in all movement techniques. (Refer to ATP 3-21.8 for more information.)

FORMATIONS

6-1. Movement formations include elements and Rangers arranged in relation to each other. Fire teams, squads, and platoons use several formations. Formations give the leader control, based on a METT-TC analysis. Leaders position themselves where they can best command and direct the formations, which are shown in figure 6-1. Typical formations are the line, vee, echelon, diamond, wedge, and file.

6-2. Formations allow the fire team leader to lead by example. (Follow me and do as I do!) All Rangers in the team must be able to see their leader.

6-3. Formations also reflect fire team formations. Squad formations are very similar with more Rangers. Squads can operate in lines and files similar to fire teams. When squads operate in wedges or in echelon, the fire teams use those formations and simply arrange themselves in a column or with one team behind the other. Squads may also use the vee, where one team forms the lines of the vee with the squad leader at the front (at the point of the vee) for mission command. Platoons use the same formations as squads. When the unit operates as a platoon, the PL carefully selects the location for the machine guns in the movement formation.

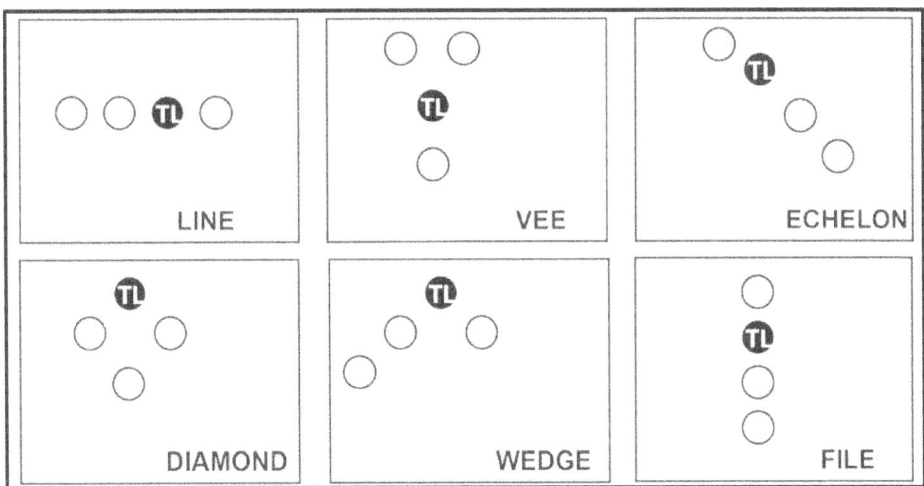

Figure 6-1. Formations

LEGEND
TL – team leader

Chapter 6

MOVEMENT TECHNIQUES

6-4. Selecting a movement technique is based on the likelihood of enemy contact and the relative need for speed. Specifically, the factors to consider include control, dispersion, speed, and security. Movement techniques are neither fixed nor are they formations. Instead, movement techniques are distinguished by a set of criteria such as distance between individual Rangers and between teams or squads.

6-5. Movement techniques are not fixed formations. They refer to the distances between Soldiers, teams, and squads that vary based on mission, enemy, terrain, visibility, and any other factor that affects control. There are three movement techniques: traveling, traveling overwatch, and bounding overwatch. The selection of a movement technique is based on the likelihood of enemy contact and the need for speed. (See table 3-5 on page 3-8.) Individual movement techniques include high and low crawl, and three to five second rushes from one covered position to another.

6-6. From these movement techniques, leaders are able to conduct actions on contact, make natural transitions to fire and movement, as well as to conduct tactical mission tasks. When analyzing the situation, some enemy positions are known. However, most of the time enemy positions will only be templated positions. Templated positions are the leader's "best guess" based on analyzing the terrain and his knowledge of the enemy. Throughout the operation, leaders are continuously trying to confirm or deny both the known positions as well as the likely positions. (See table 6-1.)

Table 6-1. Characteristics of movement techniques

MOVEMENT TECHNIQUES	WHEN NORMALLY USED	CHARACTERISTICS			
		Control	Dispersion	Speed	Security
Traveling	Contact not likely	More	Less	Fastest	Least
Traveling Overwatch	Contact possible	Less	More	Slower	More
Bounding Overwatch	Contact expected	Most	Most	Slowest	Most

6-7. Movement techniques vary depending on METT-TC. However, Rangers must always be able to see their fire team leaders. The platoon leader should be able to see the lead SL. Leaders control movement with hand and arm signals, and use radios only when needed. Leaders match the movement technique to the situation as follows:

a. **Traveling**. Use this technique when enemy contact is not likely but speed is necessary. Leave 10 m between Rangers, and 20 m between squads. Traveling is:
- More control than traveling overwatch but less controlled than bounding overwatch.
- Minimum dispersion.
- Maximum speed.
- Minimum security.

Movement

b. **Traveling overwatch.** Use this technique when enemy contact is possible. This is the most often used movement technique. Leave 20 m between Rangers, and 50 m between teams. Considerations for this technique include—
- Only the lead squad should use traveling overwatch. However, in cases where greater dispersion is desired, all squads may use it.
- In other formations, all squads use traveling overwatch unless specified not to by the PL. Traveling overwatch offers good control, dispersion, speed, and security forward.
- The lead squad is far enough ahead of the rest of the platoon to detect or engage any enemy before the enemy observes or fires on the main body. However, the lead squad stays between 50 and 100 m in front of the platoon so the platoon can support them with small arms fires. This distance is dependent on terrain, vegetation, and light and weather conditions.

c. **Bounding overwatch.** Use when enemy contact is likely or when crossing a danger area. Both squad and platoon have bounding and overwatch elements. The bounding element moves while the other one occupies a position where it can overwatch by fire the bounding element's route. The bounding element remains within firing range of the overwatching element at all times. Considerations for this technique include:

- **Characteristics.** Bounding overwatch offers maximum control, dispersion, and security with minimum speed.
- **Types of bounds:**
  - Successive bounds. One element moves to a position, and then the overwatching element moves to a position generally online with the first element.
  - Alternating bounds. One element moves into position, and then the overwatching element moves to a position in front of the first element.
- **Length.** The length of a bound depends on the terrain, visibility, and control.
- **Instructions.** Before a bound, the leader gives the following instructions to subordinates:
  - Direction of the enemy, if known.
  - Position of overwatch elements.
  - Next overwatch position.
  - Route of the bounding element.
  - What to do after the bounding element reaches the next position.
  - How the elements receive follow-on orders.
- **Squad bounding overwatch.** Rangers leave about 20 m between them. The distance between teams and squads varies. (See figure 6-2 on page 6-4.)
- **Platoon bounding overwatch.** When platoons use bounding overwatch (see figure 6-3 on page 6-5), one squad bounds, a second squad overwatches, and a third awaits orders. Rangers leave about 20 m between them. The distance between teams and squads varies. Forward observers stay with the overwatching squad to call-for-fire. Platoon leaders normally stay with the overwatching squad, which uses machine guns and attached weapons to support the bounding squad. Another technique is to have one squad use bounding overwatch while the other two use traveling or traveling overwatch. When deciding where to move the bounding element, consider the:
  - Enemy's likely action.
  - Mission.
  - Routes to the next overwatch position.
  - Weapon ranges of the overwatching unit.
  - Responsiveness of the rest of the unit.
  - Fields of fire at the next overwatch position.

Chapter 6

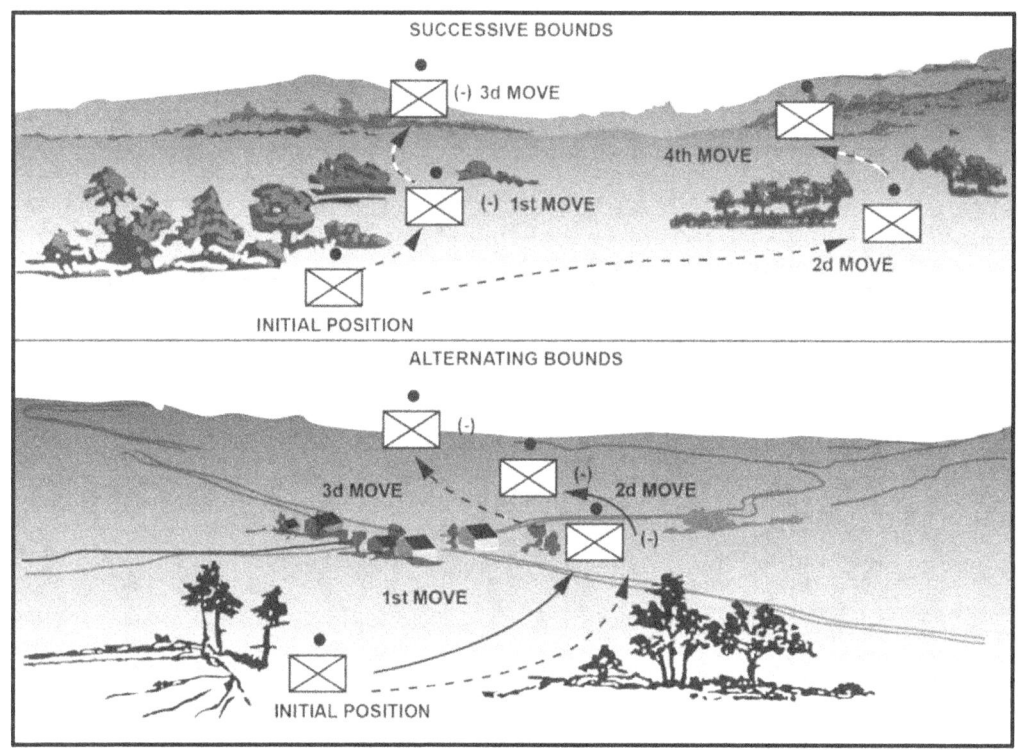

Figure 6-2. Squad bounding overwatch

Movement

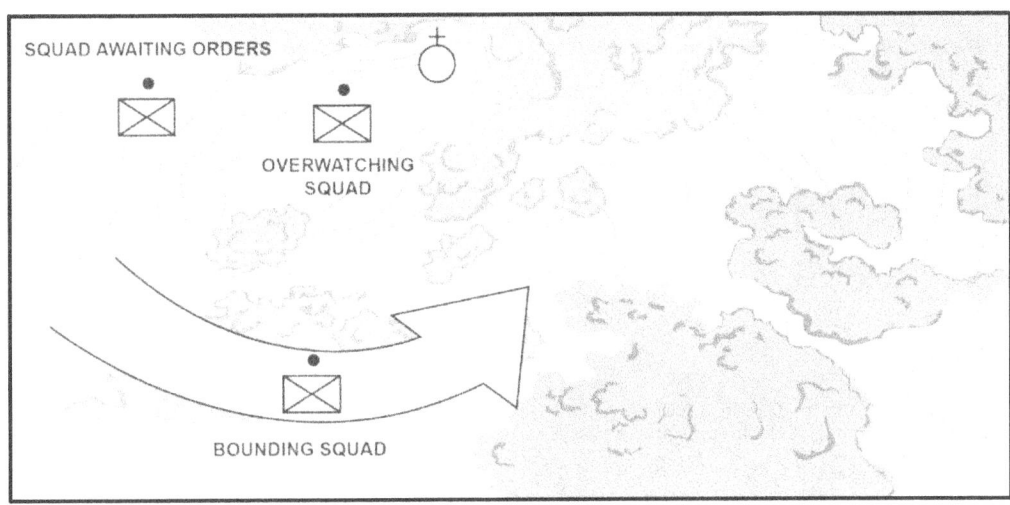

Figure 6-3. Platoon bounding overwatch

STANDARDS

6-8. Unit moves on designated route or arrives at specified location according to the OPORD, maintaining accountability of all assigned and attached personnel. Unit uses the movement formation and technique ordered by the leader (based on METT-TC). Leaders remain oriented (within 200 m) and follow a planned route, unless METT-TC dictates otherwise.

6-9. During movement, the unit maintains 360-degree security and remains 100 percent alert. During halts, unit maintains 360-degree security and remains at least 75 percent alert. If the unit makes contact with the enemy, they do so with the smallest element possible. The unit uses control measures during movement such as head counts, rally points, or phase lines.

FUNDAMENTALS

6-10. Mission accomplishment depends on successful land navigation. The patrol should use stealth and vigilance to avoid chance contact. Designate a primary and alternate compass and pace man for each patrol. All leaders except fire team leaders move inside their formations to best control the platoon.

Note: The point man is never tasked to perform compass or pace duties; the sole responsibility is forward security for the element.

6-11. Patrols use stealth, and cover and concealment of the terrain to its maximum advantage. Whenever possible, the patrol should move during limited visibility to maximize the technological advantages of night vision devices (NVDs) and hinder the enemy's ability to detect the patrol. They exploit the enemy's weaknesses, and try to time movements to coincide with other operations that distract the enemy.

Chapter 6

6-12. The patrol continues to use active and passive security measures. The leader assigns subunit responsibilities for security at danger areas, PBs, and in the objective area. The leader plans fire support such as mortars, artillery, tactical air, attack helicopter, and naval gunfire.

6-13. The enemy threat and terrain determines which of the three movement techniques to use. Fire teams maintain visual contact, but there is enough distance between them so that the entire patrol does not become engaged if it makes contact. Fire teams can spread their formations to gain better observation to the flanks, as needed. Although widely spaced, Rangers retain their relative positions in their wedge and follow their team leader. Only in extreme situations should the file be used. The lead squad secures the front, and is responsible for navigation. For a long movement, the PL may rotate lead squad responsibilities. The fire team or squad in the rear is charged with rear security. Vary movement techniques to meet the changing situation.

6-14. The patrol achieves 360-degree security, high and low. Within a fire team, squad, and so on, the leader assigns appropriate sectors of fire to subordinates. This ensures the battlefield is covered. This includes trees, multiple storied structures, tunnels, sewers, and ditches.

TACTICAL MARCHES

6-15. Platoons conduct two types of marches with the company: foot marches and motor (road) marches. A foot march is successful when troops arrive at the destination at the prescribed time, physically able to execute their tactical mission.

6-16. To meet the standard, the unit crosses the start point and release point at the times specified in the order, and follows the prescribed route, rate of march, and interval, without deviation unless required by enemy action or higher headquarters action. The fundamentals of tactical marches include effective control, detailed planning, and rehearsals. Considerations include:

a. **METT-TC.** This stands for—
- Mission—task and purpose.
- Enemy—intentions, capabilities, and course of action.
- Terrain and weather—road condition, trafficability, and visibility.
- Troops and equipment—condition of Rangers and their loads, number and types of weapons and radios.
- Time—start time, release time, rate of march, and time available.
- Civilians—movement through populated areas, refugees, and OPSEC.

b. **Task organization:**
- Security—advance and trail teams.
- Main body—the two remaining line squads and weapons squad.
- Headquarters mission command.
- Control measures.

c. **Control measures:**
- Start point and release point (given by higher HQ).
- Check points—at checkpoints report to higher HQ to remain oriented.
- Rally or rendezvous points—used when elements become separated.
- Locations of leaders—where they can best control their elements.
- Communications plan—location of radios, frequencies, call signs, and OPSKEDs.
- Dispersion between Rangers:
  - 3–5 m by day.
  - 1–3 m by night.

Movement

d. A **march order** may be issued as an OPORD, FRAGORD, or an annex to either type of order (an operational overlay or strip map is used for this). The march order includes:
- Formations and order of movement.
- Route of march, assembly area, start point, release point, rally points, check points, and break or halt points.
- Start point time, release point time, and rate of march.
- March interval for squads, teams, and individuals.
- Actions on enemy contact—air and ground.
- Actions at halts.
- Detailed plan of fire support for the march.
- Water supply plan.
- Medical evacuation (MEDEVAC) plan.

6-17. Platoons are made up of several Rangers fulfilling different positions and responsibilities. They include the platoon leader, platoon sergeant, squad leaders, security squad, assault squads, medic, and individuals. Their duties before, during, and after the movement are detailed below.

a. **Platoon leader:**
- Before—issues WARNORD, OPORD, or FRAGORD; inspects and supervises movement preparations.
- During—ensures unit makes movement time, maintains interval, and remains oriented; maintains security; checks condition of Rangers; spot checks water discipline and field sanitation.
- After—ensures Rangers are prepared to accomplish their mission, supervises SLs, and ensures Rangers receive medical coverage, as needed.

b. **Platoon sergeant:**
- Before—helps PL, makes recommendations; enforces uniform and packing lists; and obtains accountability of Rangers prior to start point time.
- During—controls stragglers, assists PL maintaining proper interval and security.
- At halts—maintains accountability, enforces security and the welfare of Soldiers, enforces field sanitation and litter discipline, performs preventive medicine, and confirms head count prior to leaving the halt.
- After—coordinates for water, rations, and medical supplies; recovers any casualties.

c. **Squad leaders:**
- Before—provides detailed instruction to TLs; inspects boots and socks for serviceability and proper fit, adjustment of equipment, full canteens, and equal distribution of loads.
- During—controls squad, maintains proper interval between Rangers and equipment, enforces security, and remains oriented.
- At halts—ensures security is maintained, provides Rangers for water resupply, as detailed. Physically checks the Rangers in his squad; ensures they drink water and change socks, as necessary, and rotates heavy equipment. (Units should plan the latter in detail to avoid confusion before, during, and after halts.)
- After—occupies squad sector assembly area, conducts foot inspection and reports condition of Soldiers to PL, and prepares them to accomplish the mission.

d. **Security squad:**
- Serves as point element for platoon, reconnoiters route to the start point (SP). Calls in checkpoints, provides frontal security, and early warning. Maintains rate of march, moves 10 to 20 m in front of main body.
- Trail team provides rear security, and moves 10 to 20 m behind main body.

e. **Assault squads:**
- Provides left, right, and rear security for the platoon.

Chapter 6

- Is prepared to provide additional combat power should the security squat receive contact, and moves 10-20 m behind the security squad.

f. **Forward observer/Radiotelephone operator:**
- Maintains constant communications with higher headquarters, moves with the PL, and sends all reports.
- Builds field-expedient antennas, as needed.

g. **Medic:**
- Assesses and treats march casualties.
- Advises chain of command on evacuation and transportation of casualties.

h. **Individual:**
- Maintains interval and follows TLs examples.
- Relays hand and arm signals.
- Remains alert during movement and at halts.

MOVEMENT DURING LIMITED VISIBILITY CONDITIONS

6-18. During hours of limited visibility, the platoon uses surveillance, target acquisition, and night observation (STANO) devices to enhance effectiveness. Leaders must be able to control, navigate, maintain security, and move during limited visibility.

6-19. When visibility is poor, methods that aid in control include moving leaders closer to the front, reducing the platoon speed, and using luminescent tape on equipment. Leaders also reduce the intervals between Soldiers and elements, and executes head counts.

6-20. While navigating during limited visibility, the unit uses the same techniques as in daylight, but leaders exercise more care to keep the patrol oriented. To maintain security, strict noise and light discipline is enforced, and radio listening silence and camouflage is used. Using the terrain to avoid detection by enemy surveillance or NVDs, Rangers make frequent halts to stop, look, listen, and smell (SLLS). Whenever possible, mask the sounds of movement. Rain, wind, and flowing water disguise movement sounds very efficiently.

6-21. Leaders plan detail actions to be taken at rally points. All elements maintain communications at all times. The two techniques for actions at rally points are:
- **Minimum force.** Patrol members assemble at the rally point, and the senior leader assumes command. When the minimum force (designated in the OPORD) is assembled and organized, the patrol continues the mission.
- **Time available.** The senior leader determines if the patrol has enough time remaining to accomplish the mission.

6-22. During halts, the unit posts security and covers all approaches into the sector with key weapons. The positions used are:
- **Short halt.** This typically takes one to two minutes. Rangers seek immediate cover and concealment and take a knee. Leaders assign sectors of fire.
- **Long halt.** This typically takes more than two minutes. Rangers assume the prone position behind cover and concealment. Leaders ensure Rangers have clear fields of fire, and assign sectors of fire.

DANGER AREAS

6-23. A danger area is any place on a unit's route where the leader determines the unit may be exposed to enemy observation or fire. Some examples of danger areas are open areas, roads and trails, urban terrain, enemy positions, and natural and manmade obstacles. **Bypass danger areas whenever possible.**

6-24. There are standards, fundamentals, and techniques for crossing danger areas. Rangers take the following steps:
a. **Standards.** The unit—
- Prevents the enemy from surprising the main body.

Movement

- Moves all personnel and equipment across the danger area.
- Prevents decisive engagement by the enemy.

b. **Fundamentals:**
- Designate nearside and farside rally points.
- Secure nearside, left and right flank, and rear security.
- Reconnoiter and secure the far side.
-
- Cross the danger area.
- Plan for fires on all known danger areas.

c. **Techniques for crossing danger areas:**
- **Linear danger area (LDA) actions for a squad** (see figure 6-4 on page 6-10):
  - **STEP 1.** The Alpha team leader observes the LDA and sends the hand and arm signal to the SL, who determines to bound across.
  - **STEP 2.** SL directs the Alpha team leader to move the team across the LDA far enough to fit the remainder of the squad on the farside of the LDA. Bravo team moves to the LDA to the right or left to provide an overwatch position prior to the A team crossing.
  - **STEP 3.** SL receives the hand and arm signal that it is safe to move the rest of the squad across (B team is still providing overwatch).
  - **STEP 4.** SL moves with radiotelephone operator (RTO) and B team across the LDA. (A team provides overwatch for squad missions.)
  - **STEP 5.** The A team assumes original azimuth at SLs command or by hand and arm signal.
- **LDA crossing for a platoon:**
  - The lead squad halts the platoon and signals danger area.
  - Platoon leader moves forward to the lead squad to confirm the danger area, and then decides if current location is suitable for crossing.
  - Platoon leader confirms danger area or crossing site and establishes nearside and farside rally points.
  - On the PLs signal, the trail squad moves forward to establish left and right nearside security.
  - Once nearside security is established, the A team of the lead squad with the SL, moves across to confirm there is enough room to fit the rest of the platoon on the farside of the LDA.
  - Once he conducts SLLS, the SL signals the PL, "ALL CLEAR." During daylight, the signal can be something such as thumbs up. At night, use something such as an infrared or red lens.
  - The PL then directs the B team of the lead squad to bound across by team and link up with the A team of the lead squad and pick up a half step while the rest of the platoon crosses.
  - Platoon leader then crosses with RTO, FO, WSL, and one gun team.
  - Once across, PL signals the second squad in movement to cross.
  - PSG with medic and one gun team crosses after second squad is across (sterilizing central crossing site).
  - PSG signals security squad to cross at their location.
  - PSG calls PL via FM radio to confirm all elements are across.
  - PL directs lead squad to pick up normal rate of movement.

Note: Platoon leader plans fires on all known larger danger area crossing sites. Nearside security in overwatch sterilizes signs of the patrol.

Chapter 6

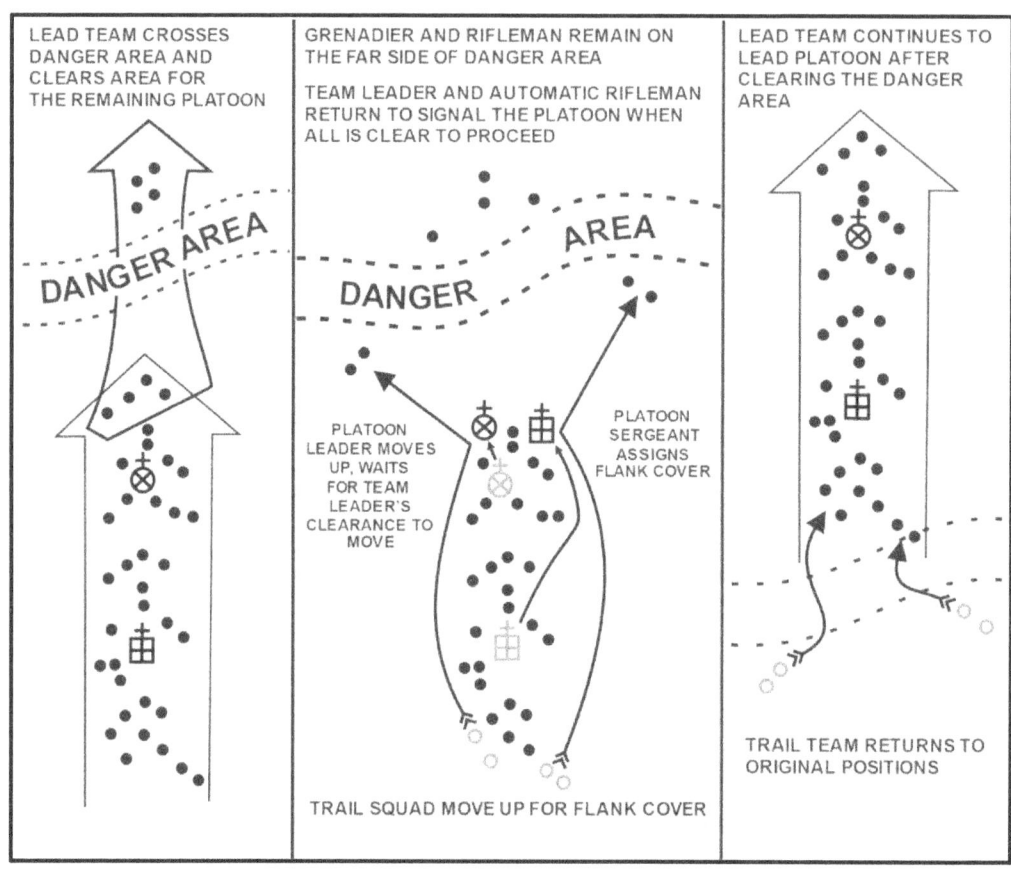

Figure 6-4. Linear danger area

- Danger area (small or open):
  - The lead squad halts the platoon and signals "danger area."
  - PL moves forward to the lead squad to confirm the danger area.
  - Platoon leader confirms danger area and establishes nearside and farside rally points.
  - PL designates lead squad to bypass danger area using the detour bypass method.
  - Paceman suspends current pace count and initiates an interim pace count. Alternate pace and compass man moves forward and offsets compass 90 degrees left or right as designated, and moves in that direction until clear of danger area.
  - After moving set distance (as instructed by PL), lead squad assumes original azimuth and primary paceman resumes original pace.

Movement

- After the open area, the alternate pace and compass Soldier offsets the compass 90 degrees left or right, and leads the platoon or squad the same distance (in meters) back to the original azimuth.
- **Danger areas (series):** two or more danger areas within an area that can be either observed or covered by fire. This includes:
  - Double linear danger area (LDA technique and cross as one).
  - Linear or small open danger area (use bypass or contour technique (See figure 6-5 on page 6-12.)
  - Linear or large open danger area (use platoon wedge when crossing).

Note: A series of danger areas are crossed using the technique that provides the most security.

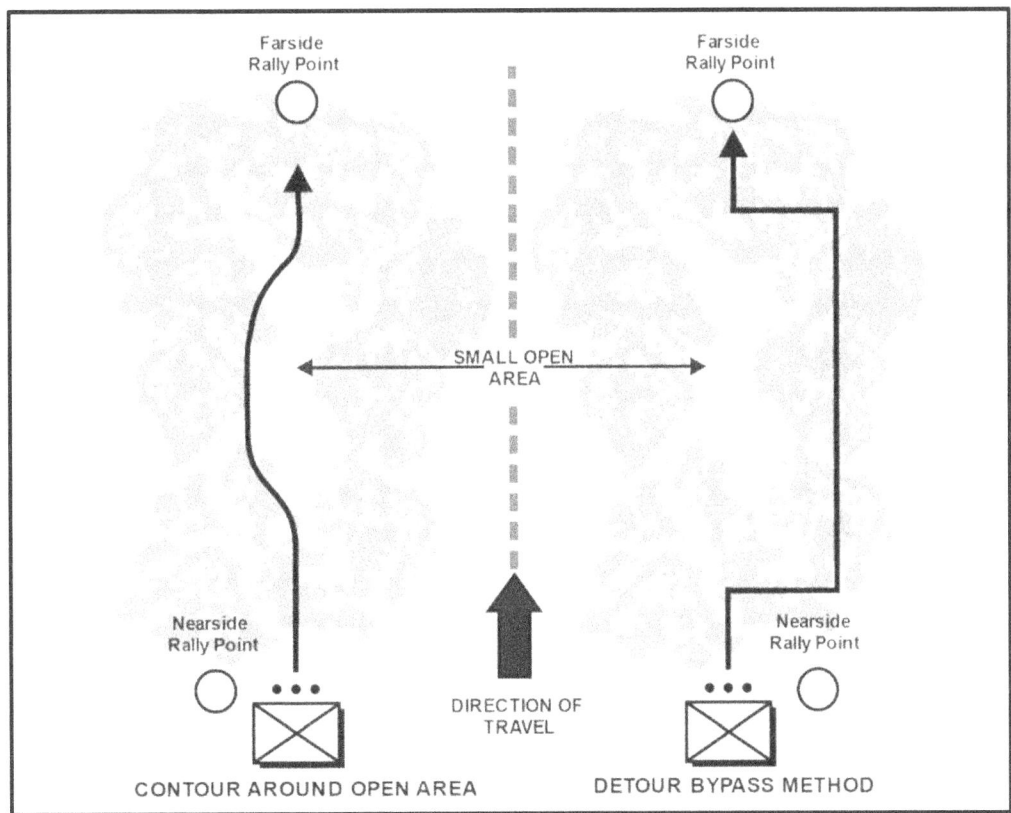

Figure 6-5. Small open danger area

Chapter 6

- **Danger area (large):**
  - Lead squad halts the platoon, and signals danger area.
  - Platoon leader moves forward with RTO and FO and confirms danger area.
  - Platoon leader confirms danger area and establishes near and farside rally points.
  - PL designates direction of movement.
  - PL designates change of formation as necessary to ensure security.

Note: Platoon leader plans for all larger danger area crossing sites. Nearside security in overwatch sterilizes signs of the patrol. Before point man steps into danger area, PL and FO adjust targets to cover movement. If farside of danger area is within 250 m, PL esta6blishes overwatch and designates lead squad to clear wood line on far side.

Chapter 7

Patrols

To survive on the battlefield, stealth, dispersion, and security is enforced in all tactical movements. The leader is skilled in all movement techniques. (Refer to ATP 3-21.8 for more information.)

PRINCIPLES

7-1. All patrols are governed by five principles: planning, reconnaissance, security, control, and common sense. In brief, each principle involves—
- **Planning:** quickly make a simple plan and effectively communicate it to the lowest level. A great plan that takes forever to complete and is poorly disseminated is not a great plan. Plan and prepare to a realistic standard and rehearse everything.
- **Reconnaissance:** your responsibility as a Ranger leader is to confirm what you think you know, and to learn that which you do not already know.
- **Security:** preserve your force as a whole. Every Ranger and every rifle counts, either one could be the difference between victory and defeat.
- **Control:** clarify the concept of the operation and commander's intent, coupled with disciplined communications, to bring every Soldier and weapon available to overwhelm the enemy at the decisive point.
- **Common sense:** use all available information and good judgment to make sound, timely decisions.

PLANNING

7-2. Planning considerations common to most patrols include task organization, initial planning and coordination, completion of the plan, and contingency planning. A patrol is a detachment sent out by a larger unit to conduct a specific mission.

7-3. Patrols operate semi-independently and return to the main body upon completion of their mission. Patrolling fulfills the Infantry's primary function of finding the enemy to engage them or report their disposition, location, and actions. Patrols act as the eyes and ears of the larger unit; and as a fist to deliver a sharp, devastating jab and then withdraw before the enemy can recover. Some definitions associated with patrols are—
- **Patrol:** sent out by a larger unit to conduct a specific combat, reconnaissance, or security mission. A patrol's organization is temporary and specifically matched to the immediate task. Because a patrol is an organization, not a mission, it is not correct to speak of giving a unit a mission to "patrol."
- **Patrolling or conducting a patrol:** the semi-independent operation conducted to accomplish the patrol's mission. A patrol requires a specific task and purpose.
- **Employment:** the commander sends a patrol out from the main body to conduct a specific tactical task with an associated purpose. Upon completion of that task, the patrol leader returns to the main body, reports to the commander and describes the events that took place, the status of the patrol's members, and equipment, and any observations that were made.
- **Leadership:** if a patrol is made up of an organic unit such as a rifle squad, the SL is responsible. If a patrol is made up of mixed elements from several units, an officer or noncommissioned officer is designated as the patrol leader. This temporary title defines their role and responsibilities for that mission. The patrol leader may designate an assistant, normally the next senior Soldier in the patrol, and any subordinate element leaders required.

Chapter 7

- **Size:** a patrol can be a unit as small as a fire team. Squad-size and platoon-size patrols are normal. Sometimes, for combat tasks such as a raid, the patrol can consist of most of the combat elements of a rifle company. Unlike operations in which the Infantry platoon or squad is integrated into a larger organization, the patrol is semi-independent and relies on itself for security.

7-4. There are common elements of all patrols and teams. Elements and teams for platoons conducting patrols include—

- **Reconnaissance and surveillance (R&S) teams:** normally used in a zone reconnaissance, but may be useful in any situation when it is impractical to separate the responsibilities for reconnaissance and security. When these responsibilities are separate, the security element provides security at danger areas, secures the ORP, isolates the objective, and supports the withdrawal of the rest of the platoon once the reconnaissance is complete. The security element may have separate security teams, each with an assigned task or sequence of tasks.
- **Assault element:** seizes and secures the objective and protects special teams as they complete their assigned actions on the objective.
- **Security element:** provides security at danger areas, secures the ORP, isolates the objective, and supports the withdrawal of the rest of the patrol once actions on the objective are complete. The security element may have separate security teams, each with an assigned task or sequence of tasks.
- **Support element:** provides direct and indirect fire support for the unit. Direct fires include machine guns, medium and light anti-armor weapons, and small recoilless rifles. The available indirect fires may include mortars, artillery, CAS, and organic M320 weapon systems.
- **Demolition team:** responsible for preparing and detonating the charges to destroy designated equipment, vehicles, or facilities on the objective.
- **Enemy prisoner of war (EPW) and search teams:** the assault element may provide two-Ranger (buddy teams) or four-Ranger (fire team) search teams to search bunkers, buildings, or tunnels on the objective. These teams search the objective or kill zone for any PIR that may give the PL an idea of the enemy concept for future operations. Primary and alternate teams may be assigned to ensure enough prepared personnel are available on the objective.
- **Breach element:** conducts initial penetration of enemy obstacles to seize a foothold and allow the patrol to enter an objective. This is typically done according to METT-TC and the steps outlined in the Conduct the Initial Breach of a Mined Wire Obstacle in Chapter 8 of this publication.
- **Reconnaissance teams:** reconnoiter the objective area from various vantage points once the security teams are in position. Normally, reconnaissance teams are two-soldier teams, to reduce the possibility of detection.

INITIAL PLANNING AND COORDINATION

7-5. Leaders plan and prepare for patrols using the troop leading procedures and estimation of the situation, as described in Chapter 2. Through an estimate of the situation, leaders identify required actions on the objective (mission analysis) and plan backward to departure from friendly lines and forward to reentry of friendly lines. Because patrolling units act independently, move beyond the direct fire support of the parent unit and operate forward of friendly units. Coordination is thorough, detailed, and continuous throughout planning and preparation. PLs use checklists to avoid omitting any items vital to the accomplishment of the mission.

7-6. Coordination with higher HQ includes intelligence, operations, and fire support. This initial coordination is an integral part of the troop leading procedures (Step 3, make a tentative plan). The leader also coordinates the unit's patrol activities with the leaders of other units that will be patrolling in adjacent areas at the same time.

COMPLETION OF PLAN

7-7. As the PL completes the plan, specified and implied tasks are considered. The PL ensures that all specified tasks to be performed on the objective, at rally points, danger areas, security or surveillance locations, along the route(s), and at passage lanes are assigned. These make up the maneuver and tasks to maneuver units subparagraphs of the execution paragraph.

Patrols

7-8. The PL also considers key travel and execution times. The leader estimates time requirements for movement to the objective, leader's reconnaissance of the objective, establishment of security and surveillance, completion of all assigned tasks on the objective, and passage through friendly lines. Some planning factors are—
- **Movement:** average of one kilometer per hour (kph) during daylight hours in woodland terrain and average limited visibility of 1/2 kph. Add additional time for restrictive or severely restrictive terrain, such as mountains, swamps, or thick vegetation.
- **Leader's reconnaissance:** not later than (NLT) 1.5 hours.
- **Establishment of security and surveillance:** 0.5 hour.

7-9. The leader selects primary and alternate routes to and from the objective. The return routes should differ from the routes to the objective. The PL may delegate route selection to a subordinate, but is ultimately responsible for the routes selected.

7-10. The leader should consider the use of special signals. These include hand and arm signals, flares, voice, whistles, radios, and infrared equipment. Primary and alternate signals must be identified and rehearsed so that all Rangers know their meaning. Password systems to consider are—
- **Odd number system.** The leader specifies an odd number. The challenge can be any number less than the specified number. The password is the number that must be added to it to equal the specified number. For example, the number is seven, the challenge is three, and the password is four.
- **Running password.** This code word alerts a unit that friendly Rangers are approaching in a less than organized manner and possibly under pressure. The number of Rangers approaching follows the running password. For example, if the running password is "Ranger," and five friendly Rangers are approaching, they would say, "Ranger five."

7-11. Always know the location of leaders. The PL considers where the PSG and other key leaders are located during each phase of the mission. The PL is positioned to best control the actions of the patrol. The PSG is normally located with the assault element during a raid or attack to help the PL control the use of additional assaulting squads, and assist with securing the objective (OBJ). The PSG locates at the casualty collection point (CCP) to facilitate casualty treatment and evacuation. During a reconnaissance mission, the PSG stays behind in the ORP to facilitate the transfer of intelligence to the higher HQ, and control the reconnaissance elements movement into and out of the ORP.

7-12. Unless required by the mission, the unit avoids enemy contact. The leader's plan addresses actions on enemy contact at each phase of the patrol mission. The unit's ability to continue depends on how early contact is made, whether the platoon is able to break contact successfully (so that its subsequent direction of movement is undetected), and whether the unit receives any casualties due to the contact. The plan addresses the handling of seriously wounded Rangers and those KIA. The plan also addresses the handling of prisoners who are captured because of chance contact but are not part of the planned mission.

7-13. The leader leaves the unit for many reasons throughout the planning, coordination, preparation, and execution of the patrol mission. Each time the leader departs the patrol main body, a five-point contingency plan is issued to the leader left in charge of the unit. The patrol leader issues additional specific guidance stating what tasks are to be accomplished in his absence. The contingency plan is remembered using the memory aid GOTWA:

> **G** Where the leader is **GOING**.
>
> **O** **OTHERS** he is taking with him.
>
> **T** **TIME** the leader plans to be gone.
>
> **W** **WHAT** to do if the leader does not return in time.
>
> **A** **ACTIONS** by the unit in the event contact is made while the leader is gone. (A stay-behind leader is designated until returning.)

Chapter 7

7-14. The leader considers the use and location of rally points. A rally point is a place designated by the leader where the unit moves to reassemble and reorganize if it becomes dispersed. Rangers know which rally point to move to at each phase of the patrol mission should they become separated from the unit. They also know what actions are required there and how long they are to wait at each rally point before moving to another. The most common types of rally points include initial, en route, objective, and nearside and farside rally points. Rally points must be—
- Easily identifiable in daylight and limited visibility.
- Show no signs of recent enemy activity.
- Covered and concealed from the ground and air.
- Away from natural lines of drift and high-speed avenues of approach.
- Defendable for short periods of time.

7-15. The objective rally point typically lies 200 to 400 m from the objective, or at a minimum, one major terrain feature away. Actions at the ORP include—
- Conducting SLLS and pinpoint location.
- Conducting a leader's reconnaissance of the objective.
- Issuing a FRAGORD, if needed.
- Making final preparations before continuing operations. (For example, recamouflage; prepare demolitions; line up rucksacks for quick recovery; prepare EPW bindings, first aid kits, and litters; and inspect weapons.
- Accounting for Rangers and equipment after completing actions on the objective.
- Disseminating information from reconnaissance, if no contact was made.

7-16. The plan includes a leader's reconnaissance of the objective once the platoon or squad establishes the ORP. Before departing, the leader issues a five-point contingency plan. During reconnaissance, the leader pinpoints the objective; selects reconnaissance, security, support, and assault positions for the elements; and adjusts the plan based on observation of the objective.

7-17. Each type of patrol requires different tasks during the leader's reconnaissance. The leader brings different elements with him. These are discussed separately under each type of patrol. The leader plans time to return to the ORP, complete the plan, disseminate information, issue orders and instructions, and allow the squads to make any additional preparations. During the Leader's Reconnaissance for a Raid or Ambush, the PL leaves surveillance on the objective. Each type of patrol also requires different actions on the objective. Actions on the objective are discussed under each type of patrol.

RECONNAISSANCE PATROLS

7-18. Area and zone reconnaissance patrols provide timely and accurate information on the enemy and terrain, and confirm the leader's plan before it is executed. Units on reconnaissance operations collect specific information (PIR) or general information based on the instructions from their higher commander.

7-19. In order to have a successful area reconnaissance, the PL applies the fundamentals of the reconnaissance to the plan while conducting the operation. To obtain required information, the parent unit tells the patrol leader what information is needed. This is in the form of the information requirement and priority intelligence requirements. The platoon's mission is then tailored to what information is required. During the entire patrol, members continuously gain and exchange all information gathered, but cannot consider the mission accomplished unless all PIR has been gathered.

7-20. A patrol avoids letting the enemy know that it is in the objective area. If the enemy knows they are being observed, they may move, change plans, or increase security measures. Methods of avoiding detection are—
- Minimizing movement in the objective area (area reconnaissance).
- Moving no closer to the enemy than necessary.
- Using long-range surveillance or NVDs, if possible.

Patrols

- Using camouflage, stealth, and noise and light discipline.
- Minimizing radio traffic.

7-21. A patrol must be able to break contact and return to the friendly unit with what information is gathered. If necessary, they break contact and continue the mission. Leaders emplace security elements where they can overwatch the reconnaissance elements. They suppress the enemy so the reconnaissance element can break contact.

7-22. When the platoon leader receives the order, the mission is analyzed to ensure it is understood what is to be done. The PL task-organizes the platoon to accomplish the mission according to METT-TC. A reconnaissance is typically a squad-sized mission. The types of reconnaissance are—

- **Area reconnaissance:** the area reconnaissance patrol collects all available information on PIR and other intelligence not specified in the order for the area. The patrol completes the reconnaissance and reports all information by the time specified in the order. The patrol is not compromised.
- **Zone reconnaissance:** the zone reconnaissance patrol determines all PIR and other intelligence not specified in the order for its assigned zone. The patrol reconnoiters without detection by the enemy. The patrol completes the reconnaissance and reports all information by the time specified in the order.

ACTIONS ON THE OBJECTIVE, AREA RECONNAISSANCE

7-23. The element occupies the ORP as discussed in the section on occupation of the ORP. (See figure 7-1 [page 7-6].) The RTO reports to higher HQ that the unit has occupied the ORP. The leader confirms the location on the map while subordinate leaders make necessary perimeter adjustments. The PL organizes the platoon in one of two ways: separate reconnaissance and security elements, or combined reconnaissance and security elements.

7-24. The PL takes subordinate leaders and key personnel on a leader's reconnaissance to confirm the objective and plan. The PL also—

- Issues a five-point contingency plan before departure.
- Establishes a suitable RP that is beyond sight and sound of the objective if possible, but that is definitely out of sight. The RP should also have good rally point characteristics.
- Allows all personnel to become familiar with the RP and surrounding area.
- Identifies the objective and emplaces surveillance, designates a surveillance team to keep the objective under surveillance, and issues a contingency plan to the senior Soldier remaining with the surveillance team. The surveillance team is positioned with one Soldier facing the objective, and one facing back in the direction of the RP.
- Takes subordinate leaders forward to pinpoint the objective, establish a limit of advance (LOA), and choose vantage points.
- Maintains communications with the platoon throughout the leader's reconnaissance.

7-25. The PSG maintains security, supervises priorities of work in the ORP, and reestablishes security at the ORP. The PSG also disseminates the PLs contingency plan, oversees preparation of reconnaissance personnel (Soldiers recamouflaged, NVDs and binoculars prepared, weapons on safe with a round in the chamber, and other duties).

7-26. The PL and reconnaissance party return to the ORP. The PL confirms the plan or issues a FRAGORD, and allows subordinate leaders time to disseminate the plan.

7-27. The patrol conducts the reconnaissance by long-range observation and surveillance, if possible. R&S element moves to observation points that offer cover and concealment and are outside of small arms range, establishes a series of OPs if information cannot be gathered from one location, and gathers all PIR using the SALUTE format.

7-28. If necessary, the patrol conducts its reconnaissance by short-range observation and surveillance (see figure 7-1), moves to an observation post near the objective, passes close enough to the objective to gain information, and gathers all PIR using the SALUTE format.

Chapter 7

7-29. R&S teams move using a technique such as the cloverleaf method to move to successive OPs. (See figure 7-1.) In this method, R&S teams avoid paralleling the objective site, maintain extreme stealth, do not cross the LOA, and maximize the use of available cover and concealment.

7-30. During the conduct of the reconnaissance, each R&S team returns to the RP when any of the following occurs:
- All their PIR is gathered.
- The LOA is reached.
- The allocated time to conduct the reconnaissance has elapsed.
- Enemy contact is made.

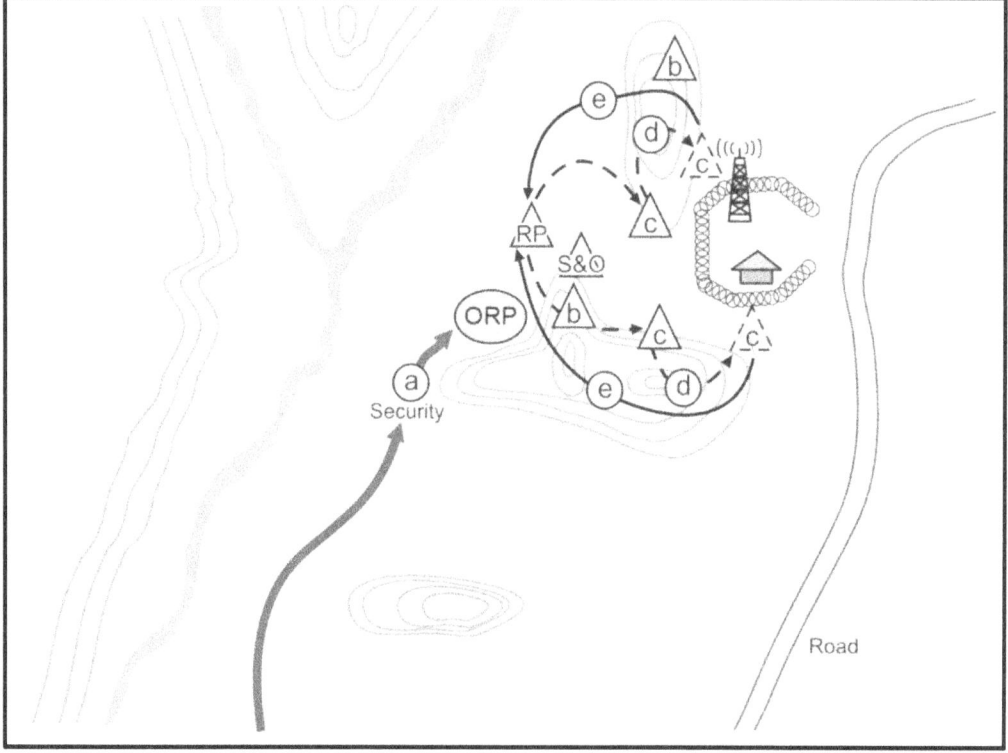

Figure 7-1. Area reconnaissance

LEGEND
ORP – objective rally point; RP – release point; S&O – surveillance and observation

Patrols

7-31. At the RP, the leader analyzes what information has been gathered and determines if it meets the PIR requirements. If the leader determines that insufficient information to meet the PIR requirements has been gathered, or if the information he and the subordinate leader gathered differs drastically, R&S teams may be sent back to the objective site. In this case, R&S teams alternate areas of responsibilities. For example, if one team reconnoitered from the 6–3–12, then that team now reconnoiters from the 6–9–12.

7-32. The R&S element returns undetected to the ORP by the specified time, and disseminates information to all patrol members through key leaders, or moves to a position at least one terrain feature or one kilometer away to disseminate. To disseminate, the leader has the RTO prepare three sketches of the objective site based on the leader's sketch and provides copies to the subordinate leaders to assist in dissemination. The R&S element reports any information requirements or any information requiring immediate attention to higher HQ, and departs for the designated area.

7-33. If contact is made, the R&S element moves to the RP. The reconnaissance element tries to break contact and return to the ORP, secure rucksacks, and quickly move out of the area. Once they have moved a safe distance away, the leader informs higher HQ of the situation and takes further instructions from them:

    a. While emplacing surveillance, the reconnaissance element withdraws through the RP to the ORP, and follows the same procedures as above.

    b. While conducting the reconnaissance, the compromised element returns a sufficient volume of fire to allow them to break contact. Surveillance can fire an AT-4 at the largest weapon on the objective. All elements pull off the objective and move to the RP. The senior leader quickly accounts for all personnel and returns to the ORP. Once in the ORP, leadership follows the procedures previously described. The critical tasks for a patrol are:

- Secure and occupy the ORP.
- Conduct a leader's reconnaissance of the objective:
  - Estimate RP.
  - Pinpoint objective.
  - Emplace surveillance with the surveillance and observation (S&O) team. Position security element, if used.
- Conduct reconnaissance by long-range surveillance, if possible.
- Conduct reconnaissance by short-range surveillance, if necessary.
- Teams:
  - Move as necessary to successive observation posts.
  - On order, return to RP.
  - Once PIR is gathered, return to ORP.
- Patrol:
  - Linkup as directed in the ORP.
  - Disseminate information before moving.

ACTIONS ON THE OBJECTIVE, ZONE RECONNAISSANCE

7-34. The element occupies the initial ORP, as discussed. The radio operator calls in "SPARE" for occupation of the ORP. The leader confirms the location on the map while subordinate leaders make necessary perimeter adjustments.

    a. **Organization.** The reconnaissance team leaders organize their reconnaissance elements:

- Designate security and reconnaissance elements.
- Assign responsibilities (point man, pace man, en route recorder, and rear security), if not already assigned.
- Designates easily recognizable rally points.
- Ensures local security at all halts.

Chapter 7

b. **Actions.** The patrol reconnoiters the zone:
- Moves tactically to the ORPs.
- Occupies designated ORPs.
- Follows the method designated by the PL (fan, converging routes, or box method). (See table 7-1.)
- The reconnaissance teams reconnoiter:
    - During movement, the squad gathers all PIR specified in the order.
    - Reconnaissance team leaders ensure sketches are drawn or digital photos are taken of all enemy hard sites, roads, and trails.
    - Return to the ORP or linkup at the rendezvous point on time.
    - When the squad arrives at a new rendezvous point or ORP, the reconnaissance team leader's report to the PL with all gathered information.
- The PL continues to control the reconnaissance elements:
    - Moves with the reconnaissance element that establishes the rendezvous point.
    - Changes reconnaissance methods, as required.
    - Designates times for the elements to return to the ORP, or to linkup.
    - Collects all information and disseminates it to the entire patrol. PL briefs all key subordinate leaders on information gathered by other squads; establishing one consolidated sketch, if possible; and allows team leaders time to brief their teams.
    - With the PSG, accounts for all personnel.
- The patrol continues the reconnaissance until all designated areas have been reconnoitered, and returns undetected to friendly lines.

Table 7-1. Reconnaissance methods

FAN METHOD	CONVERGING ROUTES METHOD	BOX METHOD
Uses a series (fan) of ORPs. Patrol establishes security at first ORP. Each reconnaissance element moves from the ORP along a different fan-shaped route. Route overlaps with that of other reconnaissance elements. This ensures reconnaissance of the entire area. Leader maintains a reserve at the ORP. When all reconnaissance elements return to the ORP, PL collects and disseminates all information before moving to the next ORP.	PL selects routes from the ORP through zone to a rendezvous point at the farside of the zone from the ORP. Each reconnaissance element moves and reconnoiters along a specified route. They converge (linkup) at one time and place.	PL sends reconnaissance elements from the first ORP along routes that form a box, and then sends other elements along routes throughout the box. All teams linkup at the farside of the box from the ORP.
	LEGEND ORP – objective rally point; PL – platoon leader	

COMBAT PATROLS

7-35. Combat patrols are the second type of patrol. Combat patrols are further divided into raids, ambushes, and security patrols. Units conduct combat patrols to destroy or capture enemy soldiers or equipment; destroy installations, facilities, or key points; or harass enemy forces. Combat patrols also provide security for larger units.

Patrols

7-36. In planning a combat patrol, the PL considers the following:

a. **Tasks to maneuver units.** Normally the platoon HQ element controls the patrol on a combat patrol mission. The PL makes every effort to maintain squad and fire team integrity, as tasks are assigned to subordinate units.

- The PL considers the requirements for assaulting the objective, supporting the assault by fire, and security of the entire unit throughout the mission.
  - For the assault on the objective, the PL considers the required actions on the objective, the size of the objective, and the known or presumed strength and disposition of the enemy on and near the objective.
  - The PL considers the weapons available, and the type and volume of fires required to provide fire support for the assault on the objective.
  - The PL considers the requirement to secure the platoon at points along the route, at danger areas, at the ORP, along enemy avenues of approach into the objective, and elsewhere during the mission.
  - The PL also designates engagement and disengagement criteria.
- The PL assigns additional tasks to the squads for demolition, search for EPWs, guarding EPWs, treatment and evacuation (litter teams) of friendly casualties, and other tasks required for successful completion of patrol mission (if not already in the SOP)
  - The PL determines who controls any attachments of skilled personnel or special equipment.

b. **Leader's reconnaissance of the objective.** In a combat patrol, the PL has additional considerations for the conduct of the reconnaissance of the objective from the ORP.

- **Composition of the leader's reconnaissance party.** The platoon leader normally brings the following personnel:
  - Squad leaders, including the weapons squad leader.
  - Surveillance team.
  - Forward observer.
  - Security element (depending on time available)
- **Conduct of the leader's reconnaissance.** In a combat patrol, the PL considers the following additional actions in the conduct of the leader's reconnaissance of the objective:
  - The PL designates an RP about halfway between the ORP and the objective, using the same characteristics as a rally point. The PL then issues a five-point contingency plan to the security element and the PL, FO, and surveillance team move to pinpoint the OBJ and emplace the surveillance team with eyes-on the OBJ. PL then moves back to the RP and emplaces the security element.
  - The PL confirms the location of the objective or kill zone, notes the terrain, and identifies where Claymore mines are emplaced to cover dead space. Any change to the plan is issued to the SLs (while overlooking the objective, if possible).
  - If the objective is the kill zone for an ambush, the leader's reconnaissance party should not cross the objective: to do so leaves tracks that may compromise the mission.
  - The PL confirms the suitability of the assault and support positions, and routes from them back to the ORP.
  - The PL issues a five-point contingency plan to the surveillance team before returning to the ORP.

AMBUSH

7-37. An ambush is a surprise attack from a concealed position on a moving or temporarily halted target. Ambushes are categorized as hasty or deliberate, divided into two types—point or area, and the formation is linear or L-shaped.

7-38. The leader considers various key factors in determining the ambush category, type, and formation; and from these decisions, develops the ambush plan.

- **Key factors** include:
  - Coverage (ideally, the whole kill zone) by fire.
  - METT-TC.

Chapter 7

- Existing or reinforcing obstacles, including Claymore mines, to keep the enemy in the kill zone.
- Security teams usually have hand-held antitank weapons such as AT-4s or light antitank weapons (LAWs), Claymore mines, and various means of communication.
- Security elements or teams to isolate the kill zone.
- Protection of the assault and support elements with Claymore mines or explosives.
- Assault through the kill zone to the LOA.

Note: The assault element must be able to move quickly through its own protective obstacles.

- Time the actions of all elements of the platoon to preclude loss of surprise. In the event any member of the ambush is compromised, the leader may immediately initiate the ambush.
- When the ambush is manned for a long time, use only one squad to conduct the entire ambush and determine movement time of the rotating squads from the ORP to the ambush site.
- **Categories** include:
  - **Hasty.** A unit conducts a hasty ambush when it makes visual contact with an enemy force and has time to establish an ambush without being detected. The actions for a hasty ambush are well rehearsed so that Rangers know what to do on the leader's signal. They also know what actions to take if the unit is detected before it is ready to initiate the ambush.
  - **Deliberate.** A deliberate ambush is conducted at a predetermined location against any enemy element that meets the commander's engagement criteria. The leader requires the following detailed information in planning a deliberate ambush: size and composition of the targeted enemy, and weapons and equipment available to the enemy.
- **Types** are:
  - **Point.** In a point ambush, Rangers deploy to attack an enemy in a single kill zone.
  - **Area.** In an area, Rangers deploy in two or more related point ambushes.
  - **Anti-armor.** An anti-armor ambush focuses on moving or temporarily halted enemy armored vehicles.
- **Formations** (see figure 7-2 on page 7-11):
  - **Linear.** In an ambush using a linear formation, the assault and support elements deploy parallel to the enemy's route. This positions both elements on the long axis of the kill zone and subjects the enemy to flanking fire. This formation can be used in close terrain that restricts the enemy's ability to maneuver against the platoon or in open terrain, provided a means of keeping the enemy in the kill zone can be effected.
  - **L-shaped.** In an L-shaped ambush, the assault element forms the long leg parallel to the enemy's direction of movement along the kill zone. The support element forms the short leg at one end and at a right angle to the assault element. This provides both flanking (long leg) and enfilading (short leg) fires against the enemy. The L-shaped ambush can be used at a sharp bend in a trail, road, or stream. It should not be used where the short leg would have to cross a straight road or trail.

Patrols

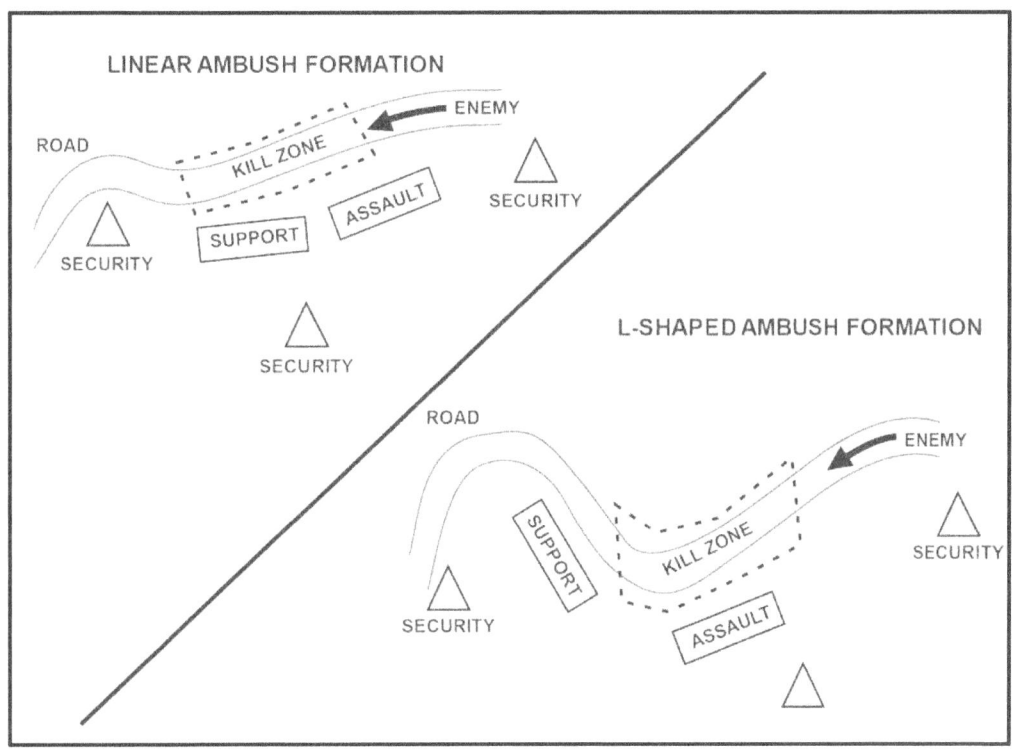

Figure 7-2. Ambush formations

HASTY AMBUSH

7-39. The platoon moves quickly to concealed positions. The ambush is not initiated until the majority of the enemy is in the kill zone. The unit does not become decisively engaged but surprises the enemy. The patrol captures, kills, or forces the withdrawal of the entire enemy within the kill zone.

7-40. On order, the patrol withdraws all personnel and equipment in the kill zone from observation and direct fire. The unit does not become decisively engaged by follow-on elements. The platoon continues follow-on operations. Actions on the objective are as follows (see figure 7-3 on page 7-12):

- Using visual signals, any Ranger alerts the unit that an enemy force is in sight. The Ranger continues to monitor the location and activities of the enemy force until the team or squad leader relieves him, and gives the enemy location and direction of movement.
- The platoon or squad halts and remains motionless:
  - The PL gives the signal to conduct a hasty ambush, taking care not to alert the enemy of the patrol's presence.
  - The leader determines the best nearby location for a hasty ambush and uses arm and hand signals to direct the unit members to covered and concealed positions.

Chapter 7

- The leader designates the location and extent of the kill zone.
- Teams and squads move silently to covered and concealed positions, ensuring positions are undetected and have good observation and fields of fire into the kill zone.
- Security elements move out to cover each flank and the rear of the unit. The leader directs the security elements to move a given distance, set up, and then rejoin the unit on order or after the ambush (the sound of firing ceases). At squad level, the two outside buddy teams normally provide flank security, as well as fires into the kill zone. At platoon level, fire teams make up the security elements.
- The PL assigns sectors of fire and issues any other commands necessary, such as control measures.
- The PL initiates the ambush, using the greatest casualty-producing weapon available, when the largest percentage of enemy is in the kill zone. The PL:
    - Controls the rate and distribution of fire.
    - Employs indirect fire to support the ambush.
    - Orders cease-fire.
    - If the situation dictates, orders the patrol to assault through the kill zone.
- The PL designates personnel to conduct a hasty search of enemy personnel, and process enemy prisoners and equipment.
- The PL orders the platoon to withdraw from the ambush site along a covered and concealed route.
- The PL gains accountability, reorganizes as necessary, disseminates information, reports the situation, and continues the mission as directed.

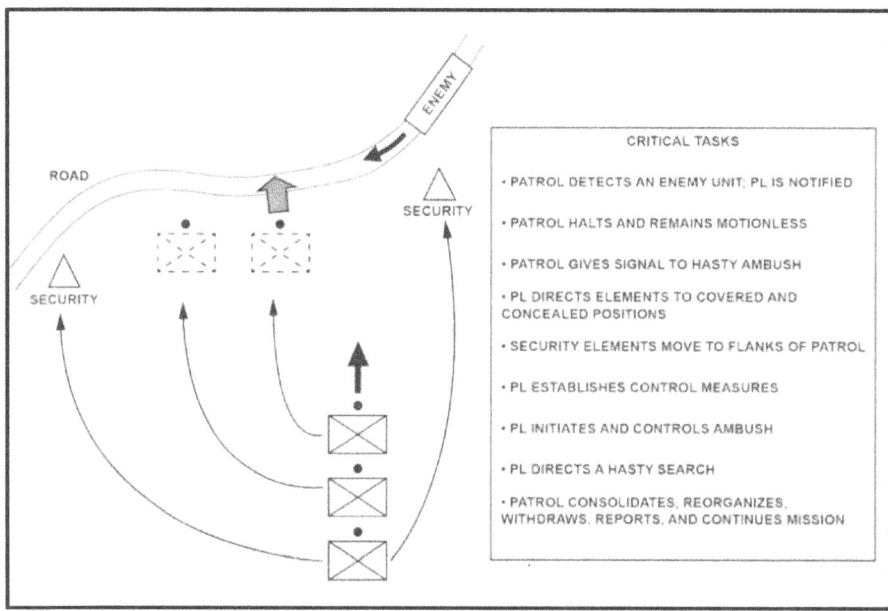

Figure 7-3. Hasty ambush

LEGEND
PL – platoon leader

Patrols

DELIBERATE (POINT/AREA) AMBUSH

7-41. The ambush is emplaced NLT the time specified in the order. The patrol surprises the enemy and engages their main body, killing or capturing all enemy in the kill zone, and destroying equipment based on the commander's intent. The patrol withdraws all personnel and equipment from the objective, on order, within the time specified in the order. The patrol obtains all available PIR from the ambush and continues follow-on operations. Actions on the objective are as follows (see figure 7-4 on page 7-15):

 a. The PL prepares the patrol for the ambush in the ORP.
 b. The PL prepares to conduct a leader's reconnaissance:
- Designates the members of the leader's reconnaissance party (SLs, surveillance team, FO, and possibly the security element).
- Issues a contingency plan to the PSG.

 c. The PL conducts the leader's reconnaissance:
- Ensures the leader's reconnaissance party moves undetected.
- Confirms the objective location and suitability for the ambush.
- Selects a kill zone.
- Posts the surveillance team at the site and issues a contingency plan.
- The security team(s) occupy prior to the PL reconnoitering the assault or support-by-fire positions, securing the flanks of the ambush site and providing early warning. A security team remains in the ORP if the patrol plans to return to the ORP after actions on the objective. If the ORP is abandoned, a rear security team should be emplaced " after "plan" and before "confirms." The OBJ must be secure prior to initiating reconnaissance of the assault or SBF positions.
- Confirms suitability of assault and support positions, and routes from the positions to the ORP.
- Selects position for each weapon system in SBF position, and then designates sectors of fire.
- Identifies all offensive control measures to be used. Identifies the probable line of deployment (PLD), the assault position, LOA, any boundaries or other control measures. If available, the PL can use infrared aiming devices to identify these positions on the ground.

 d. The PL adjusts the plan based on information from the reconnaissance:
- Assigns positions.
- Designates withdrawal routes.

 e. The PL confirms the ambush formation.
 f. Support element leader assigns sectors of fire:
- Emplaces Claymore mines and obstacles, as designated.
- Identifies sectors of fire and emplaces limiting stakes, or uses metal-to-metal contact with the machine gun tripod to prevent fratricide on the objective.
- Overwatches the movement of the assault element into position.

 g. Once the support element is in position, or on the PLs order, the assault element:
- Departs the ORP and moves into position.
- Upon reaching the assault position, leaders identify individual sectors of fire, as assigned by the PL. Emplaces aiming stakes.
- Emplaces Claymore mines to help destroy the enemy in the kill zone.
- Camouflages positions.

 h. The security element spots the enemy and notifies the PL with reports on the direction of movement, size of the target, and any special weapons or equipment carried. The security element also keeps the PL informed if any enemy forces are following the lead force.

 i. The PL alerts other elements and determines if the enemy force is too large, or if the ambush can engage the enemy successfully.

Chapter 7

j. The PL initiates the ambush using the highest casualty-producing device. The PL may use a command-detonated Claymore mine, and plans a backup method for initiating the ambush, in case the primary means fails. This should also be a casualty-producing device, such as an individual weapon. This information is passed to all Rangers, and practiced during rehearsals.

k. The PL ensures that the assault and support elements deliver fire with the heaviest, most accurate volume possible on the enemy in the kill zone. In limited visibility, the PL may use infrared lasers to define specific targets in the kill zone.

l. Before assaulting the target, the PL gives the signal to lift or shift fires.

m. The assault element:
- Assaults before the remaining enemy can react.
- Kills or captures enemy in the kill zone.
- Uses individual movement techniques or bounds by fire teams to move.
- Upon reaching the LOA, halts and establishes security. If needed, it reestablishes the chain of command and key weapon systems. All Rangers load a fresh magazine or drum of ammunition using the buddy system. LACE reports are submitted through the chain of command. The PL submits an initial contact report to higher HQ.

n. The PL directs special teams (EPW search, aid and litter, demolition) to accomplish their assigned tasks once the assault element has established its LOA:
- Once the kill zone is clear, collect and secure all EPWs and move them out of the kill zone before searching their bodies. Coordinate for an EPW exchange point to linkup with higher HQ to extract all EPWs and treat them according to secure, search, segregate, safeguard, and speed (5-Ss).
- Search from one side to the other and mark bodies that have been searched to ensure the area is thoroughly covered. Units should use the clear out/search in the intelligence technique (clear from the center of the objective and out), ensuring the area is clear of all enemy combatants. Then search all enemy personnel towards the center of the objective. Search all dead enemy personnel using the two-Ranger search technique:
    - As the search team approaches a dead enemy soldier, one Ranger guards while the other Ranger searches. First, kick the enemy's weapon away.
    - Second, roll the body over (if on the stomach) by lying on top and when given the go-ahead by the guard (who is positioned at the enemy's head), the searcher rolls the body over on him. This is done for protection in case the enemy soldier has a grenade with the pin pulled underneath him.
    - The searchers then conduct a systematic search of the dead soldier from head to toe, removing all papers and anything new (different type rank, shoulder boards, different unit patch, pistol, weapon, or NVD). They note if the enemy has a fresh or shabby haircut and the condition of the uniform and boots. They note the radio frequency, and then they secure the SOI, maps, documents, and overlays.
    - Once the body has been thoroughly searched, the search team continues in this manner until all enemy personnel in and near the kill zone have been examined.
- Identify, collect, and prepare all equipment to be carried back or destroyed.
- Evacuate and treat friendly wounded first, then enemy wounded if time permits.
- The demolition team prepares dual primed explosives or incendiary grenades and awaits the signal to initiate. This is normally the last action performed before the unit departs the objective and may signal the security elements to return to the ORP.
- All actions on the objective with stationary assault line are the same, with the exception of the search teams. To provide security within the teams to the farside of the kill zone during the search, they work in three-Ranger teams. Before the search begins, the Rangers move all KIAs to the near side of the kill zone.

o. If enemy reinforcements try to penetrate the kill zone, the flank security engages to prevent the assault element from being compromised.

p. The PL directs the units' withdrawal from the ambush site:
- Elements normally withdraw in the reverse order when they established their positions.

Patrols

- The elements may return to the RP or directly to the ORP, depending on the distance between elements.
- The security element of the ORP is alert to assist the platoon's return to the ORP. It maintains security for the ORP while the rest of the platoon prepares to leave.
- If possible, all elements should return to the location where they separated from the main body. This location should usually be the RP.

q. The PL and PSG direct actions at the ORP, including accountability of personnel and equipment, and recovery of rucksacks and other equipment left at the ORP during the ambush.

r. The PL disseminates information, or moves the platoon to a safe location (no less than one kilometer or one terrain feature away from the objective) and disseminates information.

s. As required, the PL and FO execute indirect fires to cover the platoon's withdrawal.

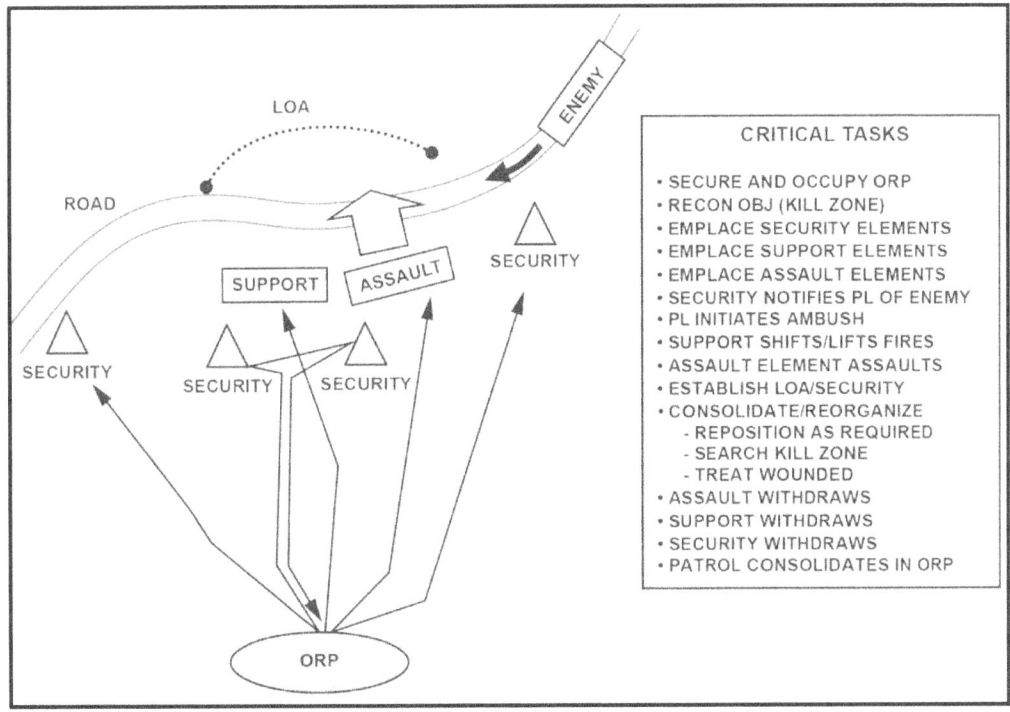

Figure 7-4. Deliberate ambush

LEGEND
LOA – limit of advance; OBJ – objective; ORP – objective rally point; PL – platoon leader; RECON - reconnoiter

Chapter 7

PERFORMING A RAID

7-42. The patrol initiates the raid NLT the time specified in the order, surprises the enemy, assaults the objective, and accomplishes its assigned mission within the commander's intent. The patrol does not become decisively engaged en route to the objective. The patrol obtains all available PIR from the raid objective and continues follow-on operations:

   a. **Planning considerations.** A raid is a form of attack, usually small scale, involving a swift entry into hostile territory to secure information, confuse the enemy, or destroy installations, followed by a planned withdrawal. Squads do not conduct raids. The sequence of platoon actions for a raid is similar to those for an ambush. Additionally, the assault element of the platoon may have to conduct a breach of an obstacle. It may have additional tasks to perform on the objective, such as demolition of fixed facilities. Fundamentals of the raid include:
   - Surprise and speed. Infiltrate and surprise the enemy without being detected.
   - Coordinated fires. Seal off the objective with well-synchronized direct and indirect fires.
   - Violence of action. Overwhelm the enemy with fire and maneuver.
   - Planned withdrawal. Withdraw from the objective in an organized manner, maintaining security.

   b. **Actions on the objective (raid).** (See figure 7-5 on page 7-18.)

   (1) The patrol moves to and occupies the ORP according to the patrol SOP. The patrol prepares for the leader's reconnaissance.

   (2) The PL, SLs, and selected personnel conduct a leader's reconnaissance:
   - PL leaves a five-point contingency plan with the PSG.
   - PL establishes the RP; pinpoints the objective; contacts the PSG to prepare Soldiers, weapons, and equipment; emplaces the surveillance team to observe the objective; and verifies and updates intelligence information. Upon emplacing the surveillance team, the PL provides a five-point contingency plan.
   - Security teams are brought forward on the leader's reconnaissance and emplaced before the leader's reconnaissance leaves the RP.
   - Leader's reconnaissance verifies location of and routes to security, support, and assault positions.
   - Leader's conduct the reconnaissance without compromising the patrol.
   - Leader's normally conduct a reconnaissance of SBF position first, then the assault position.

   (3) The PL confirms, denies, or modifies the plan and issues instructions to SLs.
   - Assigns positions and withdrawal routes to all elements.
   - Designates control measures on the objective (element objectives, lanes, limits of advance, target reference points, and assault line).
   - Allows SLs time to disseminate information and confirm that their elements are ready.

   (4) Security elements occupy designated positions, moving undetected into positions that provide early warning and can seal off the objective from outside support or reinforcement.

   (5) The support element leader moves the support element to designated positions. The support element leader ensures the element can place well-aimed fire on the objective.

   (6) The PL moves with the assault element into the assault position. The assault position is normally the last covered and concealed position before reaching the objective. As it passes through the assault position, the platoon deploys into its assault formation. Its squads and fire teams deploy to place the bulk of their firepower to the front as they assault the objective. They also:
   - Make contact with the surveillance team to confirm any enemy activity on the objective.
   - Ensure that the assault position is close enough for immediate assault, if the element is detected early.
   - Move into position undetected, and establish local security and fire control measures.

   (7) Element leaders inform the PL when their elements are in position and ready.

   (8) The PL initiates the raid and directs the support element to fire.

Patrols

(9) Upon gaining fire superiority, the PL directs the assault element to move towards the objective:
- Assault element holds fire until engaged, or until ready to penetrate the objective.
- PL signals the support element to lift or shift fires. The element lifts or shifts fires as directed, shifting fire to the flanks of targets or areas as directed in the order.

(10) The assault element attacks and secures the objective. The assault element may be required to breech a wire obstacle. As the platoon or its assault element moves onto the objective, it increases the volume and accuracy of fires. Squad leaders assign specific targets or objectives for their fire teams. Only when these direct fires keep the enemy suppressed can the rest of the unit maneuver. As the assault element gets closer to the enemy, there is more emphasis on suppression and less on maneuver. Ultimately, all but one fire team may be suppressing to allow that one fire team to break into the enemy position. Throughout the assault, Rangers use proper individual movement techniques, and fire teams retain their basic shallow wedge formation. The platoon does not get "on line" to sweep across the objective:
- Assault element assaults through the objective to the designated LOA.
- Assault element leaders establish local security along the LOA, and consolidate and reorganize as necessary. They provide LACE reports to the PL and PSG. The platoon establishes security, reorganizes key weapons, provides first aid, and prepares wounded Rangers for MEDEVAC. They redistribute ammunition and supplies, and relocate selected weapons to alternate positions if leaders believe that the enemy may have pinpointed them during the attack. They adjust other positions for mutual support. The squad and team leader provide LACE reports to the PL. The PL or PSG reorganizes the patrol based on the contact:
  - On order, special teams accomplish all assigned tasks under the supervision of the PL, who is positioned to control the patrol.
  - Special team leaders report to PL when assigned tasks are complete.

(11) On order or signal of the PL, the assault element withdraws from the objective. Using prearranged signals, the assault line begins an organized withdrawal from the objective site, maintaining control and security throughout the withdrawal. The assault element bounds back near the original assault line and begins a single file withdrawal through the antipersonnel land mine choke point. All Rangers move through the choke point for an accurate count. Once the assault element is a safe distance from the objective and the headcount is confirmed, the platoon can withdraw the support element. If the support elements were a part of the assault line, they withdraw together and security is signaled to withdraw. Once the support is a safe distance off the objective, they notify the PL, who contacts the security element and signals them to withdraw. All security teams linkup at the RP and notify the PL before moving to the ORP. Personnel returning to the ORP immediately secure their equipment and establish all-around security. Once the security element returns, the platoon moves out of the objective area as soon as possible:
- Before withdrawing, the demolition team activates devices and charges.
- Support element or designated personnel in the assault element maintain local security during the withdrawal.
- Leaders report updated accountability and status to the PL and PSG.

(12) Squads withdraw from the objective in the order designated in the order to the ORP:
- Account for personnel and equipment.
- Disseminate information.
- Redistribute ammunition and equipment as required.

(13) The PL reports mission accomplishment to higher HQ and continues the mission:
- Reports raid assessment to higher HQ.
- Informs higher HQ of any information requirements and PIR gathered.

Chapter 7

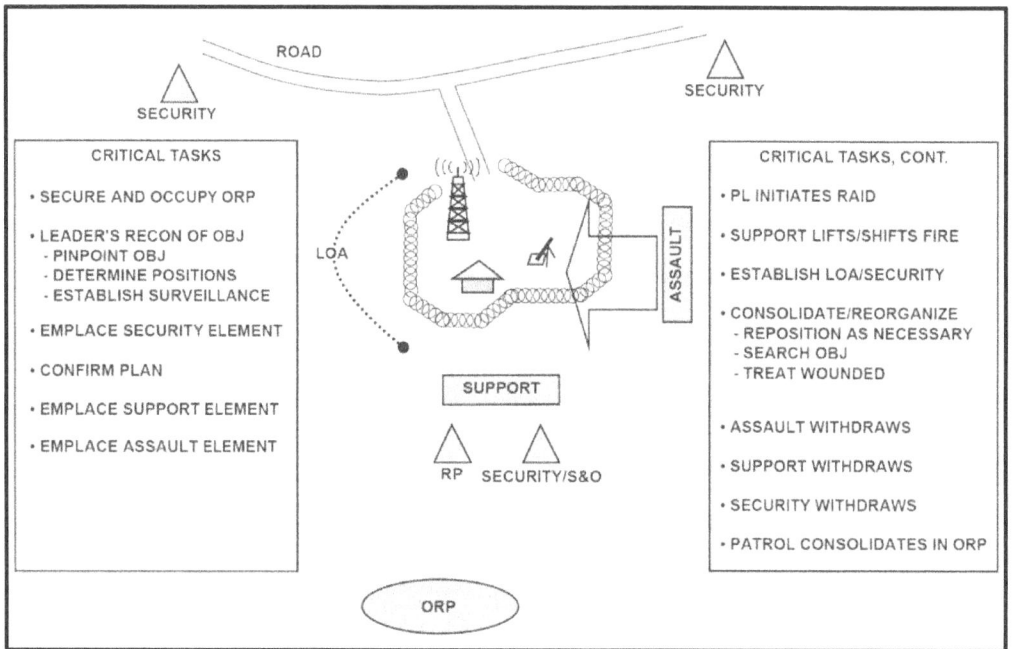

Figure 7-5. Actions on the objective, raid

LEGEND
CONT. - continued; LOA – limit of advance; OBJ – objective; ORP – objective rally point; PL – platoon leader; RECON – reconnaissance; RP – release point; S&O – surveillance and observation

SUPPORTING TASKS

7-43. Supporting tasks include a linkup, patrol debriefing, and occupation of an ORP. A linkup is a meeting of friendly ground forces. Linkups depend on control, detailed planning, communications, and stealth. This includes:

    a. **Task standard.** The unit's linkup at the time and place specified in the order. The enemy does not surprise the main bodies. The linkup units establish a consolidated chain of command.

    b. **Site selection.** The leader identifies a tentative linkup site by map reconnaissance, other imagery, or higher HQ designates a linkup site. The linkup site should have the following characteristics:

- Ease of recognition.
- Cover and concealment from ground and air.
- No tactical value to the enemy.
- Location away from natural lines of drift.
- Defendable for a short period of time.
- Multiple access and escape routes.

Patrols

c. **Execution.** Linkup procedure begins as the unit moves to the linkup point. The steps of this procedure are:
(1) The stationary unit performs linkup actions:
- Occupies the linkup rally point NLT the time specified in the order.
- Establishes all-around security, establishes communications, and prepares to accept the moving unit.
- The security team clears the immediate area around the linkup point. It then marks the linkup point with the coordinated recognition signal. The security team moves to a covered and concealed position, and observes the linkup point and immediate area around it.

(2) The moving unit:
- Performs linkup actions.
- The unit reports its location using phase lines, checkpoints, or other control measures.
- Halts at a safe distance from the linkup point in a covered and concealed position (the linkup rally point).

(3) The PL and a contact team:
- Prepare to make physical contact with the stationary unit.
- Issue a contingency plan to the PSG.
- Maintain communications with the platoon, and verifies near and far recognition signals for linkup (good visibility and limited visibility).
- Exchange far and near recognition signals with the linkup unit; conduct final coordination with the linkup unit.

(4) The stationary unit:
- Guides the patrol from its linkup rally point to the stationary unit linkup rally point.
- Linkup is complete by the time specified in the order.
- The main body of the stationary unit is alerted before the moving unit is brought forward.

(5) The patrol continues its mission according to the order.

d. **Coordination checklist.** The PL coordinates or obtains the following information from the unit that the patrol will linkup with:
- Exchange frequencies, call signs, codes, and other communication information.
- Verify near and far recognition signals.
- Exchange fire coordination measures.
- Determine command relationship with the linkup unit; plan for consolidation of chain of command.
- Plan actions following linkup.
- Exchange control measures such as phase lines and contact points, as appropriate.

7-44. Immediately after the platoon or squad returns, personnel from higher HQ conduct a thorough debrief. This may include all members of the platoon or the leaders, RTOs, and any attached personnel. Normally, the debriefing is oral. Sometimes a written report is required. Information on the written report should include:
- Size and composition of the unit conducting the patrol.
- Mission of the platoon, such as type of patrol, location, and purpose.
- Departure and return times.
- Routes. Use checkpoints and grid coordinates for each leg, or include an overlay.
- Detailed description of terrain and enemy positions that were identified.
- Results of any contact with the enemy.
- Unit status at the conclusion of the patrol mission, including the disposition of dead or wounded Rangers.
- Conclusions or recommendations.

7-45. The ORP is a point out of sight, sound, and small arms range of the objective area. It is normally located in the direction that the platoon plans to move after completion of actions on the objective. The ORP is tentative until the objective is pinpointed. This includes:

Chapter 7

a. **Occupation of the ORP.** (See figure 7-6.)
   (1) The patrol halts beyond sight and sound of the tentative ORP (200-to-400 m in good visibility, 100-to-200 m in limited visibility).
   (2) The patrol establishes a security halt according to the unit SOP.
   (3) After issuing a five-point contingency plan to the PSG, the PL moves forward with a reconnaissance element to conduct a leader's reconnaissance of the ORP.
   (4) For a squad-sized patrol, the PL moves forward with a compass man and one member of each fire team to confirm the ORP.
   - After physically clearing the ORP location, the PL leaves two Rangers at the 6 o'clock position facing in opposite directions.
   - The PL issues a contingency plan and returns with the compass man to guide the patrol forward.
   - The PL guides the patrol forward into the ORP, with one team occupying from 3 o'clock through 12 o'clock to 9 o'clock, and the other occupying from 9 o'clock through 6 o'clock to 3 o'clock.
   (5) For a platoon-sized patrol, the PL, RTO, WSL, three ammunition bearers, a team leader, a squad automatic weapon (SAW) gunner, and riflemen. Go on the leader's reconnaissance for the ORP, and position themselves at 10, 2, and 6 o'clock.
   - First squad in the order of march is the base squad, occupying from 10 to 2 o'clock. They are arranged:
   - Trail squads occupy from 2 to 6 o'clock and 6 to 10 o'clock, respectively.
   - Patrol HQ element occupies the center of the triangle.

b. **Actions in the ORP.** The unit prepares for the mission in the ORP. Once the leader's reconnaissance pinpoints the objective, the PSG generally lines up rucksacks according to unit SOP, in the center of the ORP.

Patrols

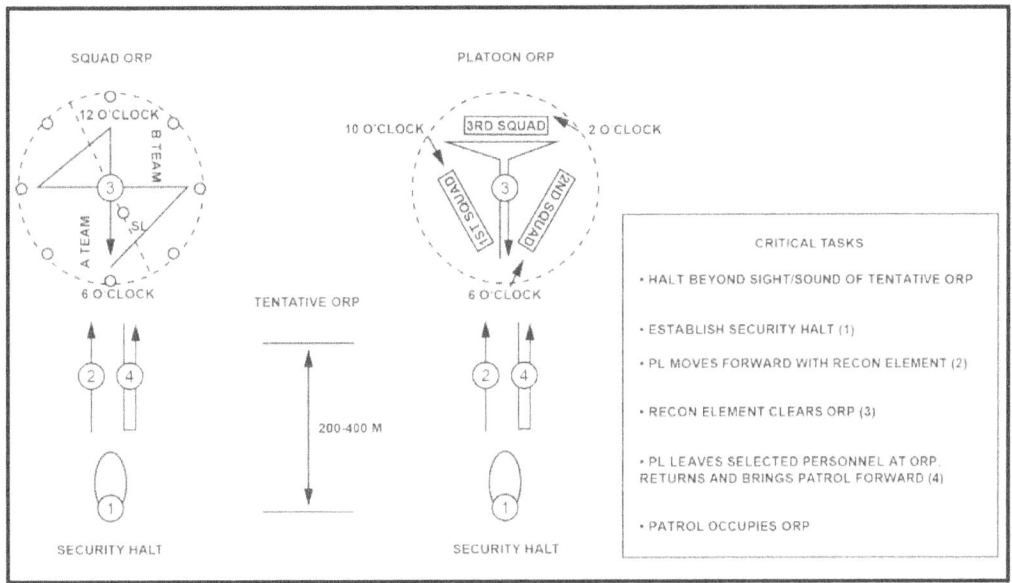

Figure 7-6. Occupation of the objective rally point

LEGEND
M – meter; ORP – objective rally point; PL platoon leader; RECON - reconnaissance

7-46. **Patrol base.** A PB is a security perimeter that is set up when a squad or platoon conducting a patrol halts for an extended period. Patrol bases should not be occupied for more than a 24-hour period (except in an emergency). A patrol never uses the same PB twice.

    a. **Use.** PBs are typically used:
- To avoid detection by eliminating movement.
- To hide a unit during a long, detailed reconnaissance.
- To perform maintenance on weapons, equipment, eat, and rest.
- To plan and issue orders.
- To reorganize after infiltrating an enemy area.
- To establish a base from which to execute several consecutive or concurrent operations.

    b. **Site selection.** The leader selects the tentative site from a map or by aerial reconnaissance. Characteristics of a PB include a site that is easily defendable for short periods of time, away from natural lines of drift, away from high-speed avenues of approach, provides cover and concealment from both ground, and provides little to no tactical advantage to the enemy. The site's suitability is confirmed and secured before the unit moves into it. Plans to establish a PB include selecting an alternate PB site. The alternate site is used if the first site is unsuitable or the patrol unexpectedly evacuates.

    c. **Planning considerations.** Leaders planning for a PB consider the mission, and passive and active security measures. A PB is located to allow the unit to accomplish its mission. This includes:
- Observation posts and communication with observation posts.

Chapter 7

- Patrol or platoon fire plan.
- Alert plan.
- Withdrawal plan from the PB, including withdrawal routes and a rally point, rendezvous point, or alternate PB.
- A security system to make sure that specific Rangers are awake at all times.
- Enforcement of camouflage, noise, and light discipline.
- The conduct of required activities with minimum movement and noise.
- Priorities of work.

d. **Security measures:**
- Select terrain the enemy would probably consider of little tactical value.
- Select terrain that is off main lines of drift.
- Select difficult terrain that would impede foot movement, such as an area of dense vegetation, preferably bushes and trees that spread close to the ground.
- Select terrain near a source of water.
- Select terrain that can be defended for a short period, and that offers good cover and concealment.
- Avoid known or suspected enemy positions.
- Avoid built up areas.
- Avoid ridges and hilltops, except as needed for maintaining communications.
- Avoid small valleys.
- Avoid roads and trails.

e. **Occupation.** (See figure 7-7.)

(1) A PB is reconnoitered and occupied in the same manner as an ORP, with the exception that the platoon typically plans to enter at a 90-degree turn. The PL leaves a two-Ranger observation post at the turn, and the patrol covers any tracks from the turn to the PB.

(2) The platoon moves into the PB. Squad-sized patrols generally occupy a cigar-shaped perimeter; platoon-sized patrols generally occupy a triangle-shaped perimeter.

(3) The PL and another designated leader inspect and adjust the entire perimeter, as necessary.

(4) After the PL has checked each squad sector, each SL sends a two-Ranger R&S team to the PL at the command post (CP). The PL issues the three R&S teams a contingency plan, reconnaissance method, and detailed guidance on what to look for (enemy, water, built-up areas or human habitat, roads, trails, or possible rally points).

Patrols

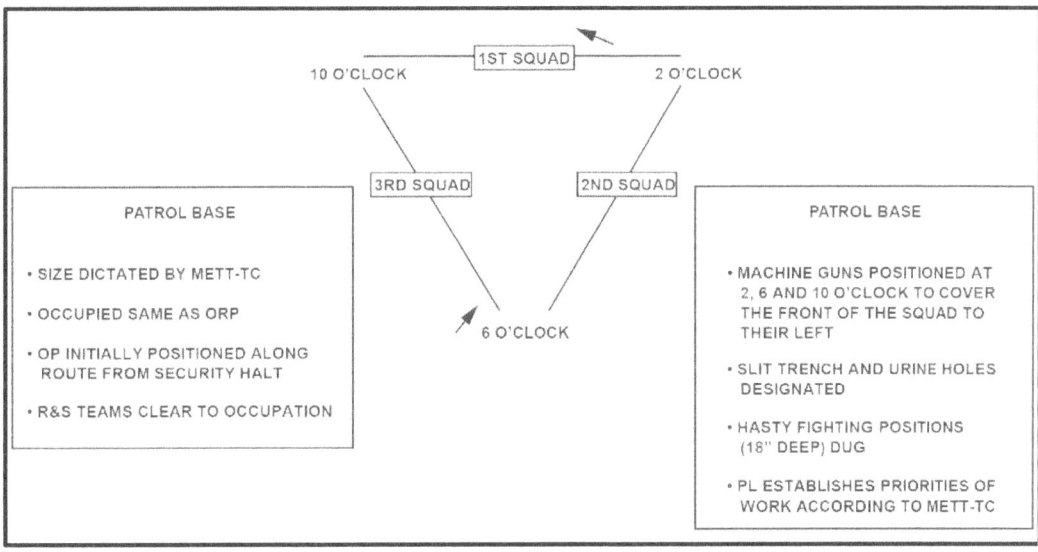

Figure 7-7. Patrol base

LEGEND
METT-TC - mission, enemy, terrain and weather, troops and support-time available, and civil considerations; OP – observation post; ORP – objective rally point; PL – platoon leader; R&S – reconnaissance and surveillance

(5) Where each R&S team departs is based on the PLs guidance. The R&S team moves a prescribed distance and direction, and reenters where the PL dictates.
- Squad-sized patrols do not normally send out an R&S team at night.
- R&S teams prepare a sketch of the area to the squad front, if possible.
- The patrol remains 100 percent alert during this reconnaissance.
- If the PL feels the patrol was tracked or followed, he may elect to wait in silence at 100 percent alert before sending out R&S teams.
- The R&S teams may use methods such as the "I," the "Box," or the "T." Regardless of the method chosen, the R&S team must be able to provide the PL with the same information.
- Upon completion of R&S, the PL confirms or denies the PB location and either moves the patrol or begins priorities of work.

f. **Passive (clandestine) patrol base (squad):**
- The purpose of a passive PB is for a squad or smaller-size element to rest.
- Unit moves as a whole and occupies in force.
- Squad leader ensures that the unit moves in at a 90-degree angle to the order of movement.
- A Claymore mine is emplaced on the route entering the PB.
- Alpha and Bravo teams sit back-to-back facing outward, ensuring that at least one individual on each team is alert and providing security.

g. **Priorities of work (platoon and squad).** Once the PL is briefed by the R&S teams and determines the area is suitable for a PB, the leader establishes or modifies defensive work priorities in order to establish the defense of the PB. Priorities of work are not a laundry list of tasks to be completed; to be effective, priorities of work consist of a task, a given time, and a measurable performance standard. For each priority of work, a clear standard is issued to guide the element in the successful accomplishment of each task. It is also designated whether the work is controlled in a centralized or decentralized manner. Priorities of work are determined according to METT-TC and may include, but are not limited to the following tasks:

(1) **Security (continuous):**
- Prepare to use all passive and active measures to cover the entire perimeter at all times, regardless of the percentage of weapons used to cover all of the terrain.
- Readjust after R&S teams return, or readjust based on current priority of work (such as weapons maintenance).
- Employ all elements, weapons, and personnel to meet conditions of the terrain, enemy, or situation.
- Assign sectors of fire to all personnel and weapons. Develop squad sector sketches and platoon fire plan.
- Confirm location of fighting positions for cover, concealment and observation, and fields of fire. SLs supervise placement of aiming stakes and Claymore mines.
- Only use one point of entry and exit, and count personnel in and out. Everyone is challenged, according to the unit SOP.
- Hasty fighting positions are prepared at least 18 inches deep (at the front) and sloping gently from front to rear, with a grenade sump, if possible.

(2) **Withdrawal plan.** The PL designates the signal for withdrawal, order of withdrawal, and the platoon rendezvous point or alternate PB.

(3) **Communication (continuous).** Communications are maintained with higher HQ, OPs, and within the unit. This may be rotated between the patrol's RTOs to allow accomplishment of continuous radio monitoring, radio maintenance, act as runners for the PL, or conduct other priorities of work.

(4) **Mission preparation and planning.** The PL uses the PB to plan, issue orders, rehearse, inspect, and prepare for future missions.

(5) **Weapons and equipment maintenance.** The PL ensures that machine guns, weapon systems, communications equipment, and night vision devices (as well as other equipment) are maintained. These items are not disassembled at the same time for maintenance (no more than 33 percent at a time), and weapons are not disassembled at night. If one machine gun is down, then security for all remaining systems is raised.

(6) **Water resupply.** The PSG organizes watering parties, as necessary. The watering party carries canteens in an empty rucksack or duffel bag, and has communications and a contingency plan prior to departure.

(7) **Mess plan.** At a minimum, security and weapons maintenance are performed prior to mess. Normally, no more than half the platoon eats at one time. Rangers typically eat one-to-three meters behind their fighting positions.

- Rest and sleep plan management. The patrol conducts rest as necessary, to prepare for future operations.
- Alert plan and stand to. The PL states the alert posture and the stand to time. The plan ensures all positions are checked periodically, OPs are relieved periodically, and at least one leader is always alert. The patrol typically conducts stand to at a time specified by the unit SOP, such as 30 minutes before and after begin evening nautical twilight (BMNT) or ending evening nautical twilight (EENT).
- Resupply. Distribute or cross-load ammunition, meals, equipment, and other items.
- Sanitation and personal hygiene. The PSG and medic ensure a slit trench is prepared and marked. All Rangers brush their teeth; wash faces; shave; and wash hands, armpits, groin, and feet. The patrol does not leave trash behind.

Patrols

MOVEMENT TO CONTACT

7-47. The movement to contact (MTC) is one of the five types of offensive operations. An MTC gains or regains contact with the enemy. Once contact is made, the unit develops the situation. Normally, a platoon conducts an MTC as part of a larger force. The two techniques for conducting an MTC are search and attack (S&A), and cordon and search.

   a. **Search and attack.** The search and attack technique is used when the enemy is dispersed, expected to avoid contact, disengaged or withdraws, or their movement in an area is denied. The search and attack technique involves the use of multiple platoons, squads, and fire teams coordinating their actions to make contact with the enemy. Platoons typically try to find the enemy, and then fix and finish them. They combine patrolling techniques with the requirement to conduct hasty or deliberate attacks once the enemy has been found.

   (1) Planning considerations:
   - Factors of METT-TC.
   - Requirement for decentralized execution.
   - Requirement for mutual support.
   - Length of operations.
   - Minimize soldier's load to improve stealth and speed.
   - Resupply and MEDEVAC.
   - Positioning key leaders and equipment.
   - Employment of key weapons.
   - Requirement for PBs.
   - Concept for entering the zone of action.
   - Concept for linkups while in contact.

   (2) Critical performance measures.
   - The platoon locates the enemy without being detected.
   - Once engaged, fixes the enemy in position and maneuvers against the enemy.
   - Maintains security throughout actions to avoid being flanked.

   b. **Cordon and search.** A technique of conducting an MTC that involves isolating a target area and searching suspect locations within that target area to capture or destroy possible enemy forces and contraband. A platoon uses the cordon and search technique as part of a larger unit. It can be tasked as the cordon or the search element for the company or battalion.

   (1) **Planning.** Immediately on receipt of the mission, the company-level commander conducts a reconnaissance of the target to be searched. Ideally, the intelligence officer (S-2) can provide maps or satellite imagery. If possible, avoid sending a patrol to the area of the target since the patrol may unnerve the target and cause them to flee prior to the search. However, depending on how much information that a commander has, he may have no alternative to conduct a reconnaissance to determine where the target is located. The patrol may not alert the target if patrols frequent the area, but the patrol does not need to loiter in the area any longer than necessary.

   (2) **Preparing.** Rehearsals are key to a successful cordon and search operations, units should develop their own requirements for what to rehearse.

   (3) **Organization** of cordon and search elements:
   - Outer cordon element: security and support element.
   - Inner cordon element: search element.
   - Assault force: detention and collection element.

   (4) **Execution.** As the unit approaches the objective, the inner cordon and assault forces make sure they have allowed enough time for the outer cordon force to set before actually arriving at the target. The locals know the sound of military vehicles, and any subversive element who may be home will likely try to flee on hearing the unit approach. While the impact may not be immediate, vehicles and foot traffic (aside from curious onlookers) around the objective will decline quickly once the outer cordon is set, facilitating the movement of other elements to the objective.

Chapter 7

TASK STANDARDS

7-48. The platoon moves NLT the time specified in the order, and makes contact with the smallest element possible. The main body is not surprised by the enemy. Once the platoon makes contact, it maintains contact. The platoon destroys squad-sized and smaller-sized elements, and fixes elements larger than a squad. The platoon maintains sufficient fighting force capable of conducting further combat operations.

7-49. Reports of enemy locations and contact are forwarded. If not detected by the enemy, the PL initiates a hasty attack. The platoon sustains no casualties from friendly fire. The platoon is prepared to initiate further movement within 25 minutes of contact, and all personnel and equipment are accounted for.

Chapter 8

Battle Drills

This chapter provides battle drills that are standard collective actions made in response to common battle occurrences. Graphic examples can be found in Appendix A. Quick reference cards can be found in Appendix B. (For more information, see ATP 3-21.8.)

A drill is a collective action (or task) is performed by a platoon or smaller element without the application of a deliberate decision-making process. It is initiated on a cue, accomplished with minimal leader orders, and performed to standard throughout like units in the Army. The action is vital to success in combat operations, or critical to preserving life. It usually involves fire or maneuver. The drill is initiated on a cue, such as an enemy action or a leader's brief order, and is a trained response to the given stimulus.

REACT TO DIRECT FIRE CONTACT (07-3-D9501)

CONDITIONS: The unit is moving or halted. The enemy initiates direct fire contact on the unit.

CUE: This drill begins when the enemy initiates direct fire contact.

STANDARDS: The element in contact returns fire immediately and seeks cover. Element in contact locates the enemy and places well-aimed fire on known enemy position(s). The leader can point out at least one-half of the enemy positions and identify the types of weapons (such as small arms, and light machine guns). Unit leader reports the contact to higher headquarters.

TASK STEPS

1. The element in contact immediately returns well-aimed fire on known enemy position(s). Vehicles move out of the beaten zone.

2. Soldiers and vehicles assume the nearest covered and concealed position. Mounted Soldiers dismount the vehicle, provide local security and add its suppressive fire against the enemy position. (See figure 8-1 on page 8-2.)

Chapter 8

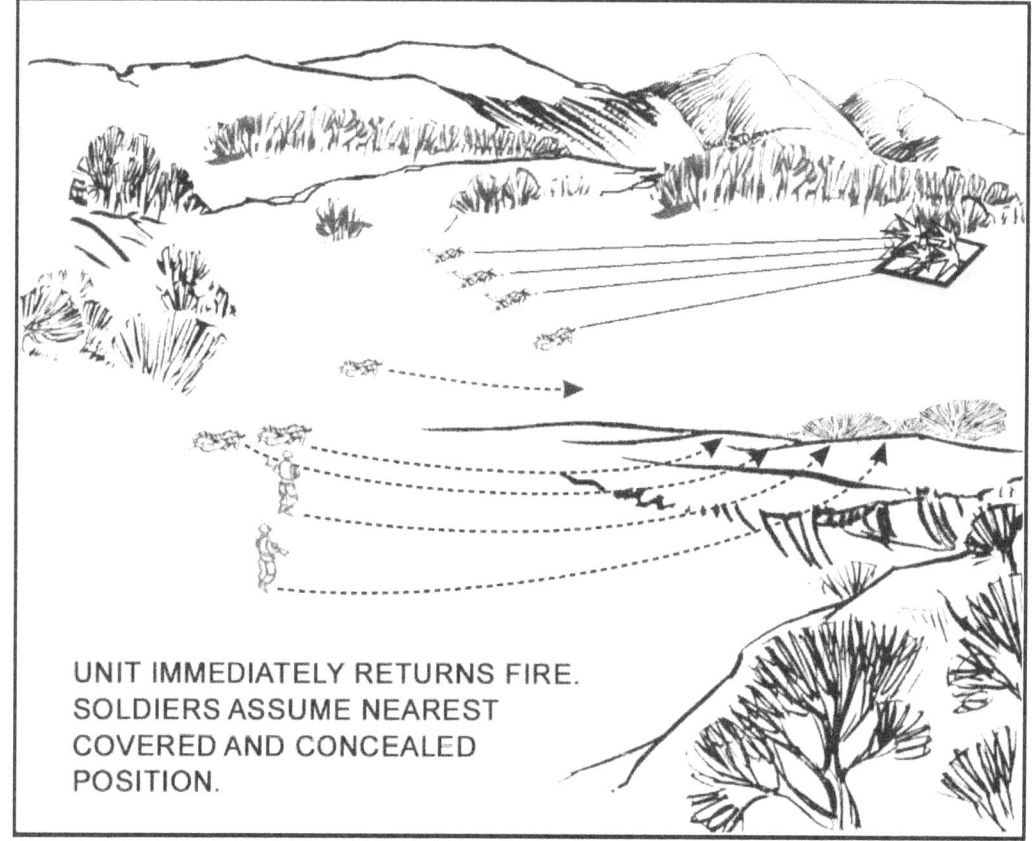

Figure 8-1. 07-3-D9501. Assuming nearest covered position

3. Element leaders locate and engage known enemy positions with well-aimed fire or battlesight fire command, and pass information to the unit leader and Soldiers.

4. Element leaders control the fire of their Soldiers by using standard fire commands (initial and supplemental) containing the following elements:

    a. Alert.

    b. Weapon or ammunition (optional).

    c. Target description.

    d. Direction.

    e. Range.

Battle Drills

    f. Method.

    g. Control (optional).

    h. Execution.

    i. Termination.

5. Soldiers and vehicle commanders maintain contact (visual or oral) with the leader, other Soldiers, and vehicles on their left or right.

6. Soldiers maintain contact with the team leader and indicate the location of enemy positions. Vehicle commanders relay all commands to the mounted Infantry squads.

7. Unit leaders (visually or orally) check the status of their personnel.

8. Element leaders maintain visual contact with the unit leader.

9. The unit leader moves up to the element in contact and links up with its leader:

    a. Unit leader brings the radiotelephone operator, forward observer, element leader of the nearest element, one crew-served weapon team (machine gun team, if available).

    b. Element leaders of the elements not in contact move to the front of their element.

    c. The platoon sergeant moves forward with the remaining crew-served weapons and links up with the unit leader, and assumes control of the support element. (See figure 8-2 on page 8-4.)

Chapter 8

Figure 8-2. 07-3-D9501. Control of the support element

10. The unit leader determines whether or not the unit must move out of the engagement area.

11. The unit leader determines whether or not the unit can gain and maintain suppressive fires with the element already in contact (based on the volume and accuracy of enemy fires against the element in contact).

12. The unit leader makes an assessment of the situation and identifies—

   a. The location of the enemy position and obstacles.

   b. The size of the enemy force engaging the unit in contact. (The number of enemy automatic weapons, the presence of any vehicles, and the employment of indirect fires are indicators of enemy strength.)

   c. Vulnerable flanks.

   d. Covered and concealed flanking routes to the enemy positions.

Battle Drills

13. The unit leader decides whether to conduct an assault, bypass (if authorized by the company commander), or break contact.

14. The unit leader reports the situation to higher headquarters and begins to maneuver the unit.

CONDUCT A PLATOON ASSAULT (07-3-D9514)

CONDITIONS: The platoon is moving as part of a larger force conducting a movement to contact or an attack. The enemy initiates direct fire contact on the lead squad.

CUE: This drill begins when the enemy initiates direct fire contact.

STANDARDS: The platoon lead squad locates and suppresses the enemy, establishes supporting fire, and assaults the enemy position using fire and maneuver. The platoon destroys or causes the enemy to withdraw, and consolidates and reorganizes.

TASK STEPS

1. The platoon conducts action on enemy contact. The squad or section in contact reacts to contact by immediately returning well-aimed fire on known enemy positions. Dismounted Soldiers assume the nearest covered positions. Vehicles move out of the beaten zone and Soldiers dismount the vehicle. The element in contact attempts to achieve suppressive fires. The element leader notifies the platoon leader of the action.

2. Platoon leader gives the command to dismount the vehicles. The platoon sergeant takes control of the vehicles.

3. The platoon leader, radiotelephone operator, forward observer, squad leader of the next squad, and one machine gun team move forward to linkup with the squad leader of the squad in contact.

4. The squad leader of the trail squad moves to the front of the lead fire team.

5. The weapons squad leader and second machine gun team move forward and linkup with the platoon leader. If directed, the weapons squad leader assumes control of the base-of-fire element and positions the machine guns to add suppressive fires against the enemy.

6. Platoon sergeant repositions vehicles, as necessary, to provide observation and supporting fire against the enemy.

7. The platoon leader assesses the situation.

8. If the squad in contact cannot achieve suppressive fire, the squad leader reports to the platoon leader.

   a. The squad in contact establishes a base of fire. The squad leader deploys the squad to provide effective, sustained fires on the enemy position. The squad leader reports the final position to the platoon leader.

   b. The remaining squads (not in contact) take up covered and concealed positions in place, and observe to the flanks and rear of the platoon.

   c. The platoon leader moves forward with the radiotelephone operator, platoon forward observer, squad leader of the nearest squad, and one machine gun team.

9. Lead squad locates the enemy.

   a. The squad leader of the squad in contact reports the enemy size and location, and any other information to the platoon leader. The platoon leader completes the squad leader's assessment of the situation.

   b. The squad continues to engage the enemy's position.

   c. The weapons squad leader moves forward with the second machine gun team and links up with the platoon leader.

Chapter 8

d. The platoon sergeant repositions vehicles, as necessary, to provide observation and supporting fire against the enemy.

10. Lead squad suppresses the enemy:

a. The platoon leader determines if the squad in contact can gain suppressive fire against the enemy based on the volume and accuracy of the enemy's return fire.

(1) If the answer is **YES**, the platoon leader directs the squad (with one or both machine guns) and vehicle element in contact to continue suppressing the enemy:

(a) The squad in contact destroys or suppresses enemy weapons that are firing most effectively against it; normally crew-served weapons.

(b) The vehicle section in contact destroys or suppresses enemy weapons that were firing most effectively against them, including vehicles and crew-served weapons.

(c) The squad in contact places screening smoke (M203/320) to prevent the enemy from seeing the maneuver element.

(2) If the answer is **NO**, the platoon leader deploys another squad, second vehicle section, and the second machine gun team to suppress the enemy position. (The platoon leader may direct the trail leader to position this squad and vehicle section, and weapons squad leader to position one or both machine gun teams in a better support-by-fire position.)

b. The platoon leader again determines if the platoon can gain suppressive fires against the enemy.

(1) If the answer is **YES**, the platoon leader continues to suppress the enemy with the two squads, two machine guns, and vehicle-mounted weapons.

(a) The trail squad leader assumes control of the base-of-fire element (squad in contact, machine gun teams, and any other squads designated by the platoon leader).

(b) The platoon sergeant assumes control of the vehicle section and base-of-fire element (squad in contact and machine gun teams designated by the platoon leader).

(c) The platoon forward observer calls for and adjusts fires based on the platoon leader's directions. (The platoon leader does not wait for indirect fires before continuing with his actions.)

(2) If the answer is still **NO**, the platoon leader deploys the last squad to provide flank and rear security; guide the rest of the company forward, as necessary; and report the situation to the company commander. Normally, the platoon becomes the base-of-fire element for the company and may deploy the last squad to add suppressive fires. The platoon continues to suppress or fix the enemy with direct and indirect fire, and responds to orders from the company commander.

11. Platoon assaults the enemy position. If the squad(s) in contact, together with the machine gun(s) and vehicle element can suppress the enemy, the platoon leader determines if the remaining squad(s) that are not in contact can maneuver.

a. The platoon leader makes the following assessment:
(1) Location of enemy positions and obstacles.
(2) Size of enemy force. (The number of enemy automatic weapons, the presence of any vehicles, and the employment of indirect fires are indicators of enemy strength.)
(3) Vulnerable flank.
(4) Covered and concealed flanking route to the enemy position.

b. If the answer is **YES**, the platoon leader maneuvers the squad(s) into the assault:
(1) Once the platoon leader has ensured that the base-of-fire element is in position and providing suppressive fires, he leads the assaulting squad(s) to the assault position.

Battle Drills

(2) If the vehicle section can effectively suppress the enemy element, the platoon leader may reposition the weapons squad or machine gun to an intermediate or local support-by-fire position to provide additional suppression during the assault.

(3) Once in position, the platoon leader gives the prearranged signal for the base-of-fire element to lift or shift direct fires to the opposite flank of the enemy position. (The assault element MUST pick up and maintain effective fires throughout the assault. Handover of responsibility for direct fires from the base-of-fire element to the assault element is critical.)

(4) The platoon forward observer shifts indirect fires to isolate the enemy position.

(5) The assaulting squad(s) fight through enemy positions using fire and maneuver. The platoon leader controls the movement of the squads, assigns specific objectives for each squad, and designates the main effort or base maneuver element. The base-of-fire element must be able to identify the near flank of the assaulting squad(s).

(6) In the assault, the squad leader determines the way in which to move the elements of the squad based on the volume and accuracy of enemy fire against the squad, and the amount of cover afforded by the terrain. In all cases, each Soldier uses individual movement techniques, as appropriate.

    (a) The squad leader designates one fire team to support the movement of the other team by fires.

    (b) The squad leader designates a distance or direction for the team to move and accompanies one of the fire teams.

    (c) Soldiers must maintain contact with team members and leaders.

    (d) Soldiers time their firing and reloading in order to sustain their rate of fire.

    (e) The moving fire team proceeds to the next covered position. Teams use the wedge formation when assaulting. Soldiers move in rushes or by crawling.

    (f) The squad leader directs the next team to move.

    (g) If necessary, the team leader directs Soldiers to bound forward as individuals within buddy teams. Soldiers coordinate their movement and fires with each other within the buddy team, and maintain contact with their team leader.

    (h) Soldiers fire from covered positions. They select the next covered position before moving and rushing forward (no more than five seconds), or use high or low crawl techniques based on terrain and enemy fires.

c. If the answer is **NO**, or the assaulting squad(s) cannot continue to move, the platoon leader deploys the squad(s) to suppress the enemy and reports to the company commander. The platoon continues suppressing enemy positions and responds to the orders of the company commander.

12. The platoon consolidates on the objective once the assaulting squad(s) seize the enemy position:

    a. Establishes local security.

    b. Platoon leader signals for the base-of-fire element to move up into designated positions.

    c. Platoon leader assigns sectors of fire for each squad and vehicle.

    d. Platoon leader positions key weapons and vehicles to cover the most dangerous avenue(s) of approach.

    e. Platoon sergeant begins coordination for ammunition resupply.

    f. Soldiers take up hasty defensive positions.

    g. Platoon leader and forward observer develop a quick fire plan.

    h. Squads place out observation points to warn of enemy counterattacks.

13. Platoon organizes by:

    a. Reestablishing the chain of command.

    b. Redistributing and resupplying ammunition.

Chapter 8

    c. Manning crew-served weapons first.

    d. Redistributing critical equipment such as radios; chemical, biological, radiological, and nuclear equipment; and night vision devices.

    e. Treating casualties and evacuating wounded.

    f. Filling vacancies in key positions.

    g. Searching, silencing, segregating, safeguarding, and speeding enemy prisoners of war to collection points.

    h. Collecting and reporting enemy information and materiel.

14. Platoon sends a situation report to the company commander.

Conduct a Squad Assault (07-4-D9515)

CONDITIONS: The squad is moving as part of the platoon conducting a movement to contact or an attack. The enemy initiates direct fire contact.

CUE: This drill begins when the enemy initiates direct fire contact.

STANDARDS: The squad locates and suppresses the enemy, establishes supporting fire, and assaults the enemy position using fire and maneuver. The squad destroys or causes the enemy to withdraw, conducts consolidation and reorganizes.

TASK STEPS

1. The team in contact immediately returns well-aimed fire on known enemy position(s) and assumes the nearest covered positions. Soldiers receiving fire take up nearest positions that afford protection from enemy fire (cover) and observation (concealment).

2. Soldiers in contact assume the nearest position that provides cover and concealment.
    a. Fire team Soldiers in contact move to positions (bound or crawl) where they can fire their weapons, position themselves to ensure that they have observation, fields of fire, cover, and concealment. They continue to fire and report known or suspected enemy positions to the fire team leader.

    b. The team leader directs fires using tracers or standard fire commands.

    c. The fire team not in contact takes covered and concealed positions in place, and observes to the flanks and rear of the squad.

    d. The squad leader reports contact to the platoon leader and moves toward the fire team in contact.

3. Lead team locates the enemy:
    a. Using sight and sound, the fire team in contact acquires known or suspected enemy positions.

    b. The fire team in contact begins to place well-aimed fire on suspected enemy positions.

    c. The squad leader moves to a position to observe the enemy and assess the situation.

    d. The squad leader requests, through the platoon leader, immediate suppression indirect fires (normally 60-mm mortars).

    e. The squad leader reports the enemy size and location, and any other information to the platoon leader. (As the platoon leader comes forward, he completes the squad leader's assessment of the situation.)

4. Team in contact suppresses the enemy.

Battle Drills

a. The squad leader determines if the fire team in contact can gain suppressive fire based on the volume and accuracy of the enemy fire.

b. If the answer is **YES**, the fire team leader continues to suppress the enemy:

(1) The fire team destroys or suppresses enemy crew-served weapons first.
(2) The fire team places smoke (M203/320) on the enemy position to obscure it.
(3) The fire team leader continues to control fires using tracers or standard fire commands. Fires must be well-aimed and continue at a sustained rate with no lulls.
(4) Buddy teams fire their weapons so that both are not reloading their weapons at the same time.

c. If the answer is **NO**, the squad leader then deploys the fire team not in contact to establish a support-by-fire position. The squad leader reports the situation to the platoon leader. Normally, the squad becomes the base-of-fire element for the platoon. The squad continues to suppress the enemy and responds to orders from the platoon leader. (The platoon leader, radiotelephone operator, forward observer, one machine gun team, squad leader of the next squad, platoon sergeant, and the other machine gun team are already moving forward according to Battle Drill 2, Platoon Assault.)

5. The unit leader maneuvers the assault elements into the assault.

   a. Squad leader adjusts fires (both direct and indirect) based on the rate of the assault element movement and the minimum safe distances of weapons systems.

   b. Once in position, the squad leader gives the prearranged signal for the supporting fire team to shift direct fires to the opposite flank of the enemy position.

   c. The assaulting fire team assumes and maintains effective fires throughout the assault. Handover of responsibility for direct fires from support element to the assault element is critical to prevent fratricide.

   d. If available, unit leader directs the forward observer to shift indirect fire (including smoke) to isolate the enemy position.

6. The assaulting element(s) fight through enemy position(s) using fire and movement.

   a. Team leader controls the movement of the team.

   b. Team leader assigns specific objectives for each buddy team and designates a base maneuver element.

   c. Base-of-fire elements maintain visual contact of the near flank of the assaulting element.

   d. The assault element conducts fire and movement based on volume and accuracy of enemy fires against his element and the amount of cover afforded by the terrain.

   (1) Assault element leader designates a distance and direction for the assault element and moves with that element.
   (2) Soldiers maintain contact with team members and leaders.
   (3) Team leaders direct Soldiers to move as individuals or teams.
   (4) Soldiers fire from covered positions. Soldiers move using 3- to 5-second rushes or the low or high crawl techniques, taking advantage of available cover and concealment.
   (5) Soldiers time their firing and reloading in order to sustain their rate of fire.
   (6) Team leaders maintain contact with the unit leader and pass signals to element members.
   (7) If the assault element cannot continue to move, the unit leader deploys the element(s) to suppress the enemy and reports to higher headquarters.

7. The squad consolidates and reorganizes.

   a. Squad leaders establish local security.
   b. The squad leader signals for the base-of-fire element to move up into designated positions.
   c. The squad leader assigns sectors of fire for each element.
   d. The squad leader positions key weapons to cover the most dangerous avenue of approach.
   e. The squad leader begins coordination for ammunition resupply.
   f. Soldiers establish hasty fighting positions.
   g. Squad leader develops a quick fire plan.

Chapter 8

    h. Squad leader place out observation posts to warn of enemy counterattacks.
    i. Reestablishes the chain of command.
    j. Redistributes and resupplies ammunition.
    k. Mans crew-served weapons, first.
    l. Redistributes critical equipment such as radios chemical, biological, radiological, and nuclear equipment; and night vision devices.
    m. Treats and evaluates wounded.
    n. Fills vacancies in key positions.
    o. Searches, silences, segregates, safeguards, speeds, and tags detainees.
    p. The unit leader consolidates ammunition, casualties' and equipment reports.

8. Squad leader reports situation to platoon leader.

BREAK CONTACT (07-3-D9505)

CONDITIONS: The unit is moving as part of a larger force, conducting a movement to contact or an attack. Following direct fire contact with the enemy, the unit leader decides to break contact.

CUE: This drill begins when the unit leader gives the command to break contact.

STANDARDS: The unit breaks contact using fire and movement, and continues to move until the enemy cannot observe or place fire on them. The unit leader reports the contact to higher headquarters.

TASK STEPS

1. The unit leader directs an element to suppress the enemy.

2. The unit leader directs the vehicles to support the disengagement of the dismounted element. (If the vehicles cannot support the disengagement of the dismounted element, the platoon leader directs one squad or fire team to suppress by fire to support the disengagement of the remainder of the element.)

3. The unit leader orders a distance and direction, terrain feature, or last rally point of the movement of the element in contact.

4. The unit leader employs indirect fires to suppress enemy position(s). (See figure 8-3.)

Battle Drills

Figure 8-3. 07-3-D9505. Employing indirect fires to suppress enemy

5. The bounding element moves to occupy the overwatch position, employs smoke (M320, grenade launchers, indirect fires, and other options) to screen movement. If necessary, employs fragmentation and concussion grenades to facilitate breaking contact.

6. The base-of-fire element continues to suppress the enemy.

7. The moving element occupies their overwatch position and engages enemy position(s). (See figure 8-4 on page 8-12.)

Chapter 8

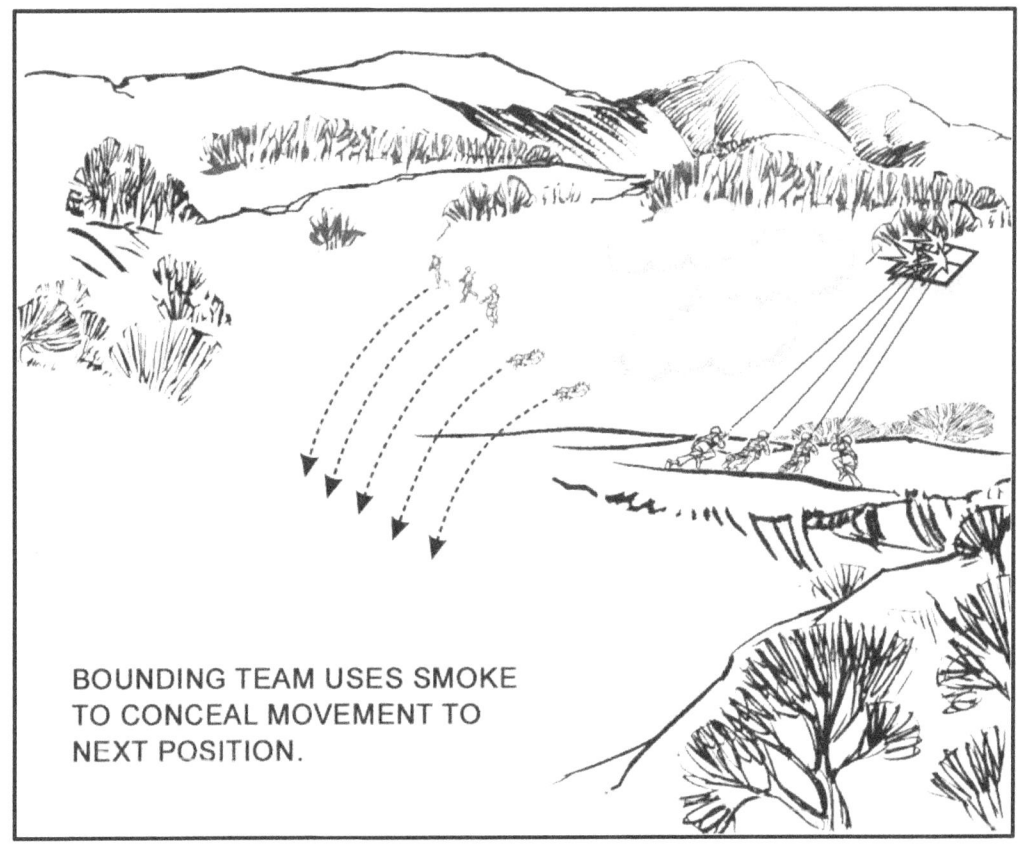

Figure 8-4. 07-3-D9505. Moving element occupies overwatch and engages enemy

8. The unit leader directs the base-of-fire element to move to its next covered and concealed position. Based on the terrain, and volume and accuracy of the enemy's fire, the moving element may need to use fire and movement techniques. (See figure 8-5.)

Battle Drills

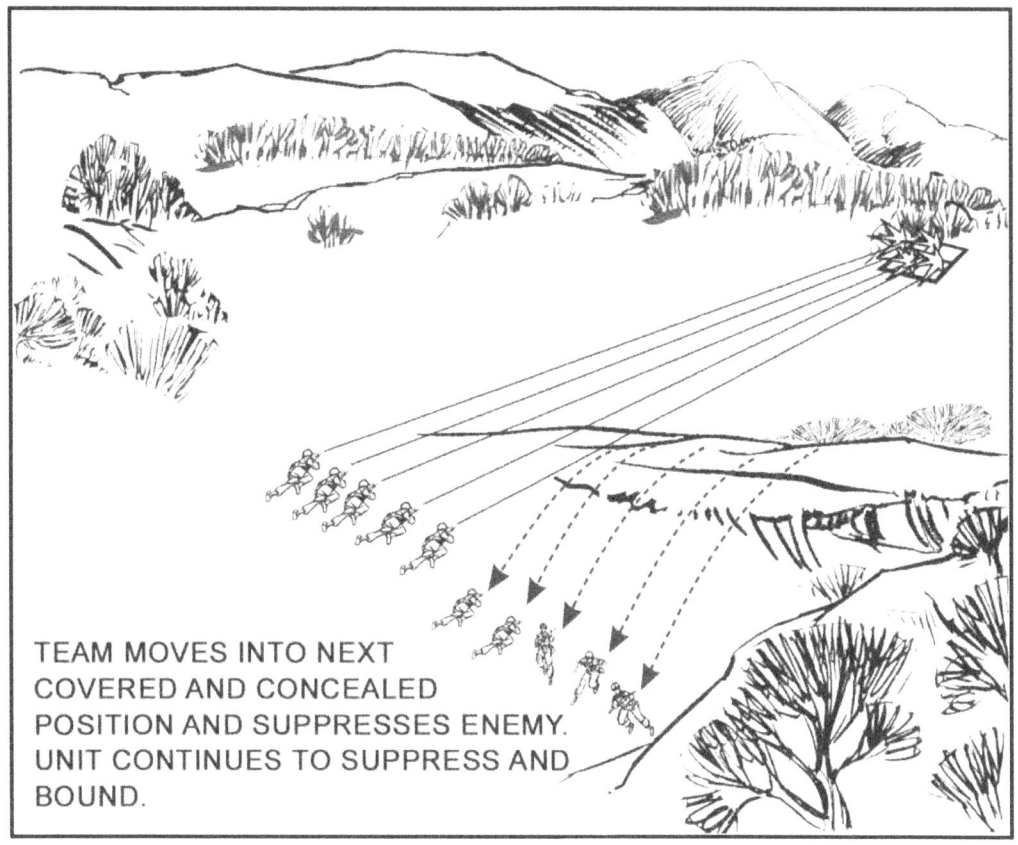

Figure 8-5. 07-3-D9505. Movement and fire technique

9. The unit continues to bound away from the enemy until:
    a. It breaks contact (the unit must continue to suppress the enemy as it breaks contact).
    b. It passes through a higher-level support-by-fire position.
    c. Its elements are in the assigned positions to conduct the next mission.

Note: For a mounted element, the platoon leader directs the vehicles to move to a rally point and linkup with the dismounted element.

Chapter 8

10. The leader should consider changing the unit's direction of movement once contact is broken. This reduces the ability of the enemy to place effective indirect fire on the unit.

11. Elements and Soldiers that become disrupted stay together and move to the last designated rally point.

12. Unit leaders account for Soldiers, reports the situation to higher leadership, reorganize as necessary, and continue the mission.

REACT TO AMBUSH (NEAR) (07-3-D9502)

CONDITIONS: (Dismounted/mounted) The unit is moving tactically, conducting operations. The enemy initiates contact with direct fire within hand grenade range. All or part of the unit is receiving accurate enemy direct fire.

CUE: This drill begins when the enemy initiates ambush within hand grenade range.

STANDARDS:

Dismounted. Soldiers in the kill zone immediately return fire on known or suspected enemy positions and assault through the kill zone. Soldiers not in the kill zone locate and place "well-aimed" suppressive fire on the enemy. The unit assaults through the kill zone and destroys the enemy.

Mounted. Vehicle gunners immediately return fire on known or suspected enemy positions as the unit continues to move out of the kill zone. Soldiers on disabled vehicles in the kill zone dismount, occupy covered positions and engage the enemy with accurate fire. Vehicle gunners and Soldiers outside the kill zone suppress the enemy. The unit assaults through the kill zone and destroys the enemy. The unit leader reports the contact to higher headquarters

TASK STEPS

1. **Dismounted unit.** (See figure 8-6.) Unit personnel take the following actions:
    a. Soldiers in the kill zone execute one of the following two actions:
       (1) Return fire immediately. If cover is not available, immediately and without order or signal, assault through the kill zone.

Battle Drills

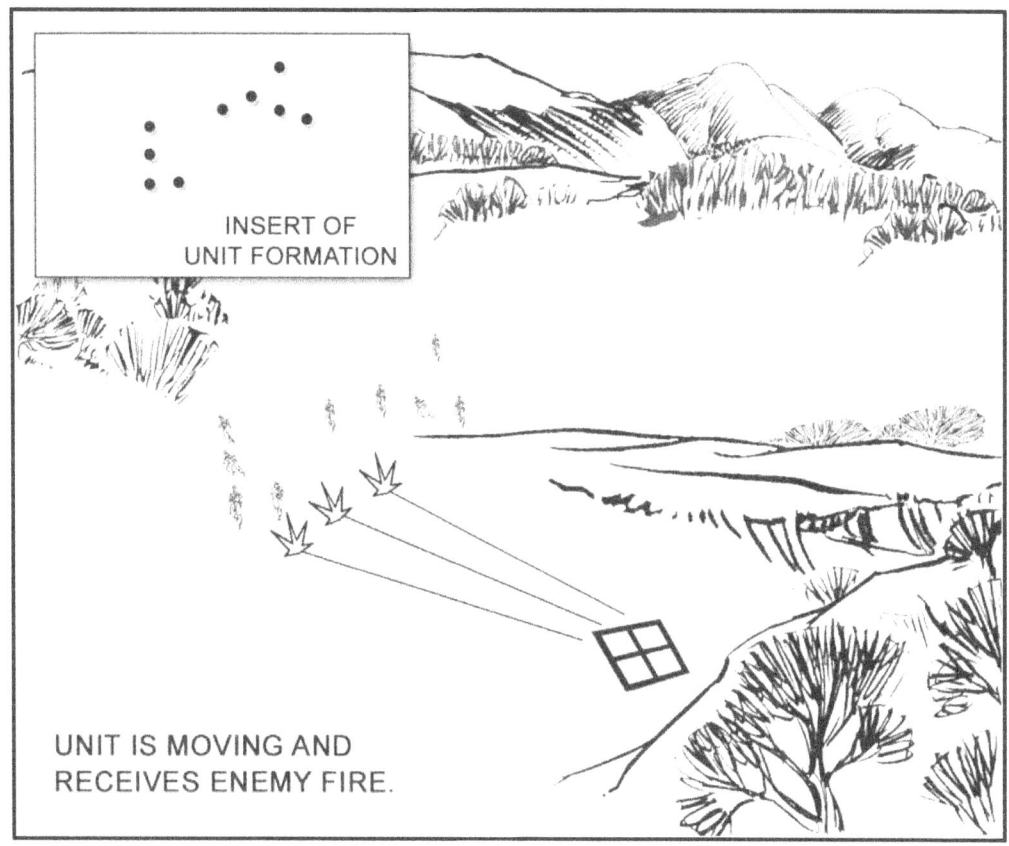

Figure 8-6. 06-3-D9502. React to ambush (near) (dismounted)

(2) Return fire immediately. If cover is not available, without order or signal, occupy the nearest covered position and throw smoke grenades. (See figure 8-7.)

Chapter 8

Figure 8-7. 07-3-D9502. Returning fire immediately

b. Soldiers in the kill zone assault through the ambush using fire and movement.

c. Soldiers not in the kill zone identify the enemy location, place "well-aimed" suppressive fire on the enemy's position and shift fire as Soldiers assault the objective

d. Soldiers assault through and destroy the enemy position. (See figure 8-8.)

e. The unit leader reports the contact to higher headquarters.

Battle Drills

Figure 8-8. 07-3-D9502. Assaulting through enemy positions

2.. **Mounted unit** takes the following actions:

    a. Vehicle gunners in the kill zone immediately return fire and deploy vehicle smoke, while moving out of the kill zone.

    b. Soldiers in disabled vehicles in the kill zone immediately obscure themselves from the enemy with smoke, dismount if possible, seek covered positions, and return fire.

    c. Vehicle gunners and Soldiers outside of the kill zone identify the enemy positions, place "well-aimed" suppressive fire on the enemy, and shift fire as Soldiers assault the objective.

    d. The unit leader calls for and adjusts indirect fire and requests close air support, according to the mission, enemy, terrain and weather, troops and support available-time available, and civil considerations (METT-TC).

    e. Soldiers in the kill zone assault through the ambush and destroy the enemy.

    f. The unit leader reports the contact to higher headquarters.

Chapter 8

ENTER AND CLEAR A ROOM (07-4-D9509)

CONDITIONS: The unit is conducting operations as part of a larger unit and has been given the mission to clear a room. Enemy personnel are believed to be in building. Noncombatants may be present in the building and are possibly intermixed with the enemy personnel. The unit has support and security elements positioned at the initial foothold and outside the building.

CUE: This drill begins on the order of the unit leader or on the command of the clearing team leader.

STANDARDS: The unit clears and secures the room by killing or capturing the enemy while minimizing friendly casualties, noncombatant casualties, and collateral damage. The team complies with rules of engagement, maintains a sufficient fighting force to repel an enemy counterattack, and continues operations.

TASK STEPS

1. The unit leader occupies a position to best control the security and clearing teams.
    a. Unit leader directs a clearing team to secure corridors or hallways outside the room with appropriate firepower.
    b. The team leader (normally, the number two Soldier) takes a position to best control the clearing team outside the room.
    c. The unit leader gives the signal to clear the room.

Note: If the unit is conducting high-intensity combat operations and grenades are being used, the unit must comply with the rules of engagement rules of engagement and consider the building structure. A Soldier of the clearing team cooks off at least one grenade (fragmentation, concussion, or stun grenade), throws the grenade into the room and announces, "FRAG OUT." The use of grenades should be consistent with the rules of engagement and building structure. Soldiers can be injured from fragments if walls and floors are thin or damaged.

2. The clearing team enters and clears the room.
    a. The first two Soldiers enter the room almost simultaneously. (See figure 8-9.)

Battle Drills

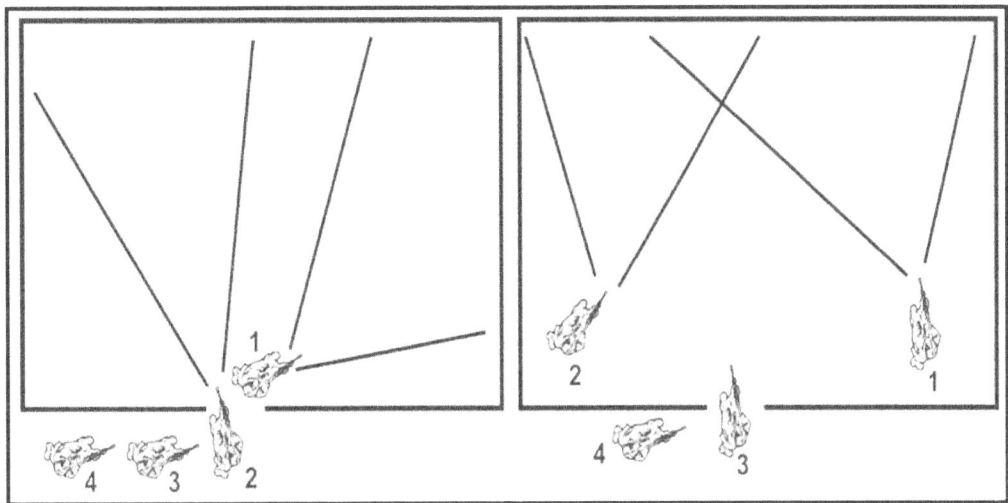

Figure 8-9. 07-4-D9509. Clear a room. First two soldiers enter simultaneously

b. The first Soldier enters the room, moves left or right along the path of least resistance to one of two corners, and assumes a position of domination facing into the room. During movement, the Soldier scans the sector and eliminates all immediate threats.

c. The second Soldier (normally the clearing team leader) enters the room immediately after the first Soldier and moves in the opposite direction of the first Soldier to his point of domination. During movement, the Soldier eliminates all immediate threats in the sector.

> *Notes:* During high intensity combat, the Soldiers enter immediately after the grenade detonates. Both Soldiers enter firing aimed bursts into their sectors, engaging all threats or hostile targets to cover their entry. If the first or second Soldier discovers the room is small or a short room (such as a closet or bathroom), he announces, "SHORT ROOM" or "SHORT." The clearing team leader informs the third and fourth Soldiers whether or not to stay outside the room or to enter.

d. The third Soldier moves in the opposite direction of the second Soldier while scanning and clearing the sector and assuming the point of domination. (See figure 8-10 on page 8-20.)

Chapter 8

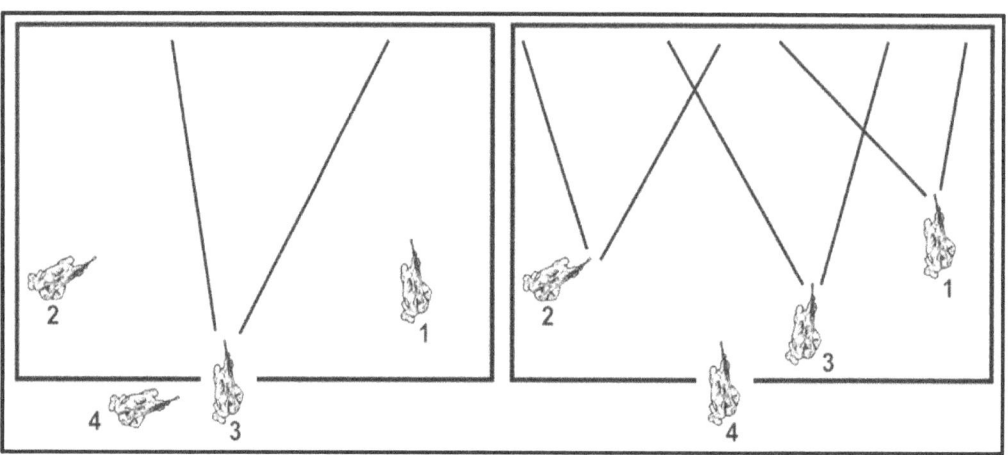

Figure 8-10. 07-4-D9509. Clear a room. Third soldier enters, clearing his sector

e. The fourth Soldier moves opposite of the third Soldier to a position dominating his sector. (See Figure 8-11.)

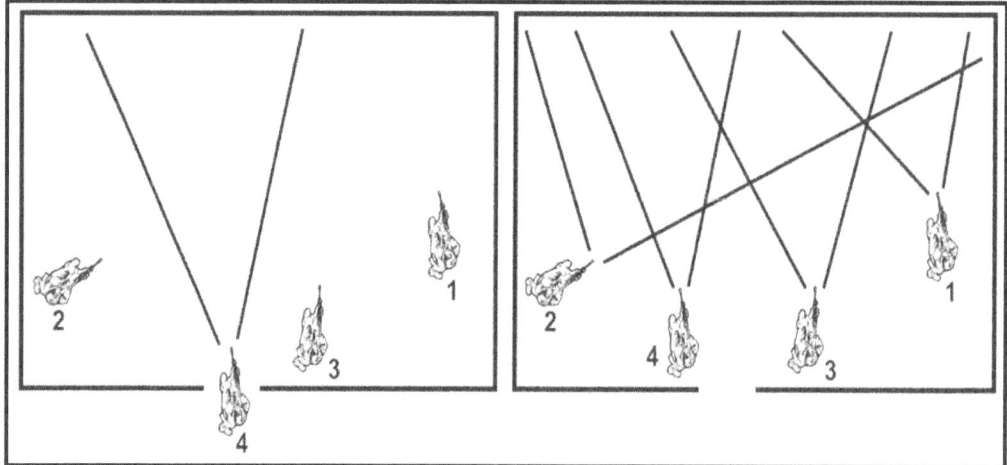

Figure 8-11. 07-4-D9509. Clear a room. Third soldier enters, dominating his sector

f. All Soldiers engage enemy combatants with precision aimed fire and identify noncombatants to avoid collateral damage.

Battle Drills

> *Note: If necessary or on order, number one and two Soldiers of the clearing team may move deeper into the room while overwatched by the other team members.*

    g. The clearing team leader announces to the unit leader when the room is CLEAR.

3. Marks the entry point according to unit SOP.

    a. Makes a quick assessment of room and threat.

    b. Determines if unit has fire power to continue clearing their assigned sector.

    c. Reports to the higher unit leader the first room is clear.

    d. Requests needed sustainment to continue clearing his sector.

    e. Marks entry point according to unit standard operating procedure (SOP).

4. The unit consolidates and reorganizes, as needed.

REACT TO INDIRECT FIRE (07-3-D9504)

CONDITIONS: Dismounted, the unit is moving, conducting operations. Any Soldier gives the alert, "INCOMING," or a round impacts nearby. Mounted, the unit is stationary or moving, conducting operations. The alert, "INCOMING," comes over the radio or intercom, or rounds impact nearby.

CUE: This drill begins when any member alerts, "INCOMING," or a round impacts.

STANDARDS:

Dismounted. Soldiers immediately seek the best available cover. The unit moves out of the area to the designated rally point after the impacts.

Mounted. When moving, drivers immediately move their vehicles out of the impact area in the direction and distance ordered. If stationary, drivers start their vehicles and move in the direction and distance ordered. Unit leaders report the contact to higher headquarters.

TASK STEPS

1. Dismounted. Unit personnel take the following actions:

    a. Any Soldier announces, "INCOMING!"

    b. Soldiers immediately assume the prone position or move to immediately available cover during initial impacts.

    c. The unit leader orders the unit to move to a rally point by giving a direction and distance.

    d. Soldiers move rapidly in the direction and distance to the designated rally point, after the impacts.

    e. The unit leaders report the contact to higher headquarters.

2. Mounted. Unit personnel take the following actions:

    a. Any Soldier announces, "INCOMING!"

    b. Vehicle commanders repeat the alert over the radio.

    c. The leaders give the direction and linkup location over the radio.

Chapter 8

d. Soldiers close all hatches, if applicable to the vehicle type; gunners stay below turret shields or get down into the vehicle.

e. Drivers move rapidly out of the impact area in the direction ordered by the leader.

f. Unit leaders report the contact to higher headquarters.

Chapter 9

Military Mountaineering

In the mountains, commanders face the challenge of maintaining their units' combat effectiveness and efficiency. Mountains exist in almost every country in the world and almost every war has included some type of mountain operations. Obstacles such as terrain, weather, and altitude must all be planned for and considered. To meet this challenge, commander's conduct training that provides Rangers with the mountaineering skills necessary to apply combat power in a rugged mountain environment, and they develop leaders capable of applying doctrine to the distinct characteristics of mountain warfare. (For more information, refer to TC 3-97.61.)

TRAINING AND PLANNING

9-1. Military mountaineering training provides units tactical mobility in mountainous terrain that would otherwise be inaccessible. Rangers are trained in the fundamental mobility and climbing skills necessary to move units safely and efficiently in mountainous terrain.

9-2. Rangers conducting combat operations in a mountainous environment should receive extensive training to prepare them for the rigor of mountain operations. Some of the areas are—
- Characteristics of the mountain environment.
- Care and use of basic mountaineering equipment.
- Mountain bivouac techniques.
- Mountain communications.
- Mountain travel and walking techniques.
- Mountain navigation, hazard recognition, and route selection.
- Rope management and knots.
- Natural and artificial anchors.
- Belay and rappel techniques.
- Installation construction and use, such as rope bridges.
- Rock climbing fundamentals.
- CASEVAC raising and lowering systems.
- Individual movement on snow and ice.
- Mountain stream crossings (including water survival techniques).

9-3. Unique planning considerations for mountain operations include movement and insertion techniques, methods that reduce the enemy's ability to observe, identification of likely enemy positions, and possible ambush and contact locations in the development of movement techniques, routes, and actions on contact in a compartmentalized and canalized terrain.

9-4. Planning also takes into consideration possible degraded reaction times from other ground units, and challenges for sustainment functions (resupply, MEDEVAC, and CASEVAC). Possible climate changes (weather, precipitation, wind, temperature) that can occur quickly and be extreme are also factored into the plans.

Chapter 9

DISMOUNTED MOBILITY

9-5. Ranger military mountaineers must be able to assess a vertical obstacle, develop a course of action to overcome the obstacle, and have the skills to accomplish the plan. Assessment of a vertical obstacle requires experience in the classifications of routes and understanding the levels of difficulty they represent. Without a solid understanding of the difficulty of a chosen route, the mountain leader can place his life and the life of other Soldiers in extreme danger. Ignorance is the most dangerous hazard in the mountain environment.

9-6. Operations in the mountains require Soldiers to be physically fit and leaders to be experienced in operations in this terrain. Problems arise in moving men and transporting loads up and down steep and varied terrain in order to accomplish the mission. Chances for success in this environment are greater when a leader has experience operating under the same conditions as his men. Acclimatizing, conditioning, and training are important factors in successful military mountaineering. Table 9-1 should be used to help determine the platoons movement formation and techniques during the planning process.

Table 9-1. Terrain classification table

CLASS	TERRAIN	MOBILITY	PLANNING CONSIDERATIONS	MOUNTAINEER SKILL LEVEL REQUIRED
1	Gentle slopes/trails	Walking	No special training required other than general environmental acclimation.	None
2	Steeper/rugged	Walking, some use of hands may be required	Environmental acclimation recommended. Unit movement/ standard operating procedure (SOP)/battle drill training on steep terrain.	Basic mountaineers helpful but not required.
3	Easy climbing/scrambling	Easy climbing, fixed ropes where exposed or fall risk	Environmental acclimation. Soldier load management. Unit movement/standard operating procedure (SOP)/battle drill training on steep terrain. Unit movement on fixed lines.	Basic mountaineers are used to install simple fixed ropes and installations.
4	Steep exposed	Fixed ropes required	Extensive environmental acclimation. Soldier load management. Unit movement/ standard operating procedure (SOP)/battle drill training on steep terrain. Unit movement on fixed lines. Negotiation of near vertical obstacles. Route selection.	Basic mountaineers. Assault climber may be required to establish anchors, fixed ropes, and hauling systems.
5	Near vertical/vertical	Technical climbing required	Extensive environmental acclimation. Extensive Soldier load management. Assault climbing. Technical rope rescue. Rope ascending/descending.	Assault climbers recommended to advise commanders and supervise complex rope systems.

Military Mountaineering

TASK ORGANIZATION

9-7. Special considerations should be taken when task organizing a Ranger platoon in a mountainous environment. Some specific teams to plan for include casualty evacuation teams, reconnaissance and installation teams, security teams specific to any obstacle that maybe encountered, who the strong swimmers and climbers are, and what climbing level-qualified mountaineers are within the platoon.

9-8. Once task organization has been completed, the platoon organizes and consolidates all rescue and mountaineering equipment.

RESCUE EQUIPMENT

9-9. Rangers use a litter system that functions like a standard basket-type litter but is more compact, lightweight, and versatile. The stretcher is made of low-density polyethylene plastic with solid brass grommets, nylon webbing, and steel buckles.

9-10. The system weighs 19 pounds (lbs.) when packed inside the carrying case (not including taglines). When packed, the case is 9 inches in diameter and 36 inches in length. The system can be used in extreme temperatures. It can be used in extremely cold weather (down to minus 120 degrees) and begins to melt at 450 degrees.

9-11. When possible, linkup with the aviation support unit to discuss unit SOPs and conduct rehearsals. Components of the litter system include:
- Stretcher.
- Nylon backpack.
- Horizontal lift slings with 10,000-pound (lb) tensile strength.
- Vertical lift sling with 5800 lb tensile strength.
- Locking steel carabiner with 9000 lb tensile strength.
- Tow strap with 300 lb tensile strength.
- Four webbing handles with 300 lb tensile strength.

9-12. There are different loading procedures for the litter system. To begin—
- Unpack and unroll the rescue litter.
- Bend litter in half backwards to make it lay flat.
- Place a patient in the litter using one of two methods:
    - Logroll method is used to load a casualty when stabilization of the patient's entire body is critical.
    - Place litter next to patient.
    - Roll patient onto side and slide litter as far under the body as possible.
    - Roll patient onto litter and carefully slide patient into center of litter.
    - Secure patient to litter.
    - Slide method is used when injuries prevent rolling the patient on the side, and in confined spaces that prevent using the logroll method.
    - Place foot end of stretcher at the head of the patient.
    - One person straddles the stretcher and supports the patient head, neck, and shoulders.
    - Two people grab straps and pull stretcher under patient while slightly lifting patient head and shoulders.

9-13. Care is taken to ensure the casualty is protected from potential environmental injuries. This includes wrapping the patient in a space blanket and sleeping bag during cold weather operations, or monitoring overheating in hot weather operations. To fasten straps and buckles—
- Lift sides of stretcher and fasten straps to buckles directly across from them.
- Feed foot straps through unused buckles at the foot of the stretcher and fasten to buckles.

Chapter 9

- After the patient is placed and strapped securely in the litter, lace the stretcher with the 30-foot lift sling of kernmantle rope. The lacing provides additional security to the litter, and the fixed loop created by the double figure-eight knot serves as the point of attachment for lifting or lowering the stretcher.

9-14. A vertical lift is used on sloping terrain and when canopy prevents a horizontal lift. When rigging this lift, remember that this type of lift forces the weight of the causality's body onto the lower extremities. To make a vertical lift—
- Create a fixed loop in the middle of the rope by tying a double figure-eight knot.
- Pass tails through grommet on either side of the head and snug knot against stretcher.
- Feed ropes through grommets along the sides, pass through the handles and through the grommets at the foot end of the stretcher, and secure with a square knot.
- Route the pigtails through the lower carrying handles (outside to inside), and secure ends. The litter may be towed by the tow handle strap hauling system, and is ready to be lifted or lowered.

9-15. When rigging for a horizontal lift, remember the head strap is four inches shorter than the foot strap. The horizontal lift position is the preferred position for lifting or lowering for several reasons. When the litter is closed and rigged correctly, it is impossible for the causality to accidently fall from the stretcher.

9-16. Horizontal is the most stable platform for the causality, as the stretcher absorbs all the weight of the causality's body. Control tag lines are also easily employed and much more efficient, reducing spin when lifting horizontally. Lifting horizontally requires the use of two (head and foot) lift straps rated at 9000 pounds each, creating redundant attaching points. The head strap is four-to-six inches shorter than the foot strap to ensure the head remains slightly raised. To make a horizontal lift—
- Insert one end of head strap through slot at head end, route under stretcher, then through slot on the opposite side.
- Repeat at the foot end with foot strap.
- Equalize weight on all straps and insert steel carabiner through sewn loops on all four straps.
- Ensure horizontal lift straps are removed if the stretcher is to be dragged. This prevents damage to the straps.

9-17. When the stretcher is rigged properly, it is ready to lift. To ascend vertical terrain with a casualty:
- Package a casualty in a stretcher for carrying and dragging.
- Package a casualty in a stretcher for helicopter evacuation horizontally and vertically.
- Task organization for a platoon for moving a casualty (carrying squad, security squads, machine guns, and key leaders). Emphasize that the PL focus on the entire tactical situation while controlling the platoon. Have a rotation of the carrying squad if they move the casualy over long distances. The PSG focuses on controlling the CASEVAC.
- Establish the primary anchor (sling rope and two opposite and opposed carabiners), and the secondary anchor for the six-to-eight wrap Prusik safety.
- Have teams move ahead to set up anchors that expedite moving the casualty up multiple pitches.

9-18. To descend vertical terrain with a casualty:
- Lower the casualty on a Munter hitch with a six-to-eight wrap Prusik safety. (The Munter hitch and Prusik safety are described in depth in this chapter under Knots and Rope.)
- Everyone else can rappel down, and the last man configures a retrievable rappel.
- Have teams move down and establish anchors to expedite the lowering if there are multiple pitches or rope lengths.

MOUNTAINEERING EQUIPMENT

9-19. Mountaineering equipment refers to all the parts and pieces that allow the trained Ranger to accomplish many tasks in the mountains. The importance of this gear to the mountaineer is no less than that of the rifle to the Infantry Soldier.

Military Mountaineering

9-20. Army mountaineering kits are made up of three separate but integrated kits with state-of-the-art, commercial equipment that meets the highest industry standards. The separate kits enable the commander to tailor the equipment to the mission environment.

- The high-angle mountaineering kit (HAMK) is designed for a minimally trained Infantry brigade combat team (BCT) platoon (40 personnel) moving through steep terrain that is void of ice or snow, on rope installations established by assault climbers. The HAMK provides each Soldier in the platoon with a harness, locking and nonlocking carabiners, sewn webbing runners, seven-mm accessory cord, and a belay/rappel device. There are also static installation ropes, a rope cutter, and a rope washer.
- The assault climber team kit (ACTK) is used by a trained assault climber team (consisting of three personnel) to establish rope installations that minimally trained Soldiers can move over using the HAMK. The ACTK provides each Soldier in the assault climber team with a harness, locking and nonlocking carabiners, sewn webbing runners, mechanical ascenders, chock pick, assault climber bag, seven-mm accessory cord, and a belay/rappel device. There are also dynamic climbing ropes and rock protection equipment, including spring-loaded camming devices and chocks.
- The snow and ice mobility kit (SIMK) is used by an Infantry platoon trained in techniques for operating in steep terrain covered by snow or ice. The kit provides each Soldier in the platoon with an avalanche transceiver, crampons, ice axe, and snowshoes. Included are also avalanche shovels, probes, and ice and snow anchors.

ROPES AND CORDS

9-21. Ropes and cords are the most important pieces of mountaineering equipment, and proper selection deserves careful thought. These items are a lifeline in the mountains, so selecting the right type and size is of utmost importance. All ropes and cord used in mountaineering and climbing today are constructed with the same basic configuration. The construction technique is referred to as *kemmantle*, which is essentially a core of nylon fibers protected by a woven sheath, similar to parachute cord.

9-22. Ropes come in two types: static and dynamic. This refers to their ability to stretch under tension. A static rope has very little stretch, perhaps as little as one-to-two percent, and is best used in rope installations. A dynamic rope is most useful for climbing and general mountaineering. Its ability to stretch up to one-third of its overall length makes it the right choice any time the user might take a fall. Dynamic and static ropes come in various diameters and lengths. For most military applications, a standard 9.5 millimeter (mm) or 11-mm X 50-m ropes are sufficient.

9-23. A short section of static rope or static cord is called a "sling rope" or "cordelette." These are critical pieces of personal equipment in mountaineering operations. The diameter usually ranges from seven mm to eight mm, and up to 21 feet long. 8 mm X 15 feet is the minimum Ranger standard.

9-24. Cordage cord or small diameter rope is indispensable to the mountaineer. Its many uses make it a valuable piece of equipment. All cord is static and constructed in the same manner as larger rope.

9-25. Rope that is used daily should be used no longer than one year. An occasionally used rope can generally be used up to five years, if properly cared for. To do this—

- Inspect ropes thoroughly before, during, and after use for cuts, frays, abrasions, mildew, and soft or worn spots.
- Never step on a rope or unnecessarily drag it on the ground.
- Avoid running rope over sharp or rough edges (pad, if necessary).
- Keep ropes away from oil, acids, and other corrosive substances.
- Avoid running ropes across one another under tension (nylon-to-nylon contact damages ropes).
- Do not leave ropes knotted or under tension longer than necessary.
- Clean in cool water, loosely coil, and hang to dry out of direct sunlight. Ultraviolet light rays harm synthetic fibers. When wet, hang rope to drip-dry on a rounded wooden peg, at room temperature. Do not apply heat.

Chapter 9

Webbing and Slings

9-26. Loops of tubular webbing or cord, called slings or runners, are the simplest pieces of equipment and some of the most useful. The uses for these simple pieces are endless, and are a critical link between the climber, the rope, carabiners, and anchors. Runners are predominately made from 9/16-inch or one-inch tubular webbing, and are tied or sewn by the manufacturer.

9-27. The carabiner is one of the most versatile pieces of equipment available in the mountains. This simple piece of gear is the critical connection between the climber, the rope, and the protection attaching the climber to the mountain. Carabiners must be strong enough to hold hard falls, yet light enough for the climber to carry a quantity of them easily. Today's highly technical metal alloys allow carabiners to meet both of these requirements. Steel carabiners are still widely used in the military, but are being replaced by lighter and stronger materials. Basic carabiner construction affords the user several different shapes.

9-28. Protection is the generic term used to describe a piece of equipment (natural or artificial) that is used to construct an anchor. Protection is used with a climber, belayer, and climbing rope to form the lifeline of the climbing team. The rope connects two climbers, and the protection connects them to the rock or ice. Figure 9-1 shows removable artificial protection, and stoppers, tri-cams, and spring-loaded cam devices. Figure 9-2 shows fixed (usually permanent) artificial protection.

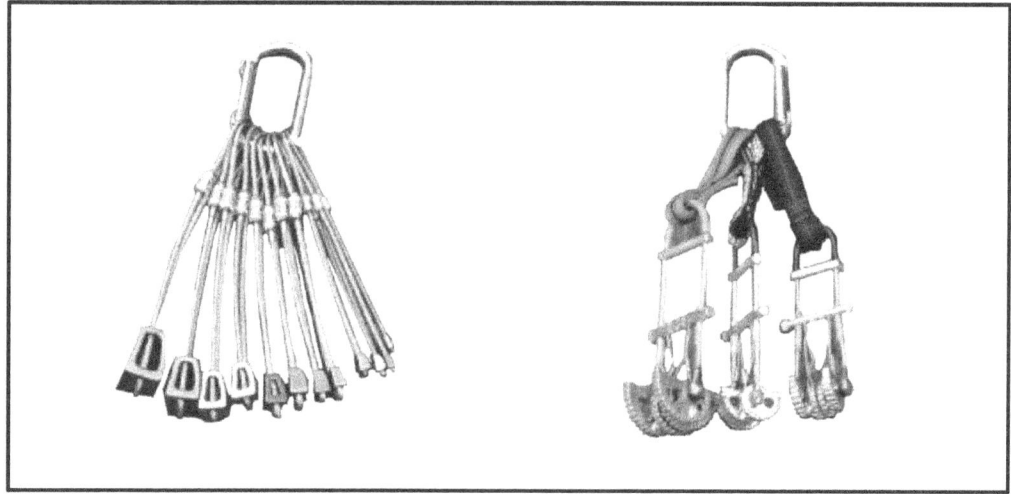

Figure 9-1. Examples of traditional (removable) protection used on rocks

Military Mountaineering

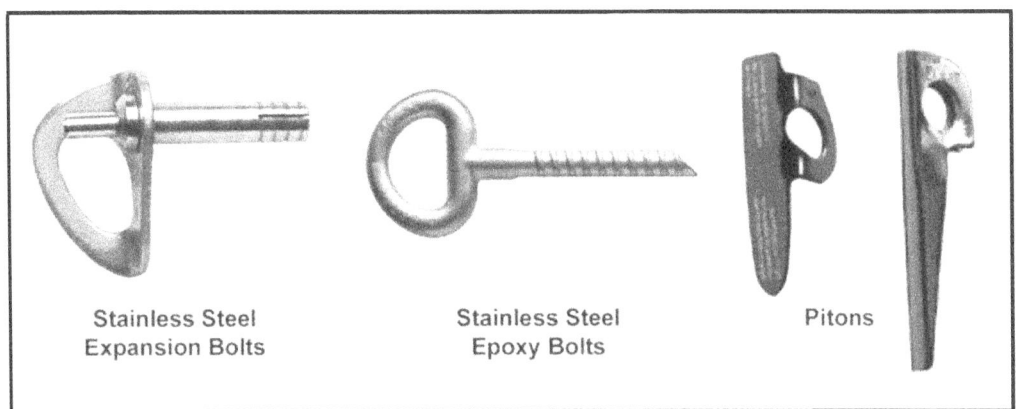

Figure 9-2. Examples of fixed (permanent or semipermanent) protection used on rocks

Equipment Inspection

9-29. Ropes should be inspected before, during, and after each use, especially when working around loose rock or sharp edges. Although the core of the kernmantle rope cannot be seen, it is possible to damage the core without damaging the sheath. Check a kernmantle rope by carefully inspecting the sheath before and after use while the rope is being coiled. When coiling, be aware of how the rope feels as it runs through the hands. Immediately note and tie off any lumps or depressions that can be felt. Carabiners and hardware should be inspected before, during, and after each use. Opening gates should open and close freely, be free of rust and corrosion, and any sharp edges or burrs can be smoothed out with a fine file.

Anchors

9-30. Anchors are the base for all installations and roped mountaineering techniques. Anchors must be strong enough to support the entire weight of the load or impact placed upon them. Several pieces of artificial or natural protection may be incorporated together to make one multipoint anchor. Anchors are classified as artificial or natural:

- **Artificial anchors** are constructed using all man-made material. The most common anchors incorporate traditional or fixed protection. (See figure 9-3 on page 9-8.)
- **Natural anchors** are usually very strong and often simple to construct using minimal equipment. Trees, shrubs, and boulders are the most common. All natural anchors simply require a method of attaching a rope. Regardless of the type of natural anchor used, the anchor must be strong enough to support the entire weight of the load. Natural anchors can be:
  - **Trees** are probably the most widely used of all anchors. In rocky terrain, trees usually have a very shallow root system. Check this by pushing or tugging on the tree to see how well it is rooted. Anchor as low as possible to prevent excess leverage on the tree. Use padding on soft, sap-producing trees to keep sap off ropes and slings.
  - **Rock projections and boulders** can be used but they must be heavy enough and have a stable enough base to support the load.
  - **Bushes and shrubs**. If no other suitable anchor is available, route a rope around the bases of several bushes. As with trees, place the anchoring rope as low as possible to reduce leverage on the anchor. Make sure all vegetation is healthy and well rooted to the ground.
  -

Chapter 9

- **Tensionless anchor** is used to anchor rope on high-load installations such as bridging. The wraps of the rope around the anchor (see figure 9-4]) absorb the tension of the installation and keep the tension off the knot and carabiner. Tie it with a minimum of four wraps around the anchor; however, a smooth anchor (small tree, pipe, or rail) may require several more wraps. Wrap the rope from top to bottom. Place a fixed loop into the end of the rope and attach loosely back onto the rope with a carabiner.

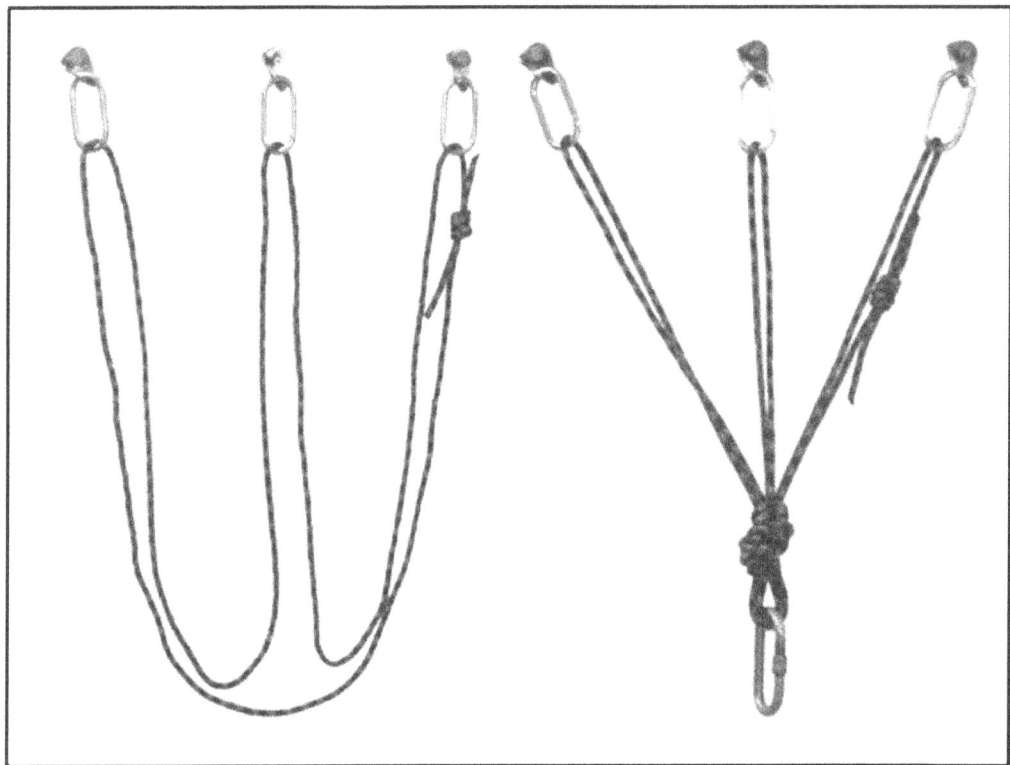

Figure 9-3. Constructing a three-point, pre-equalized anchor using fixed artificial protection

Military Mountaineering

Figure 9-4. Tensionless natural anchor

Knots

9-31. Proficiency with knots and rope is vitally important for Rangers, especially in mountaineering situations. Familiarity with the terminology associated with knots and rope is critical. (See figure 9-5 on page 9-10.) These terms include:
- **Running or working end** is the loose (or working) end of the rope.
- **Standing end** is the static, stationary, or nonworking end of the rope.
- **Bight.** Formed by placing the running end alongside the standing end, creating an open eyelet of rope.
- **Loop.** Formed by placing the running end across the standing end, creating a closed eyelet.
- **Overhand knot.** Formed by inserting the running end through the eyelet formed.
- **Half hitch.** An overhand knot tied around an object with the pigtail pulled perpendicular to the standing end.
- **Pigtail.** The pigtail is the portion of the running end of the rope between the safety knot and the end of the rope.
- **Turn.** Formed by passing the running end of a rope 360 degrees around an object.

Chapter 9

- **Round turn.** A round turn wraps around an object one and one-half times. A round turn is used to distribute the load over a small diameter anchor (three inches or less). It may also be used around larger diameter anchors to reduce the tension on the knot or provide added friction.
- **Dress** is the proper arrangement of all the knot parts, removing unnecessary kinks, twists, and slack so that all rope parts of the knot make contact.

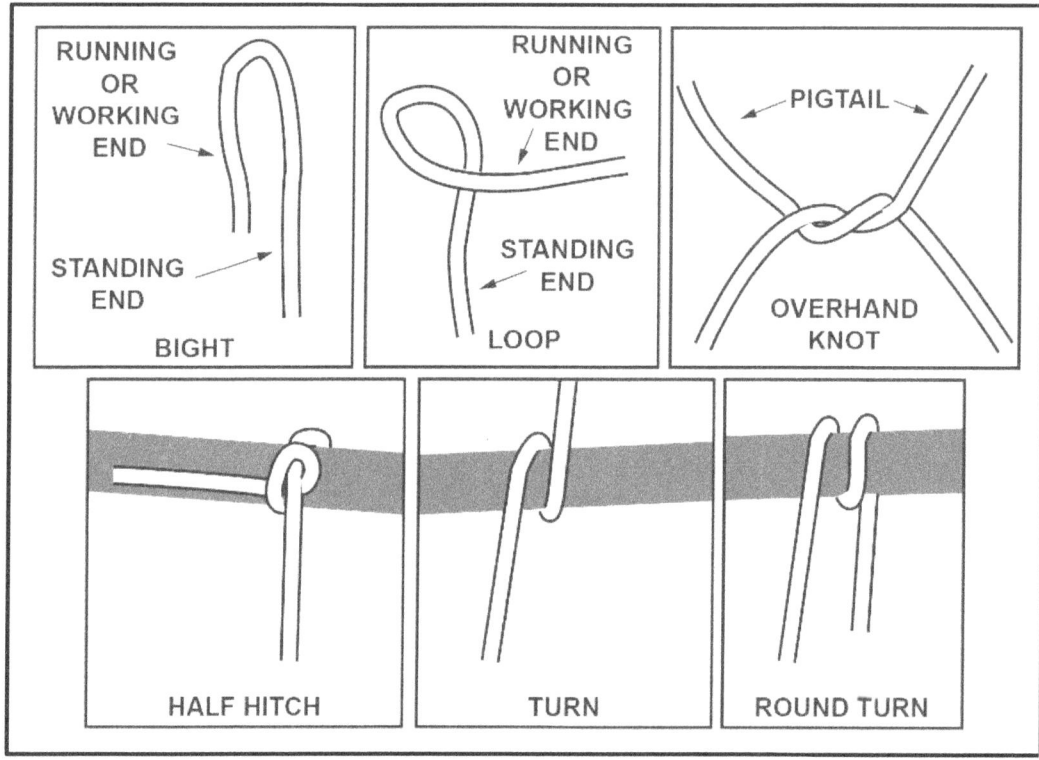

Figure 9-5. Rope terminology

9-32. A **square knot** joins two ropes of equal diameter (see figure 9-6). Two interlocking bites, running ends exit on same side of the standing portion of rope. Each tail is secured with an overhand knot on the standing end. When dressing the knot, leave at least a four-inch tail on the working end.

Military Mountaineering

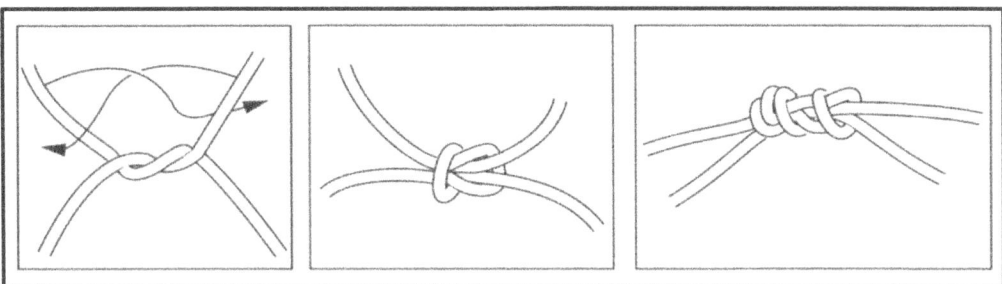

Figure 9-6. Square knot

9-33. A **round turn with two half hitches** is a constant tension anchor knot (see figure 9-7). The rope forms a complete turnaround the anchor point (where the name "round turn" comes from), with both ropes parallel and touching, but not crossing. Both half hitches are tightly dressed against the round turn, with the locking bar on top. When dressing the knot, leave at least a four-inch tail on the working end.

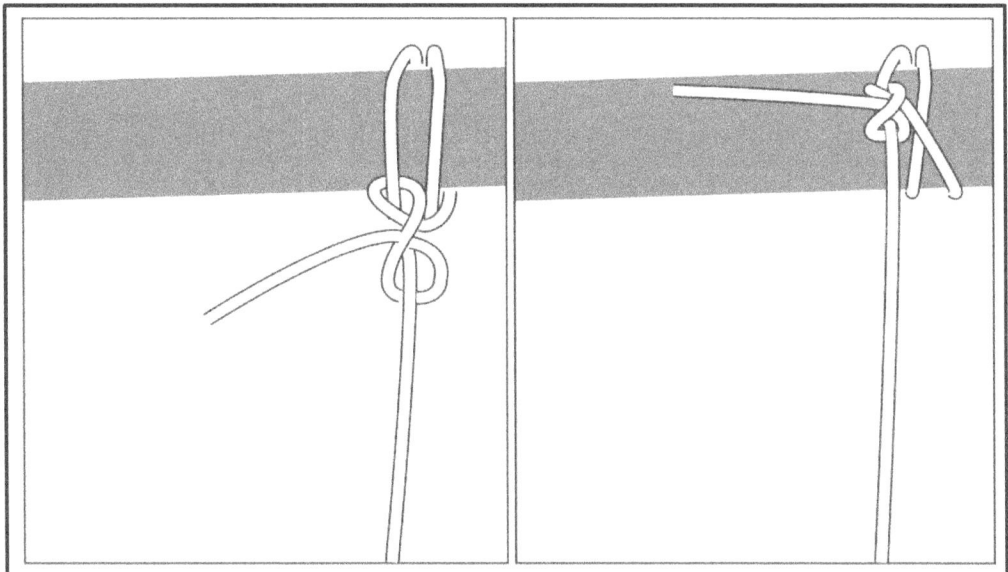

Figure 9-7. Round turn with two half hitches

Chapter 9

9-34. To make a **double figure-eight knot,** use a figure-eight loop knot (see figure 9-8) to form a fixed loop in the end of the rope. It can be tied at the end of the rope or anywhere along the length of the rope. Figure-eight loop knots are formed by two ropes parallel to each other in the shape of a figure eight, with no twists are in the figure eight. Fixed loops are large enough to insert a carabiner. When dressing the knot, leave at least a four-inch tail on the working end.

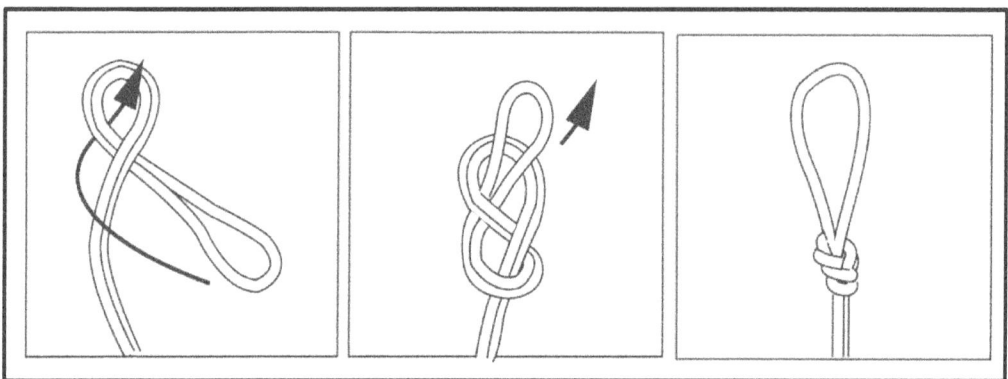

Figure 9-8. Double figure-eight knot

9-35.
9-36. The **end-of-the-rope clove hitch** is an intermediate anchor knot (see figure 9-9) that requires constant tension. Make two turns around the anchor. A locking bar runs diagonally from one side to the other. Leave no more than one rope width between turns of rope. Locking bar is opposite the direction of pull. When dressing the knot, leave at least a four-inch tail on the working end.

Military Mountaineering

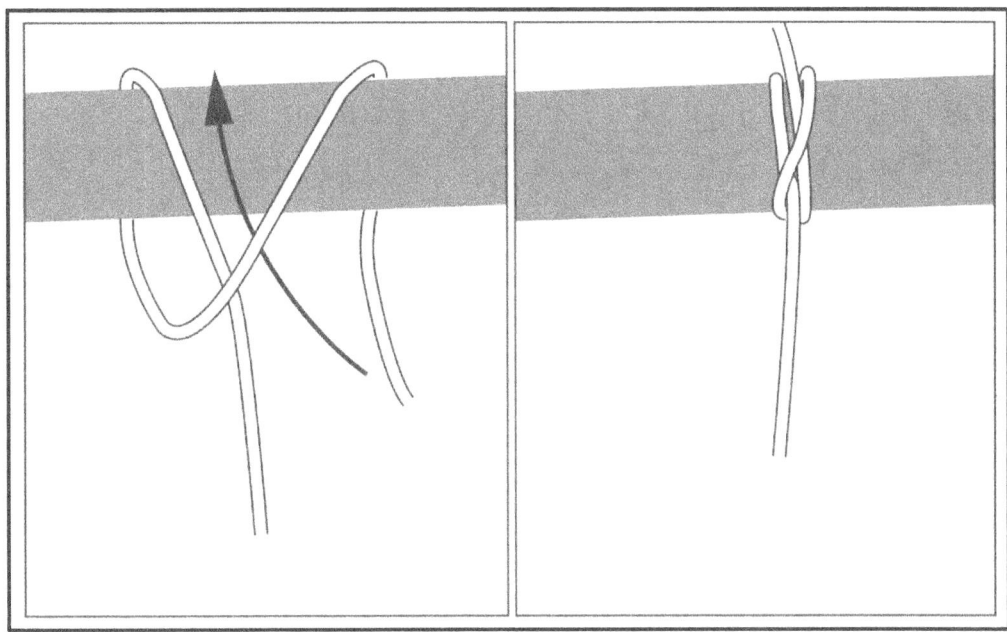

Figure 9-9. End-of-the-rope clove hitch

9-37.

9-38. The **middle-of-the-rope clove hitch** (see figure 9-10) secures the middle of a rope to an anchor. The knot forms two turns around the anchor. A locking bar runs diagonally from one side to the other. Leave no more than one rope width between turns. Ensure the locking bar is opposite the direction of pull.

Chapter 9

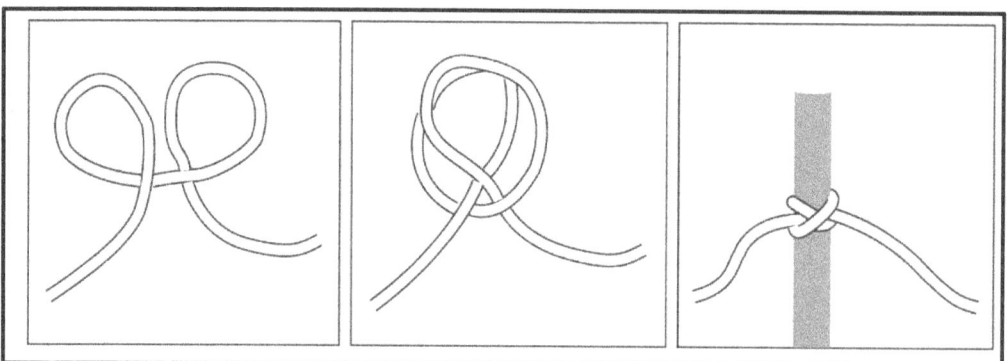

Figure 9-10. Middle-of-the-rope clove hitch

9-39. The **rappel seat** (see figure 9-11) is a rope harness used in rappelling and climbing. It can be tied for use with the left or right hand (1). Leg straps do not cross and are tightly centered on buttocks (2). Leg straps form locking half hitches on rope around waist. Square knot is properly tied on right hip (3) and finished with two overhand knots (4). Carabiner properly inserted around all ropes with opening gate opening up and away (5). Carabiner will not come in contact with square knot or overhand knot. Rappel seat is tight enough not to allow a fist to be inserted between the rappeller's body and the harness.

Military Mountaineering

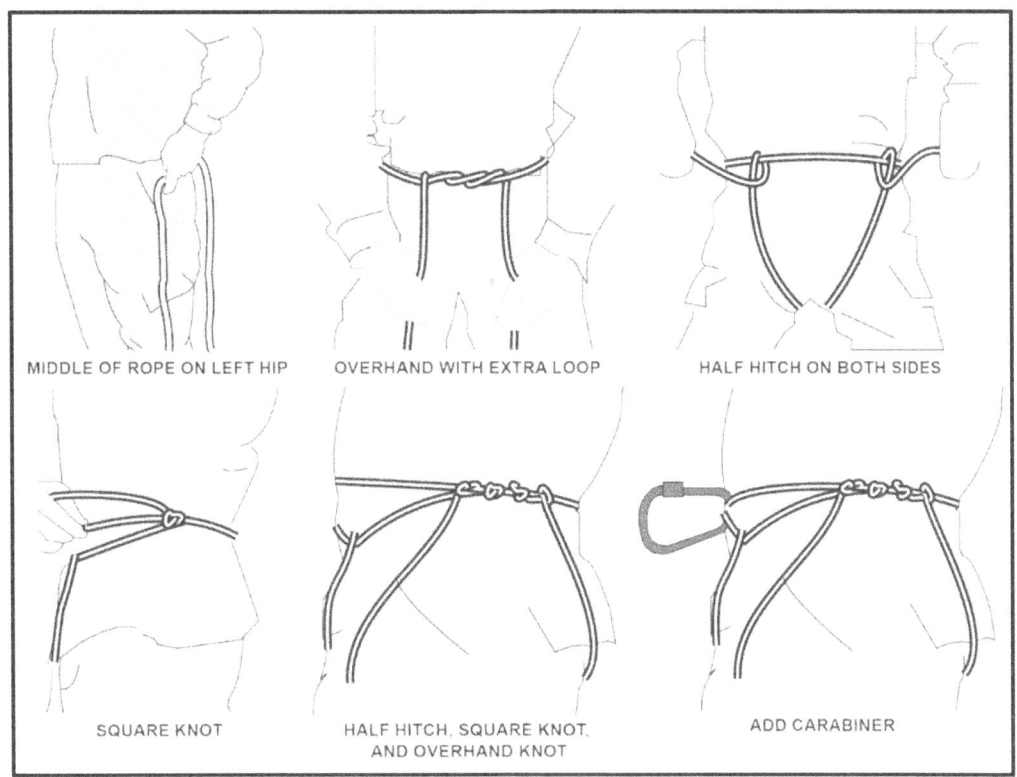

Figure 9-11. Rappel seat

9-40. The **rerouted figure-eight knot** is an anchor knot that also attaches a climber to a climbing rope. Form a figure eight in the rope and pass the working end around an anchor. Reroute the end back through to form a double figure eight. (See figure 9-12.) Tie the knot with no twists. When dressing the knot, leave at least a four-inch tail on the working end.

Chapter 9

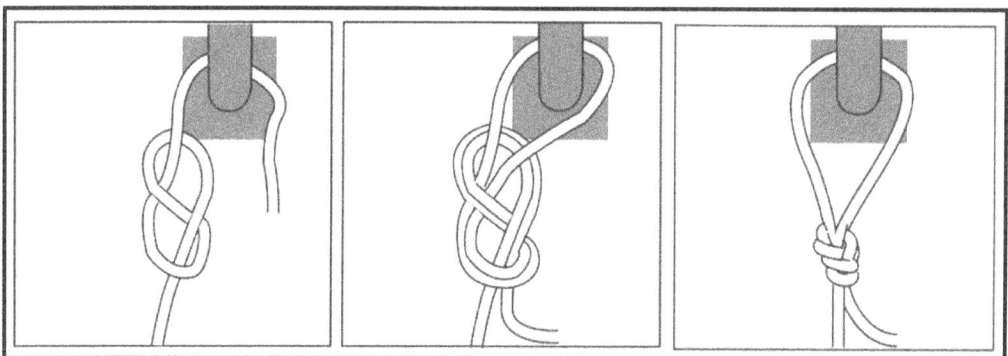

Figure 9-12. Rerouted figure-eight knot

9-41. The **figure-eight slipknot** is used to form an adjustable bight in the middle of a rope. Knot is in the shape of a figure eight. Both ropes of the bight pass through the same loop of the figure eight. The bight is adjustable by means of a sliding section. (See figure 9-13.)

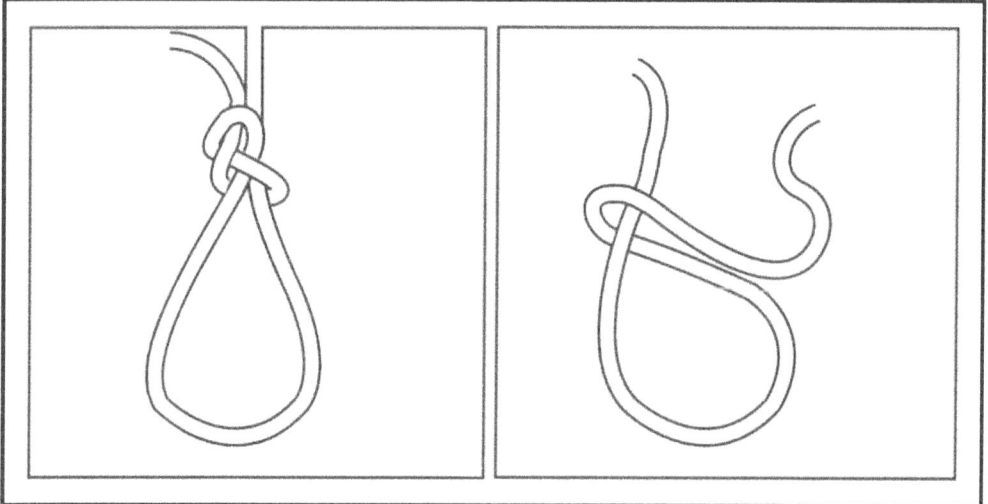

Figure 9-13. Figure-eight slipknot

9-42. The **Munter hitch** is one of the most often used belays, the Munter hitch (see figure 9-14) requires very little equipment. The rope is routed through a locking pear-shaped carabiner, then back on itself. The belayer controls the rate of descent by manipulating the working end back on itself with his brake hand.

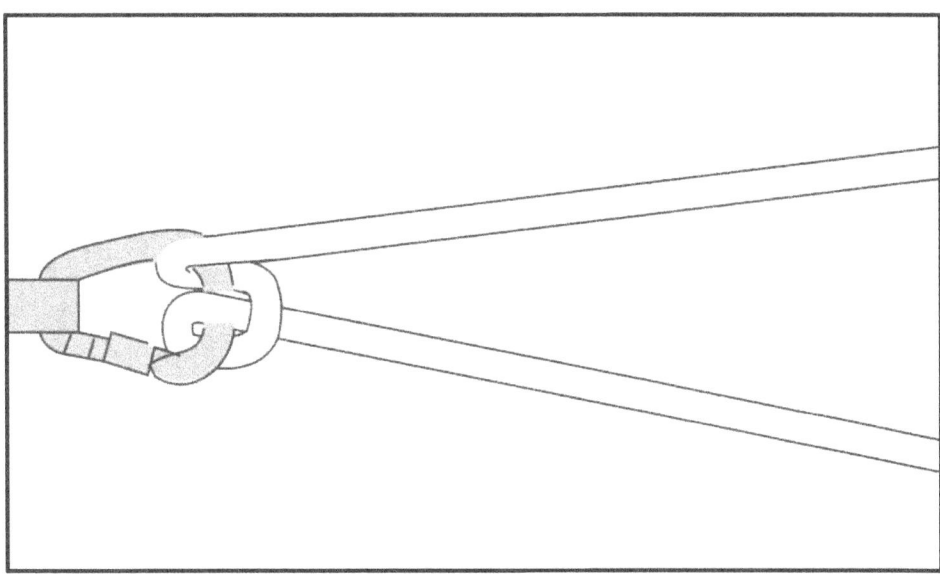

Figure 9-14. Munter hitch

9-43. The Munter mule knot (see figure 9-15 on page 9-18) is a knot that allows the user to stop movement of the rope through a Munter hitch. The Munter mule knot is tied so it can be easily released tension with the pull of a rope, allowing a smooth and controlled release. It is vital knot for many rope systems.

9-44. Start by tying a Munter hitch in the rope, ensuring it is in the loaded position. Maintaining tension, create two loops in the brake strand on either side, and behind the main strand. Pass one loop through the other on top of the main strand, and carefully pull it tight while maintaining tension. Secure the knot by tying an overhand knot with the bight of rope protruding from the knot, and include the main strand in the knot.

9-45. A **Prusik** knot (see figure 9-16 on page 9-18) attaches a movable rope to a fixed rope. The knot has two round turns, with a locking bar perpendicular to the standing end of the rope. The knot is secured with a double figure eight or bowline within six inches of the locking bar. When dressing the knot, leave at least a four-inch pigtail on the working end. The Prusik knot can be tied in the end or the middle of the rope.

Chapter 9

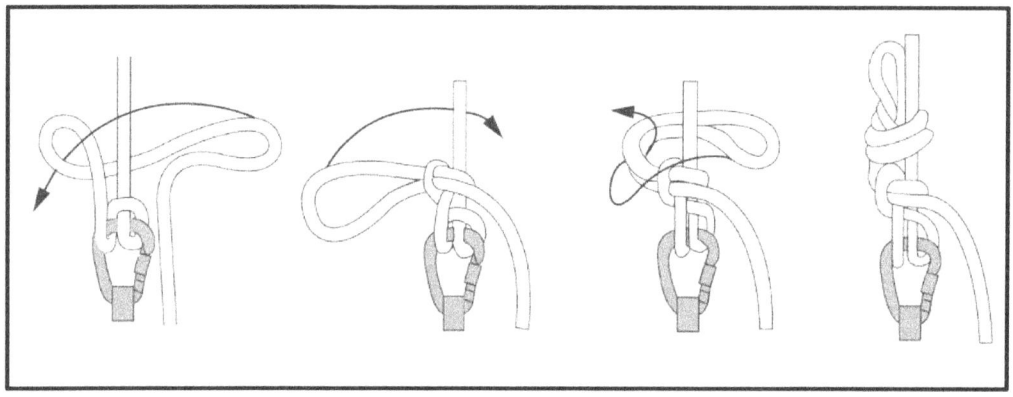

Figure 9-15. Munter mule knot

Figure 9-16. Prusik knot

9-46. The **bowline** (see figure 9-17) is used to tie the end of a rope around an anchor. It may also be used to tie a single fixed loop in the end of a rope. Bring the working end of the rope around the anchor from right to left (as the climber faces the anchor). Form an overhand loop in the standing part of the rope (on the climber's right) toward the anchor. Reach through the loop and pull up a bight. Place the working end of the rope (on the climber's left) through the bight and bring it back onto itself. Now dress the knot down. Form an overhand knot with the tail from the bight.

Military Mountaineering

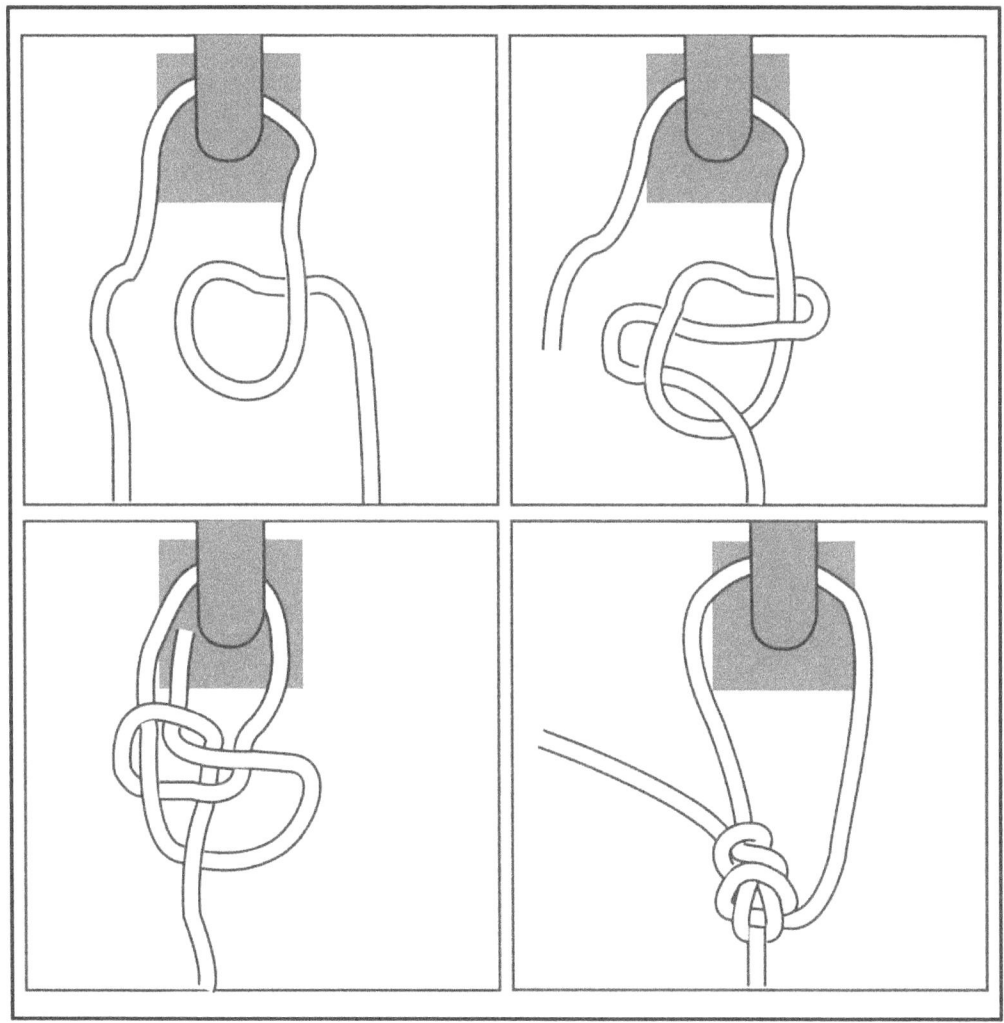

Figure 9-17. Bowline

Chapter 9

BELAYS

9-47. **Belaying** is any action taken to arrest a climber that has fallen, or to control the rate of descent of a load from a higher to lower elevation. The belayer also helps manage a climber's rope or the rate of the climber's or rappeller's descent by controlling the amount of rope that is taken in or out. The belayer is anchored in a stable position to prevent being pulled out of position and losing control of the rope. There are three types of belays: body belays, mechanical belay devices, and friction hitches.

9-48. The **body belay** (see figure 9-18) uses the belayer's body to apply friction. The belayer routes the rope around his body and uses friction to arrest a climber's fall. Care is taken because the body bears the entire weight of the load.

Military Mountaineering

Figure 9-18. Body belay

Chapter 9

9-49. The **mechanical belay** uses mechanical devices to help the belayer control the rope, as in rappelling. A variety of mountaineering devices are used to construct a mechanical belay. Most mechanical belay devices can be used as rappel devices. (See figure 9-19.)

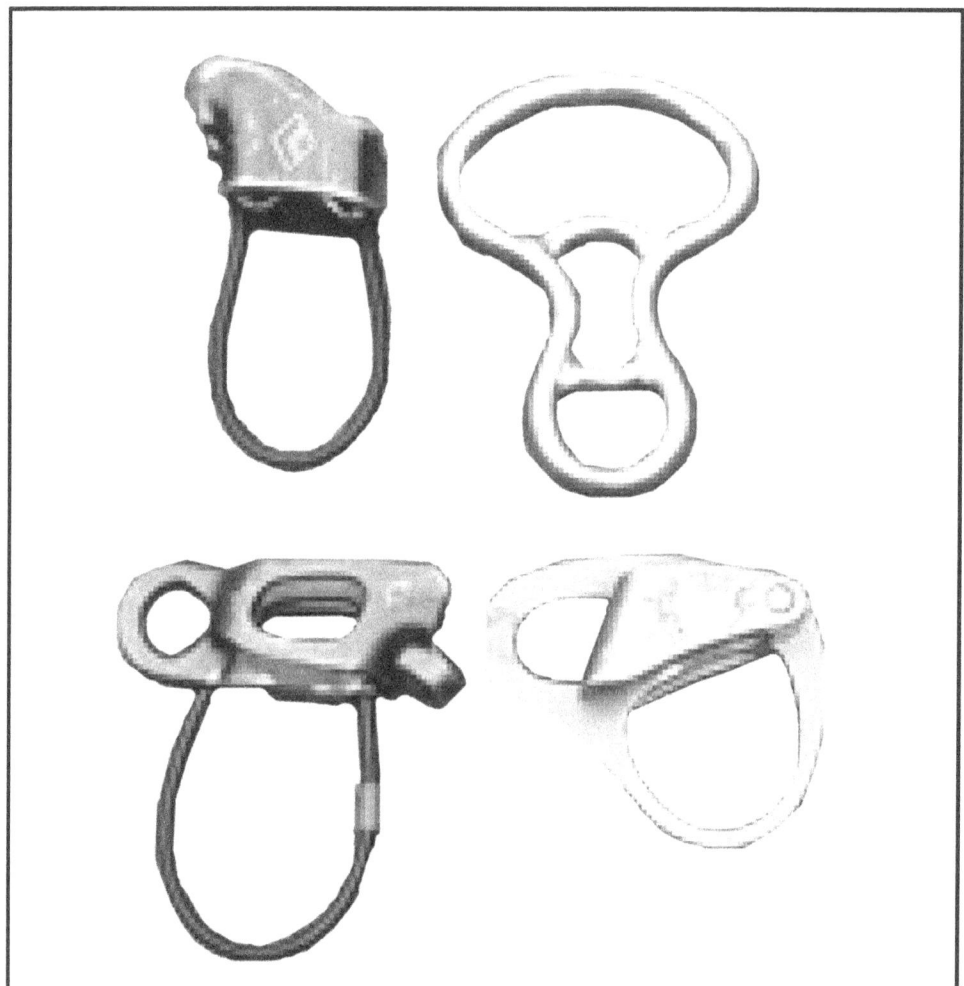

Figure 9-19. Mechanical belay devices

Military Mountaineering

9-50. The **air traffic controller (ATC)** is a locking mechanical belay device. (See figure 9-20.) It locks down on itself once tension is applied in opposite directions. This requires the belayer to apply very little force with the brake hand to control the rate of descent or to arrest a climber's fall.

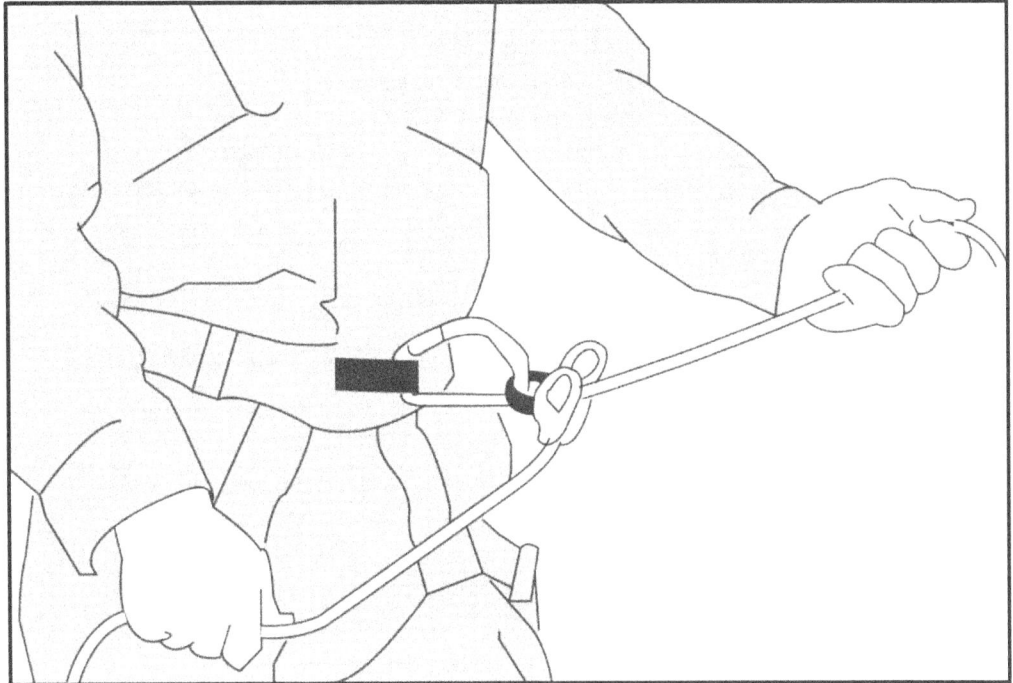

Figure 9-20. Air traffic controller

9-51. **Friction hitches** such as the Munter hitch (see figure 9-14 [page 9-16]) are excellent for belaying loads during lowering or raising loads. Table 9-2 demonstrates the sequence of commands used by climber and belayer.

Chapter 9

Table 9-2. Sequence of climbing commands

COMMAND	GIVEN BY	MEANING
BELAY ON, CLIMB	Belayer	Belay is on and climber may climb.
CLIMBING	Climber	Climber is climbing.
UP-ROPE	Climber	Belayer, remove excess slack in the rope.
BRAKE	Climber	Belayer, immediately apply brake.
FALLING	Climber	Climber is falling, immediately apply brake and prepare to arrest the fall.
TENSION	Climber	Belayer, remove all slack from climbing rope until rope is tight, apply brake, and hold position.
SLACK	Climber	Belayer, allow climber to pull slack into the climbing rope (belayer may have to assist).
ROCK	Anyone	Command given to alert everyone of an object falling near them. Belayer immediately applies the brake.
POINT	Climber	Alert belayer that the direction of pull on the climbing rope has changed in the event of a fall.
STAND BY	Climber or belayer	Hold position, stand by, I am not ready.
DO YOU HAVE ME?	Climber	Informal command to belayer to prepare for a fall, or prepare to lower me.
I HAVE YOU	Belayer	The brake is on and I am prepared for you to fall, or to lower you.
OFF-BELAY	Climber	Alert belayer that climber is safely in, or it is safe to come off belay.
THREE METERS	Belayer	Alert climber to the amount of rope between climber and belayer (may be given in feet or meters).
BELAY OFF	Belayer	I am off belay.

ROPE INSTALLATIONS

9-52. Rope installations may be constructed by teams to help units negotiate natural and man-made obstacles. Installation teams consist of a squad-sized element with two-to-four trained mountaineers. Installation teams deploy early and prepare the area of operations (AO) for safe, rapid movement by constructing various types of mountaineering installations.

9-53. Following construction of an installation, the squad, or part of it, remains on site to secure and monitor the system, assist with the control of forces across it, and adjust or repair it during use. After the unit passes, the installation team may disassemble the system and deploy to another area.

9-54. A fixed rope is anchored in place to help Rangers move over difficult terrain. Its simplest form is a rope tied off on the top of steep terrain. As terrain becomes steeper or more difficult, fixed rope systems may require intermediate anchors along the route. Planning considerations to follow include:
- Does the installation allow you to bypass the obstacle?
- (Tactical.) Can obstacle be secured from construction through negotiation, to disassembly?
- Is it in a safe and suitable location? Is it easy to negotiate? Does it avoid obstacles?

Military Mountaineering

- Are natural and artificial anchors available?
- Is the area safe from falling rock and ice?

ROPE BRIDGES

9-55. Rope bridges are employed in mountainous terrain to bridge linear obstacles such as streams or rivers where the force of flowing water may be too great or temperatures are too cold to conduct a wet crossing. The rope bridge is constructed using a static rope. The maximum span that can be bridged is half the length of the rope for a dry crossing, and three-quarters the length of the rope for a wet crossing. The ropes are anchored with an anchor knot on the farside of the obstacle and tied off at the near end with a transport-tightening system. (See figure 9-21 on page 9-26.) Rope bridge planning considerations to follow include:

- Does the installation allow you to bypass the obstacle?
- (Tactical.) Can obstacle be secured from construction through negotiation, to disassembly?
- Is it in the most suitable location, such as a bend in the river? Is it easily secured?
- Does it have nearside and farside anchors?
- Does it have good loading and off-loading platforms?
- Equipment (one-rope bridge):
  - One sling rope for every Ranger.
  - One steel locking carabiner.
  - Three steel ovals.
  - Two 120-foot static ropes.

Constructing a Rope Bridge

9-56. The first Ranger swims the rope to the far side and ties a tensionless anchor, between knee and chest level, with a minimum of four wraps. The bridge team commander (BTC) ties a transport-tightening system (see figure 9-21) to the nearside anchor point. To do this, tie a figure-eight slipknot and incorporate a locking half hitch around the adjustable bight. Insert two steel oval carabiners into the bight so the gates are opposite and opposed. The rope is then routed around the nearside anchor point at waist level and dropped into the steel oval carabiners.

- A three-man pull team moves forward from the platoon. No more than three are used to tighten the rope. Using more can over tighten the rope, bringing it near failure.
- Once the rope bridge is tight enough, the bridge team secures the transport-tightening system (see figure 9-20) using two half hitches without losing more than four inches of tension.
- Personnel cross using the commando crawl (see figure 9-22 on page 9-27) or Tyrolean traverse (see figure 9-23 on page 9-28) methods.

Chapter 9

Figure 9-21. Transport-tightening system

Military Mountaineering

Figure 9-22. Commando crawl method

Chapter 9

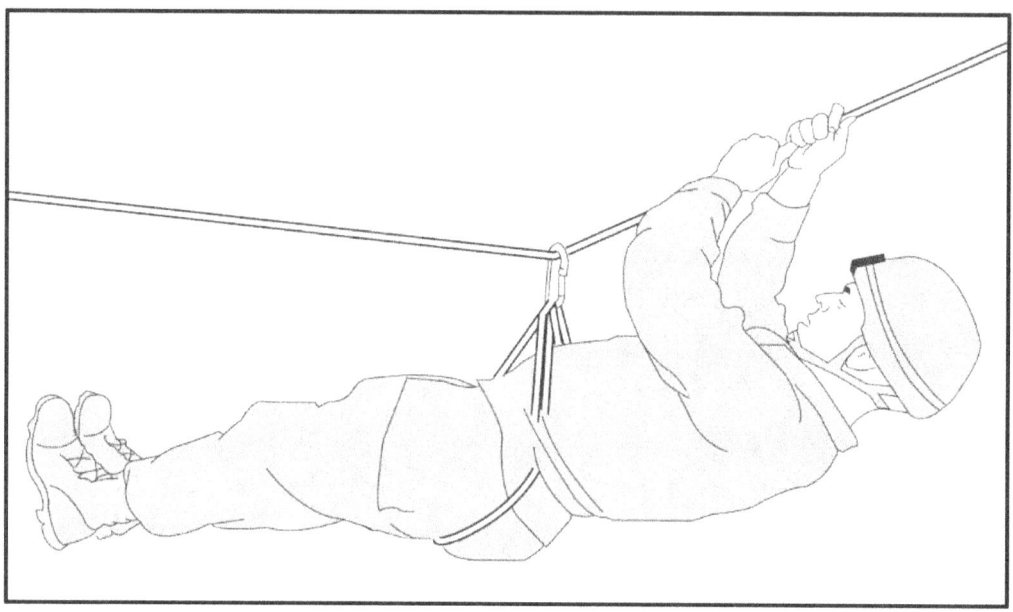

Figure 9-23. Rappel seat (Tyrolean traverse) method

9-57. The first Ranger swims the rope to the far side and ties a tensionless anchor, between knee and chest level, with a minimum of four wraps. The BTC ties a transport-tightening system to the nearside anchor point. To do this, tie a figure-eight slipknot and incorporate a locking half hitch around the adjustable bight. Insert two steel oval carabiners into the bight so the gates are opposite and opposed. The rope is then routed around the nearside anchor point at waist level and dropped into the steel oval carabiners.

Bridge Recovery

9-58. Once all except two troops have crossed the rope bridge, the BTC chooses the wet or dry method to dismantle the rope bridge. If the BTC chooses the dry method, anchor the tightening system with the transport knot. Take the following steps for bridge recovery:
- The BTC back-stacks all of the slack coming out of the transport knot, ties a fixed loop, and places a carabiner into the fixed loop.
- The next-to-last Ranger to cross should attach the carabiner to his rappel seat or harness, and then move across the bridge using the Tyrolean traverse method.
- The BTC removes all knots from the system. The far side remains anchored. The rope should now only pass around the nearside anchor.
- A three-Ranger pull team, assembled on the far side, takes the end brought across by the next-to-last Ranger, pulls, and holds the rope tight again.
- The BTC attaches himself to the rope bridge and moves across.

Military Mountaineering

- Once across, the BTC breaks down the farside anchor, removes the knots, and then pulls the rope across. If it is a wet crossing, any method can be used to anchor the tightening system.
    - During a wet crossing, all personnel cross except the BTC or the strongest swimmer.
    - The BTC then removes all knots from the system.
    - The BTC ties a fixed loop, inserts a carabiner, attaches it to his rappel seat or harness, and then manages the rope as the slack is pulled to the far side.
    - The BTC then moves across the obstacle while being belayed from the far side.

Z-Pulley System

9-59. The Z-Pulley System (see figure 9-24 on page 9-30) is a simple, easily constructed hauling system. Anchors must be sturdy and able to support the weight of the load. Site selection is governed by different factors such as tactical situation, weather, terrain, equipment, load weight, and availability of anchors. Use carabiners as a substitute if pulleys are not available. The leverage obtained using a Z-Pulley System is a three-to-one mechanical advantage. The less friction involved, the greater the mechanical advantage. Friction is caused by the rope running through carabiners, the load rubbing against the rock wall, and the rope condition. To construct a Z-Pulley System:

- Establish an Anchor Prusik System (APS).
- Place a carabiner on the runner at the anchor point, place a pulley into the carabiner, and run the hauling rope through the pulley.
- With a sling rope, tie a middle-of-rope Prusik knot secured with a figure-eight knot on the load side of the pulley. This is used as a progress capture device. A mechanical descender may be used in place of the Prusik knot.
- Take the tails exiting the figure eight and tie a Munter hitch secured by a mule knot. Ensure the Munter hitch is loaded properly before tying the mule knot.
- At an angle away from the APS, establish a Moveable Prusik System (MPS) to create a "Z" in the hauling rope.
- Tie another Prusik knot on the load side of the hauling rope. Secure it with a figure-eight knot. Using the tails, tie a double-double figure-eight knot.
- Insert a locking carabiner into the two loops formed, then place the working end into the carabiner. Mechanical ascenders should not be used as a Moveable Prusik System.
- Move the working end back on a parallel axis with the APS. Provide a pulling team on the working end with extra personnel to monitor the Prusik knots.

Chapter 9

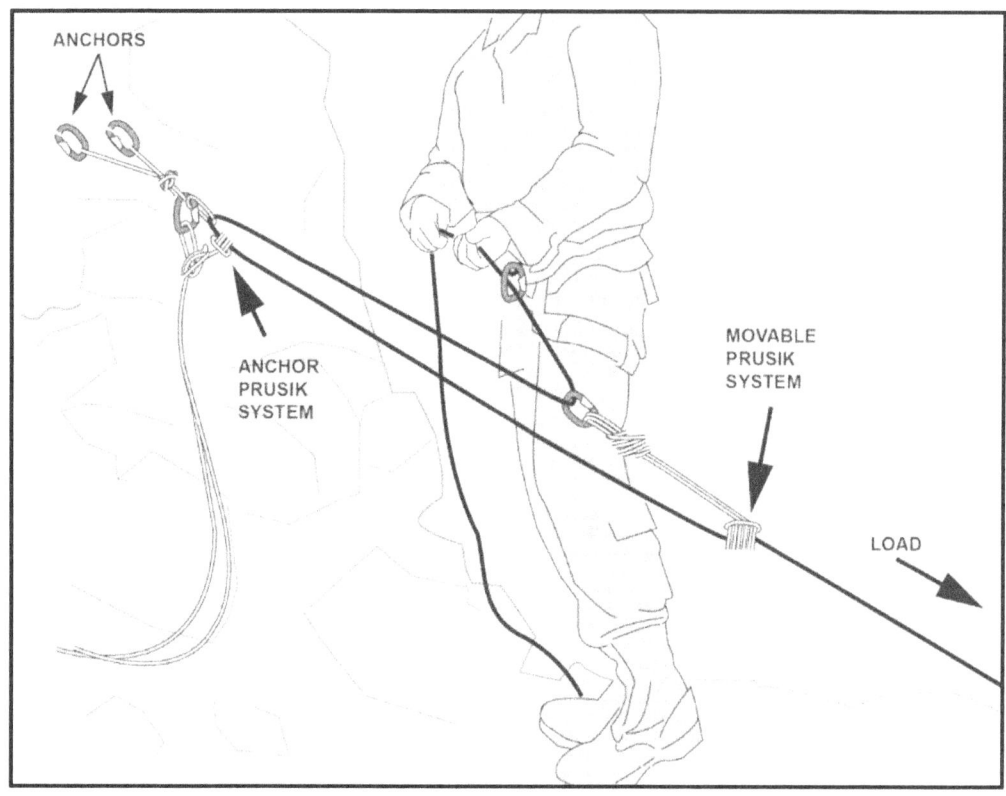

Figure 9-24. Z-Pulley system

RAPPELLING

9-60. Rappelling is a quick method of descent, but it can be extremely dangerous. Dangers include failure of the anchor or other equipment, and individual error. Anchors used in mountainous environments should be chosen carefully. Great care is taken to load the anchor slowly and avoid placing too much stress on the anchor. To ensure this, bounding rappels are prohibited—only walk-down rappels are permitted.

9-61. Hasty and body rappels are quick and easy (see figure 9-25 on page 9-31 and figure 9-26 on page 9-32) that should only be used on moderate pitches—never on vertical or overhanging terrain. Gloves are used with both rappels to prevent rope burns.

Military Mountaineering

Figure 9-25. Hasty rappel

Chapter 9

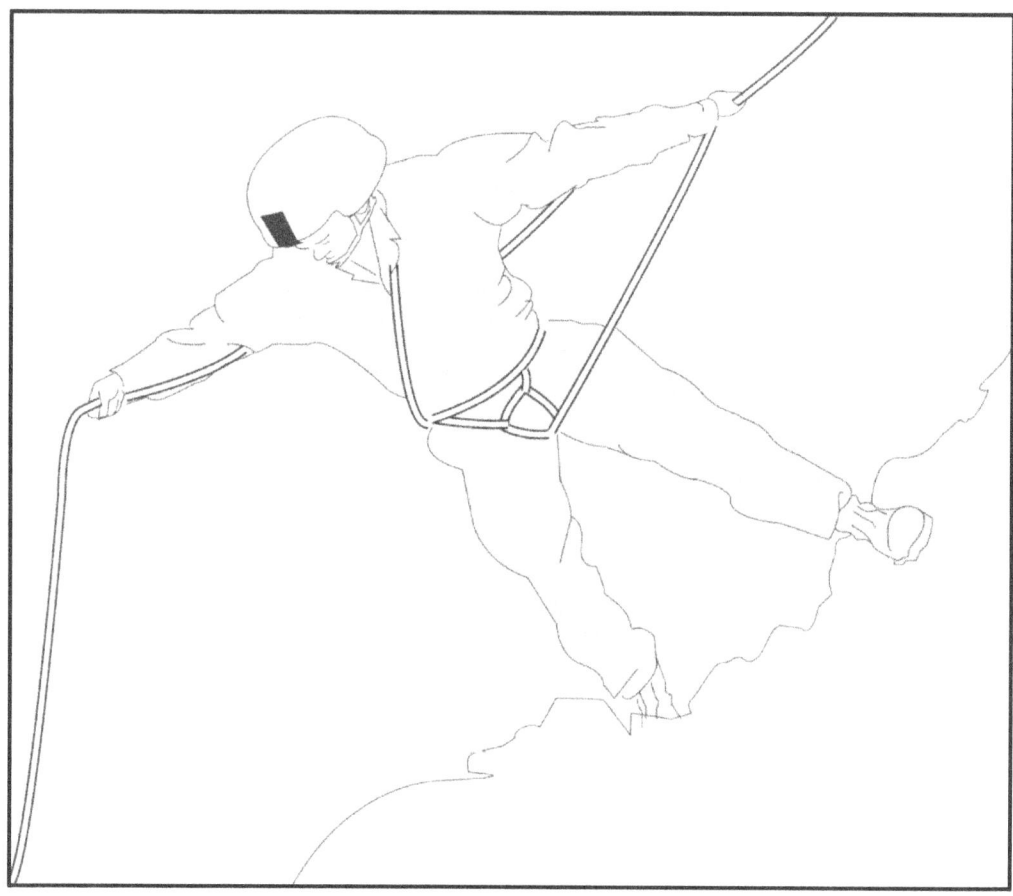

Figure 9-26. Body rappel

9-62. **Seat hip rappel** uses either a figure-eight descender (see figure 9-27) or a carabiner wrap descender (see figure 9-28). The descender is inserted in a sling rope seat and then fastened to the rappeller. This gives the Ranger enough friction for a smooth, controlled descent.

Military Mountaineering

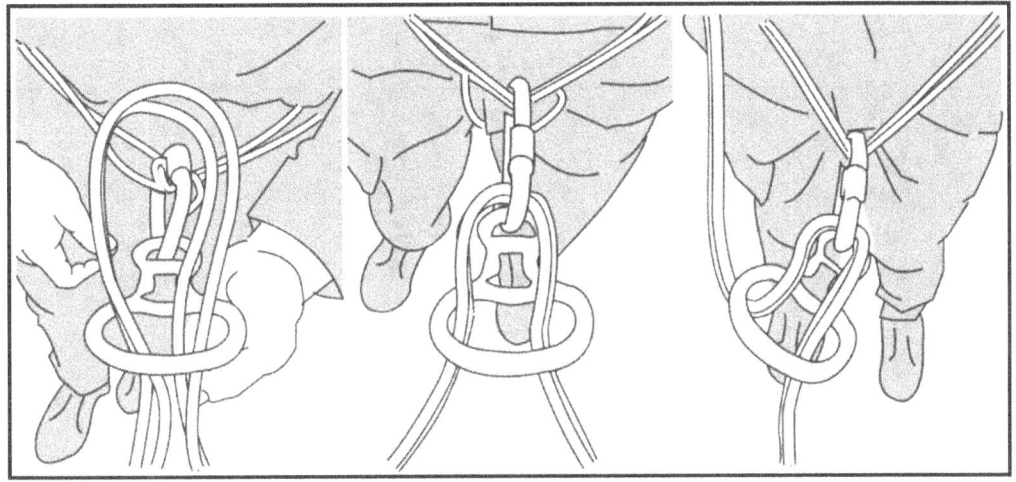

Figure 9-27. Figure-eight descender

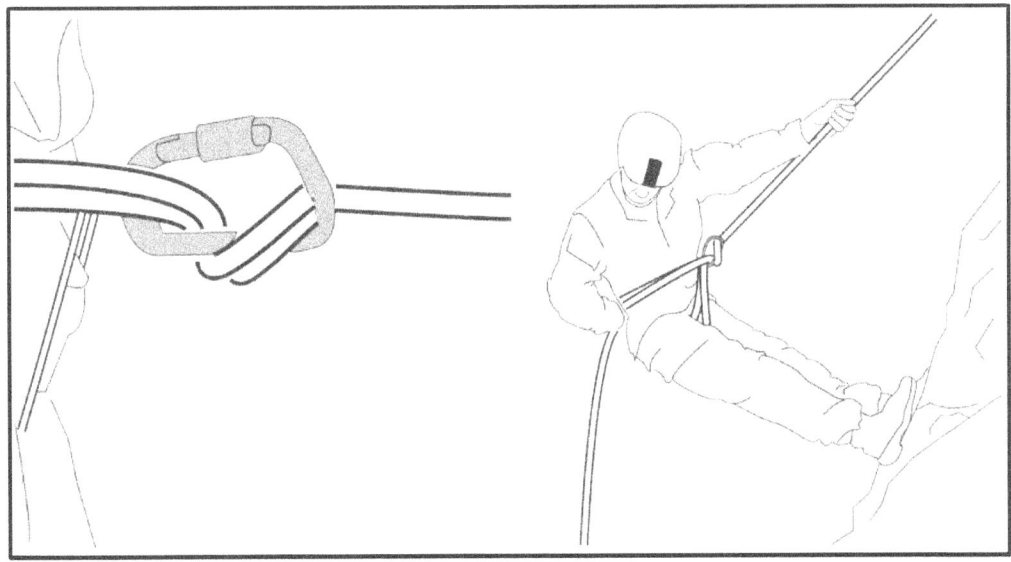

Figure 9-28. Close-up of carabiner wrap descender, and the seat hip rappel
(shown with carabiner wrap seat hip descender)

Chapter 9

9-63. For an **extended ATC** rappel, the ATC should be extended away from the body when possible. A sling rope can be used to do this (see figure 9-29). This lowers the rappeller's center of gravity, making the edge transition smoother. The rappeller uses both hands below the device to control the descent. (See figure 9-30.)

Figure 9-29. Sling extension

Military Mountaineering

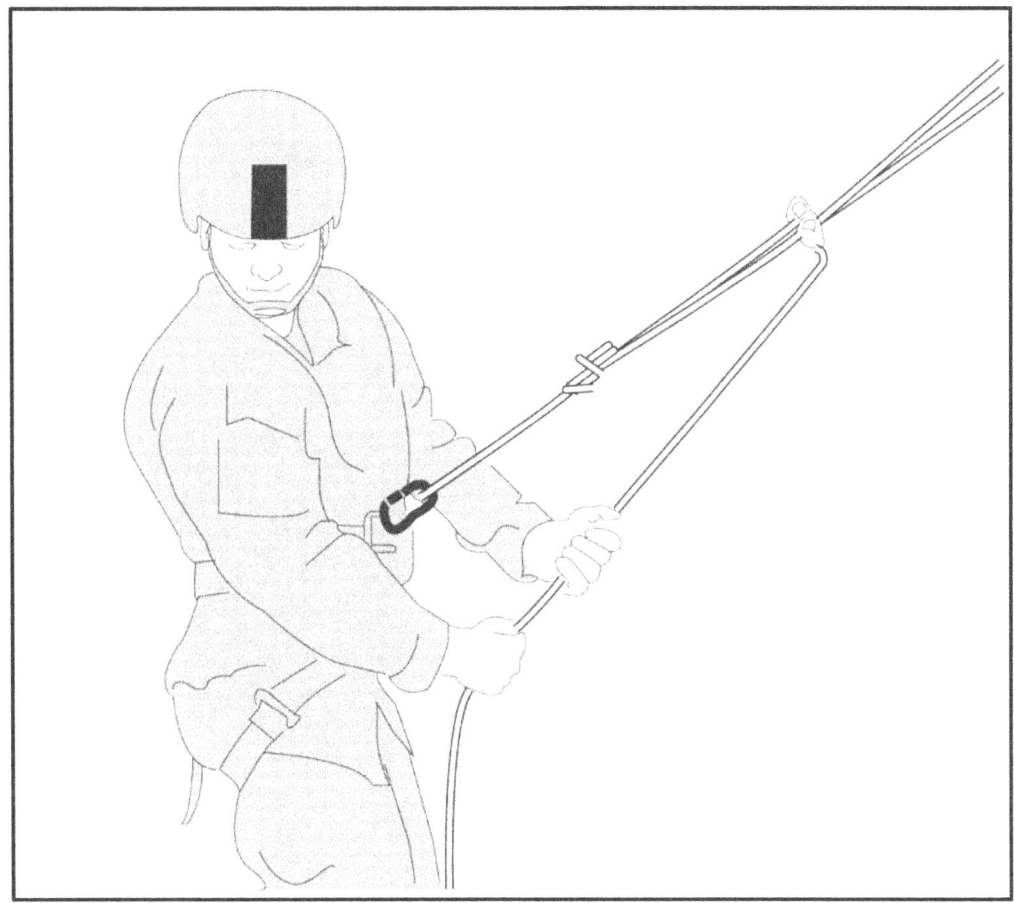

Figure 9-30. Extended ATC rappel

Chapter 10

Machine Gun Employment

Machine guns are a Ranger platoon's most effective weapons against the dismounted enemy force. Machine guns allow the Ranger unit to engage enemy forces from a greater range and with greater accuracy than individual weapons.

SPECIFICATIONS

10-1. A leader's ability to properly employ available machine guns and achieve fire superiority is often the deciding factor on the battlefield. Table 10-1 on pages 10-1 and 10-2 show references and specifications for various machine guns. Definitions associated with machine guns are found in table 10-2 on pages 10-2 and 10-3, and figure 10-1 on page 10-3, and figure 10-2 on page 10-4.

Table 10-1. Specifications of machine guns

WEAPON	M249	M240B	M2	MK19
Information	FM 3-22.68 TM 9-1005-201-10	FM 3-22.68 TM 9-1005-313-10	FM 3-22.68 TM 9-1005-213-10	FM 3-22.27 TM 9-1010-230-10
Description	5.56-mm gas-operated automatic	7.62-mm gas-operated medium	.50-caliber recoil-operated heavy	40-mm air-cooled, blowback-operated automatic GL
Weight	16.41 lbs. (gun w/barrel) 16 lbs. (tripod)	27.6 lbs. (gun w/barrel) 20 lbs. (tripod)	128 lbs. (gun w barrel and tripod)	140.6 lbs. (gun w/barrel and tripod)
Length	104 cm	110.5 cm	156 cm	109.5 cm
Max range	3600 m	3725 m	6764 m	2212 m
Max eff range	Bipod/point: 600 m Bipod/area: 800 m Tripod/area: 1000 m Grazing: 600 m	Bipod/point: 600 m Tripod/point: 800 m Bipod/area: 800 m Tripod/area:1100 m Suppression: 1,800 m Grazing: 600 m	Point: 1500 m (single shot) Area: 1830 m Grazing: 700 m	Point: 1500 m Area: 2212 m
Tracer BO	900 m	900 m	1800 m	
Sustained rate of fire	50 rpm 6 to 9 rounds 4 to 5 sec Every 10 min	100 rpm 6 to 9 rounds 4 to 5 sec Every 10 min	40 rpm 6 to 9 rounds 10 to 15 sec End of day or if damaged	40 rpm

Chapter 10

Table 10-1. Specifications of machine guns (continued)

Weapon	M249	M240B	M2	MK19
Rapid rate of fire	100 rpm 6 - 9 rounds 2 - 3 sec 2 minutes	200 rpm 10 - 13 rounds 2 - 3 sec 2 minutes	40 rpm 6 - 9 rounds 5 - 10 sec Change barrel end of day or if damaged	60 rpm
Cyclic rate of fire	850 rpm, continuous burst/ min	650 - 950 rpm, continuous burst/ min	450 - 550 rpm, continuous burst	325 – 375 rpm, continuous burst
LEGEND BO – burnout; cm – centimeter; eff – effective; FM – Field Manual; GL – grenade launcher; lbs. – pounds; m – meter; max – maximum; mm – millimeter; min – minute; rpm – revolutions per minute; sec – second; TM – Training Manual;					

Table 10-2. Machine gun terms

Line of sight	The imaginary line drawn from the firer's eye through the sights to the point of aim.
Burst of fire	A number of successive rounds fired with the same elevation and point of aim when the trigger is held to the rear. The number of rounds in a burst can vary depending on the type of fire employed.
Trajectory	The curved path of the projectile in its flight from the muzzle of the weapon to its impact. As the range to the target increases, so does the curve of trajectory.
Maximum ordinate	The height of the highest point above the line of sight the trajectory reaches between the muzzle of the weapon and the base of the target. It always occurs at a point about two-thirds of the distance from weapon to target and increases with range.
Cone of fire	The pattern formed by the different trajectories in each burst as they travel downrange. Vibration of the weapon and variations in ammunition and atmospheric conditions all contribute to the trajectories that make up the cone of fire.
Beaten zone	The elliptical pattern formed when the rounds in the cone of fire strike the ground or target. The size and shape of the beaten zone changes as a function of the range to target and slope of the target, but is normally oval or cigar shaped, and the density of the rounds decreases toward the edges. Gunners and automatic riflemen should engage targets to take maximum effect of the beaten zone. Due to the right-hand twist of the barrel, the simplest way to do this is to aim at the left base of the target.
Sector of fire	An area to be covered by fire that is assigned to an individual, a weapon, or a unit. Gunners are normally assigned a primary and a secondary sector of fire.
Primary sector of fire	The primary sector of fire is assigned to the gun team to cover the most likely avenue of enemy approach from all types of defensive positions.
Secondary sector of fire	The secondary sector of fire is assigned to the gun team to cover the second most likely avenue of enemy approach. It is fired from the same gun position as the primary sector of fire.
Final protective fire	An immediately available, prearranged barrier of fire to stop enemy movement across defensive lines or areas.

Machine Gun Employment

Table 10-2. Machine gun terms (continued)

Final protective line	A predetermined line along which grazing fire is placed to stop an enemy assault. If a final protective line (FPL) is assigned, the machine gun is sighted along it except when other targets are being engaged. An FPL becomes part of the unit's machine gun FPFs. An FPL is fixed in direction and elevation. However, a small shift for search is employed to prevent the enemy from crawling under the FPL, and to compensate for irregularities in the terrain or the sinking of the tripod legs into soft soil during firing. Fire is delivered during all conditions of visibility.
Principal direction of fire	Assigned to a gunner to cover an area that has good fields of fire, or that has a likely dismounted avenue of approach, a PDF also provides mutual support to an adjacent unit. If no FPL has been assigned, then sight machine guns using the PDF. If a PDF is assigned and other targets are not being engaged, then machine guns remain on the PDF. It is used only if an FPL is not assigned; it then becomes the machine gun's part of the unit's final protective fires.
LEGEND FPL – final protective line; FPF – final protective fire; PDF – principle direction of fire	

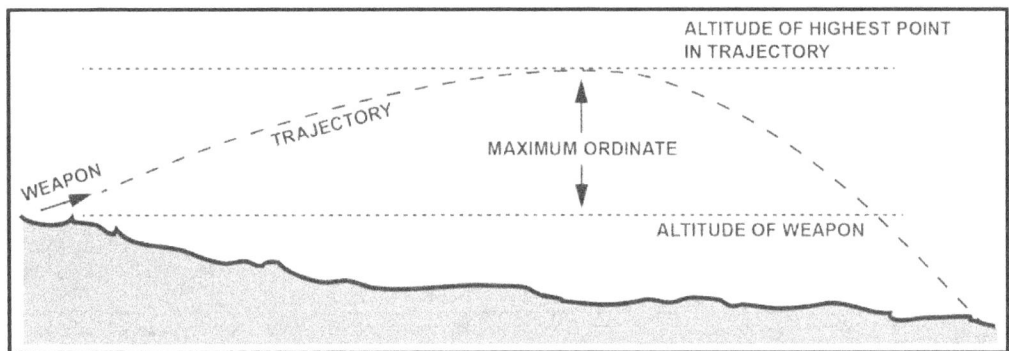

Figure 10-1. Trajectory and maximum ordinate

Chapter 10

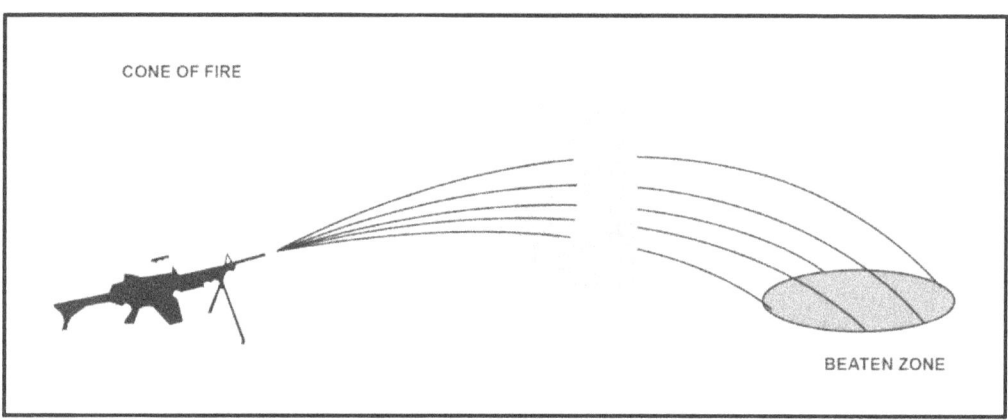

Figure 10-2. Cone of fire and the beaten zone

CLASSES OF AUTOMATIC WEAPONS FIRE

10-2. The U.S. Army classifies automatic weapon fires with respect to ground, target, and weapon. Respect to ground is detailed in table 10-3 and figure 10-3.

Table 10-3. Classes of fire, respect to ground

GRAZING FIRE	Grazing fire occurs when the center of the cone of fire rises less than one meter above ground. Grazing fire is employed in the final protective line (FPL) defense. It is possible only when the terrain is level or uniformly sloping. Any dead space encountered along the FPL is covered by indirect fire, such as from an M320. When firing over level or uniformly sloping terrain, the machine gun M240B and M249 can attain a maximum of 600 meters of grazing fire. The M2 can attain a maximum of 700 meters.
PLUNGING FIRE	Plunging fire occurs when the danger space is within the beaten zone. It occurs when weapons fire at long range, from high to low ground, into abruptly rising ground, or across uneven terrain, resulting in a loss of grazing fire at any point along the trajectory.

Machine Gun Employment

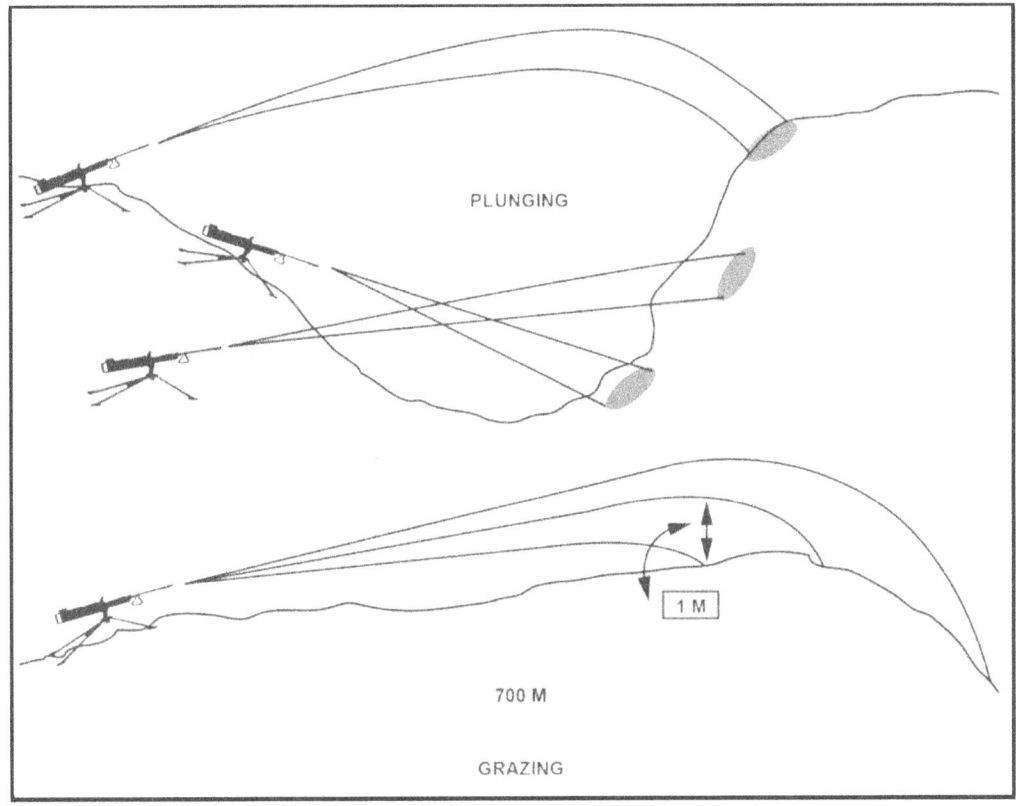

Figure 10-3. Classes of fire, respect to ground

LEGEND
M - meter

10-3. Leaders and gunners should strive at all times to position their gun teams where they can best take advantage of the machine gun's beaten zone with respect to an enemy target. Channeling the enemy by using terrain or obstacles so they approach a friendly machine gun position from the front in a column formation is one example.

10-4. In this situation, the machine gun would employ enfilade fire on the enemy column, and the effects of the machine gun's beaten zone would be much greater than if it engaged the same enemy column from the flank. Table 10-4 on page 10-6 defines and compares the four classifications of fire with respect to the target. Figures 10-4A on page 10-7 and 10-4B on page 10-8 depict these classifications.

Chapter 10

Table 10-4. Classes of fire, respect to target

ENFILADE FIRE BEST	• Occurs when long axes of beaten zone and target coincide/nearly coincide.
	• Can be frontal fire on column or flanking fire on line.
	• Most desirable class of fire with respect to the target:
	    ▪ Makes maximum use of the beaten zone.
	    ▪ Leaders and gunners should always try to position guns for enfilade fire.

FRONTAL FIRE COLUMN: YES LINE: NO	• Occurs when the long axis of the beaten zone is at a right angle to the front of the target. • Highly desirable against a column. • • Becomes enfilade fire as beaten zone coincides with long axis of target. • Less desirable against a line because most of the beaten zone normally falls below or after the enemy target.
FLANKING FIRE COLUMN: NO LINE: YES	• Delivered directly against the flank of the target. • Most desirable against a line. • Becomes enfilade fire as beaten zone coincides with the long axis of the target. • Least desirable against a column because most of the beaten zone normally falls before or after the enemy target.
OBLIQUE FIRE	• Gunners and automatic riflemen. • Occurs when long axis of beaten zone is at any angle other than a right angle to the front of the target.

Machine Gun Employment

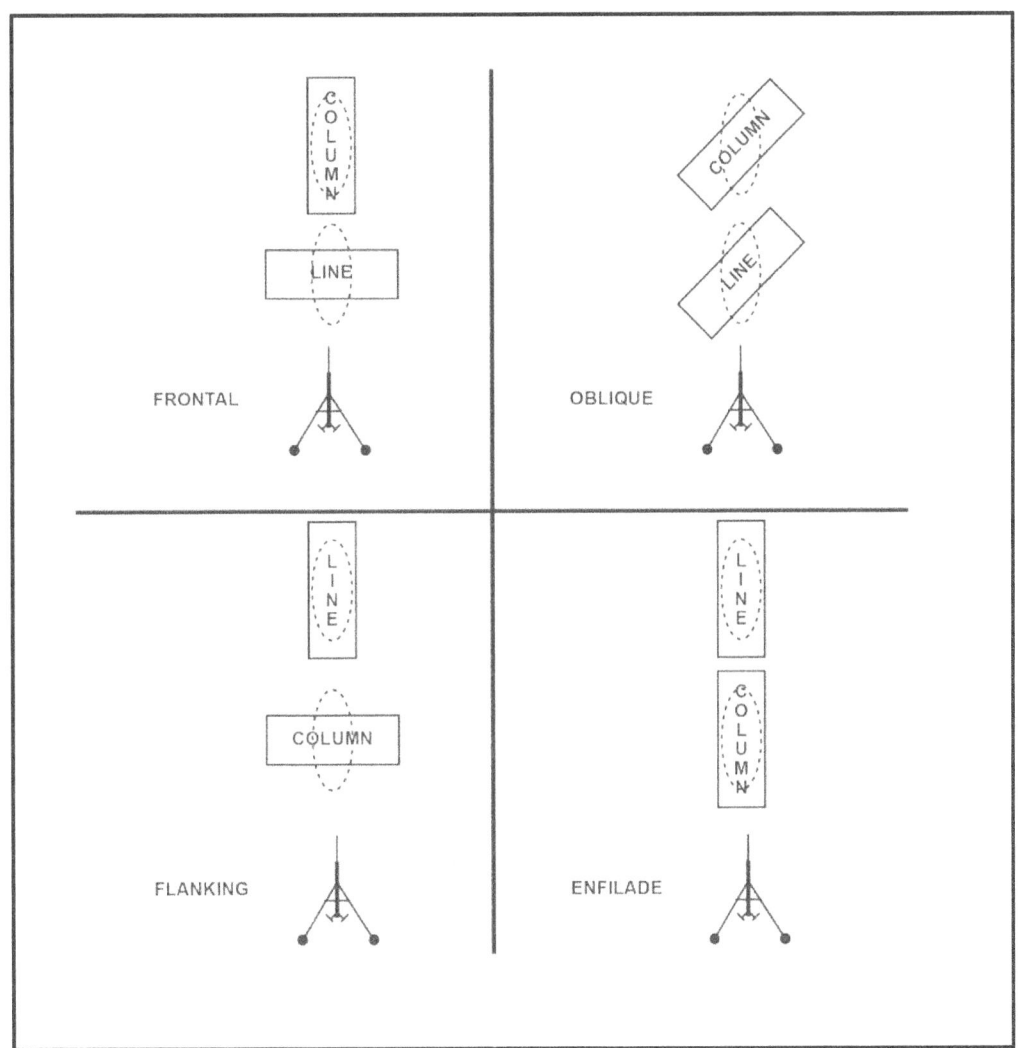

Figure 10-4A. Classes of fire, respect to target

Chapter 10

Figure 10-4B. Classes of fire, respect to target

Machine Gun Employment

10-5. Fires with respect to the machine gun include fixed, traversing, searching, traversing and searching, swinging traverse, and free gunfires. Table 10-5 describes these classifications and figure 10-5 on page 10-10 depicts them.

OFFENSE

10-6. Successful offensive operations depend on effective employment of fire and movement. They are essential, and depend on each other. For example, without the support of covering fires, maneuvering in the presence of enemy fire can produce huge losses.

10-7. Covering fires, especially those that provide fire superiority, allow maneuvering in the offense. However, fire superiority alone rarely wins battles. The primary objective of the offense is to advance, occupy, and hold the enemy position.

Table 10-5. Classes of fire, respect to gun

FIXED	Fixed fire is delivered against a stationary point target when the depth and width of the beaten zone covers the target with little or no manipulation needed. After the initial burst, the gunners follow any change or movement of the target without command.
TRAVERSING	Traversing disperses fires in width by successive changes in direction, but not elevation. It is delivered against a wide target with minimal depth. When engaging a wide target requiring traversing fire, the gunner should select successive aiming points throughout the target area. These aiming points should be close enough together to ensure adequate target coverage. However, they do not need to be so close that they waste ammunition by concentrating a heavy volume of fire in a small area.
SEARCHING	Searching distributes fires in depth by successive changes in elevation. It is employed against a deep target or a target that has depth and minimal width, requiring changes only in the elevation of the gun. The amount of elevation change depends upon the range and slope of the ground.
TRAVERSING AND SEARCHING	This class of fire is a combination in which successive changes in direction and elevation result in the distribution of fires in width and depth. It is employed against a target whose long axis is oblique to the direction of fire.
SWINGING TRAVERSE	Swinging traverse fire is employed against targets that require major changes in direction but little or no change in elevation. Targets may be dense, wide, in close formations moving slowly toward or away from the gun, or vehicles or mounted troops moving across the front. If tripod mounted, the traversing slide lock lever is loosened enough to permit the gunner to swing the gun laterally. When firing swinging traverse, the weapon is normally fired at the cyclic rate of fire. Swinging traverse consumes a lot of ammunition and does not have a beaten zone because each round seeks its own area of impact.
FREE GUN	Free gunfire is delivered against moving targets that are rapidly engaged with fast changes in direction and elevation. Examples are aerial targets, vehicles, mounted troops, or enemy soldiers in relatively close formations moving rapidly toward or away from the gun position. When firing free gun, the weapon is normally fired at the cyclic rate of fire. Free gunfire consumes a lot of ammunition and does not have a beaten zone because each round seeks its own area of impact.

Chapter 10

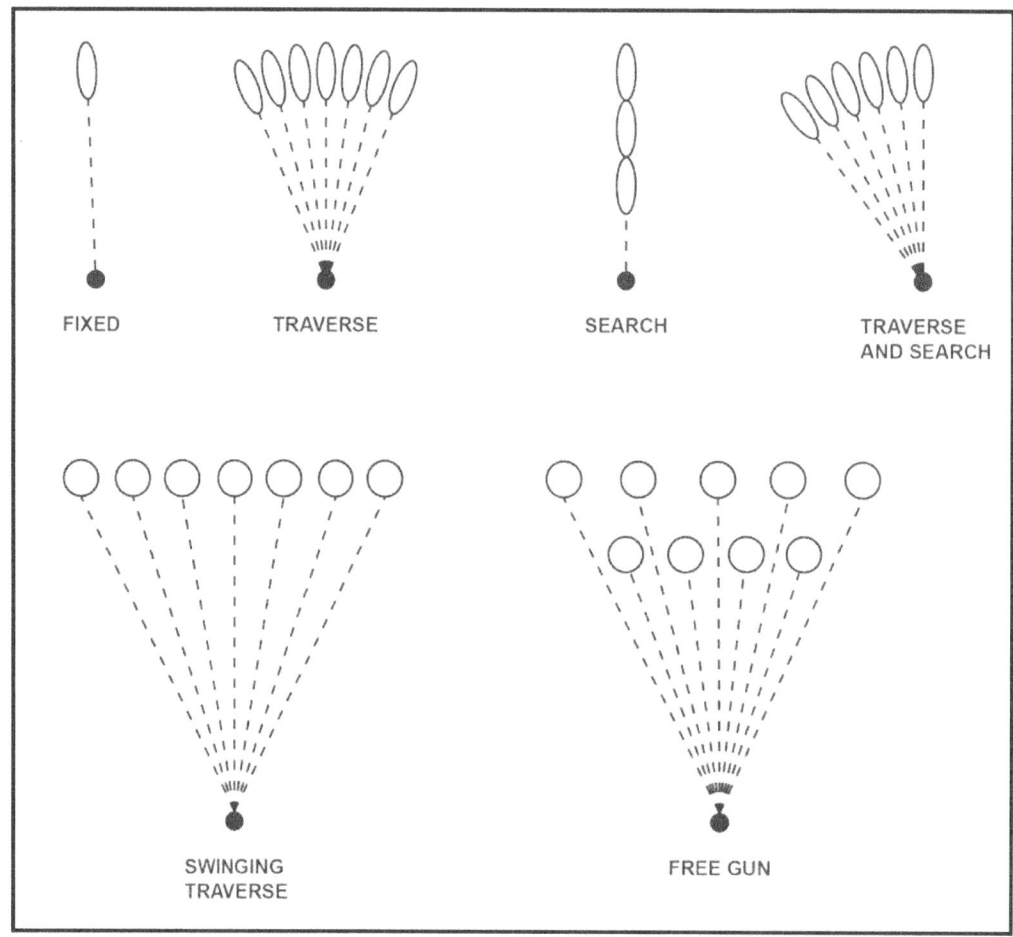

Figure 10-5. Classes of fire, respect to gun

Machine Gun Employment

MEDIUM MACHINE GUNS

10-8. In the offensive, the PL can establish a base of fire element with the M240B, the M249 light machine gun, or a combination of the weapons. When the platoon scheme of maneuver is to conduct the assault with the Infantry squads, the platoon sergeant or WSL may position this element and control its fires. The M240B machine gun is more stable and accurate at greater ranges, but takes longer to maneuver on the tripod than on the bipod. Machine gunner responsibilities include:

- Target key enemy weapons until the enemy's assault element masks the machine gunners' fires.
- Suppress the enemy's ability to return accurate fire.
- Hamper the maneuver of the enemy's assault element.
- Fix the enemy in position.
- Isolate the enemy by cutting off their avenues of reinforcement.
- Shift fire to the flank opposite the one being assaulted and continue targeting any automatic weapons providing enemy support.
- Engage enemy counterattack, if any.
- Cover the gap created between the forward element of the friendly assaulting force and terrain covered by indirect fires when the direct fires are lifted and shifted.
- On signal, displace (with the base of fire element) to join the assault element on the objective.

MK19 AND M2

10-9. As part of the base of fire element, the MK19 and M2 can help the friendly assault element. They do this by suppressing enemy bunkers and lightly armored vehicles. Even if their fire is too light to destroy enemy vehicles, well-aimed suppressive fire can keep the enemy buttoned up and unable to place effective fire on friendly assault elements.

10-10. The MK19 and M2 are particularly effective in preventing lightly armored enemy vehicles from escaping or reinforcing. Both vehicle-mounted weapons can fire from a long-range standoff position, or be moved forward with the assault element.

BASE OF FIRE

10-11. Machine gun fire from an SBF position is the minimum needed to keep the enemy from returning effective fire. Ammunition is conserved so the guns do not run out of ammunition. The WSL positions and controls the fires of all machine guns in the element.

10-12. Machine gun targets include key enemy weapons or groups of enemy targets, either on the objective or attempting to reinforce or counterattack. The nature of the terrain, desire to achieve some standoff, and the other factors of METT-TC prompt the leader to the correct tactical positioning of the base of fire element. There are distinct phases of rates of fire employed by the base of fire element:

- Initial heavy volume (rapid rate) to gain fire superiority.
- Slower rate to conserve ammunition (sustained rate) while still preventing effective return fire as the assault moves forward.
- Increased rate as the assault nears the objective.
- Lift and shift to targets of opportunity.
- Machine guns in the SBF role should be set in role and assigned a primary and alternate sector of fire, and a primary and alternate position.
- Machine guns are suppressive fire weapons used to suppress known and suspected enemy positions. Therefore, gunners cannot be allowed to empty all of their ammunition into one bunker simply because that is all they can identify at the time.
- Shift and shut down the weapon squad gun teams one at a time, not all at once. M203 and mortar or other indirect fire can be used to suppress while the machine guns are moved to where they can shoot.

Chapter 10

- Leaders take into account the surface danger zone (SDZ) of the machine guns when planning and executing the lift and or shift of the SBF guns. The effectiveness of the enemy on the objective plays a large role in how much risk should be taken with respect to the lifting or shifting of fires.
- Once the SBF line is masked by the assault element, fires are shifted, or lifted, or both to prevent enemy withdrawal or reinforcement.

MANEUVER ELEMENT

10-13. Under certain terrain conditions, and for proper control, machine guns may join the maneuver or assault unit. When this is the case, they are assigned a cover fire zone or sector. The machine guns seldom accompany the maneuver element. The gun's primary mission is to provide covering fire. The machine guns are only employed with the maneuver element when the area or zone of action assigned to the assault or company is too narrow to permit proper control of the guns. The machine guns are then moved with the unit and readied to employ on order from the leader and in the direction needing the supporting fire.

10-14. When machine guns move with the element undertaking the assault, the maneuver element brings the machine guns to provide additional firepower. These weapons are fired from a bipod, in an assault mode, from the hip, or from the underarm position. They target enemy automatic weapons anywhere on the unit's objective.

10-15. After destroying the enemy's automatic weapons, if any, the gunners distribute fire over their assigned zones or sectors. The machine gunner in the assault position engages within 300 m of the target, often at point-blank ranges.

10-16. If the platoon's organic weapons fail to cover the area or zone of action, the company commander can assign more machine guns and personnel. This might help the platoon accomplish its assigned mission. Each machine gunner is assigned a zone or a sector to cover, and they move with the maneuver element.

CONTROLLED OCCUPATION AND WITHDRAWAL OF THE SUPPORT-BY-FIRE POSITION

10-17. Controlled occupation of the support-by-fire position is one of the key elements in setting up an SBF position. To remain undetected, use stealth and control. Rangers follow these steps:
- (1) The WSL moves to and establishes a RP just short of the SBF position.
  - Order of movement for the weapons squad during movement to their position is the WSL, gun team 2 (gunner, ammunition bearer [AB], and assistant gunner [AG]), and gun team 1 (AG, gunner, AB).
- (2) The WSL then moves forward from the RP with the gun 2 gunner. The gunner gets into position and remains in bipod mode to provide security.
- (3) The WSL then brings forward the gun 1 gunner and AG.
  - The AG moves to the left of the gun and emplaces the tripod. The gunner then places the gun on the tripod.
  - The AB drops off all ammunition at the gun position and then moves to pull flank or rear security.
- (4) Once gun 1 is in place, the WSL brings the gun 2 AG and AB forward. The AG sets in the tripod then the gunner sets in gun 2 o the tripod. The AB drops off all remaining ammunition, and then pulls flank or rear security.
- (5) Once the SBF is emplaced, the WSL gets down behind both guns to ensure they cover their sectors of fire, and that everything is according to the PLs guidance.
- (6) The WSL calls the PL and notifies that the SBF position is occupied.

10-18. The PL can use controlled withdrawal of the SBF position method to cover the withdrawal of the platoon and provide security for the SBF position. Rangers follow these steps:
- Before the platoon moves off an objective, the WSL shifts the machine guns' sectors of fire to cover the objective.
- After the guns cover the objective, the WSL starts breaking down the gun positions, one at a time.

Machine Gun Employment

- After the main body of the platoon starts to move off the objective, the gun teams move one at a time into the order of movement, with the last gun breaking down as soon as the platoon is completely off the objective.
- The entire weapons squad moves tactically to linkup with the rest of the platoon.

DEFENSE

10-19. The platoon's defense centers on its machine guns. The PL positions the rifle squad to protect the machine guns against the assault of a dismounted enemy formation.

10-20. The machine gun provides the necessary range and volume of fire to cover the squad front in the defense. However, position is very important. The requirements and employment of positioning machine guns are:
- The main requirement of a suitable machine gun position in the defense is its effectiveness in accomplishing specific missions. The position should be accessible and afford cover and concealment. Machine guns are positioned to protect the front, flanks, and rear of occupied portions of the defensive position, and to be mutually supporting. Attacking troops usually seek easily traveled ground that provides cover from fire. For each machine gun, the leader chooses three positions: primary, alternate, and supplementary. This ensures they cover the sector and have protection on their flanks.
- The leader positions each machine gun to cover the entire sector or to overlap sectors with the other machine guns. The engagement range may extend from over 1000 m where the enemy begins their assault, to point-blank range. Machine gun targets include enemy automatic weapons, and command and control elements.

10-21. . Machine gun fire is distributed in width and depth in a defensive position. Machine guns are the backbone or framework of the defense, because the leader can use them to subject the enemy to increasingly devastating fire from the initial phases of the attack, and neutralize any partial enemy successes the enemy might attain by delivering intense fires in support of counterattacks. It also helps the unit hold ground due to its tremendous firepower.

MEDIUM MACHINE GUNS

10-22. In the defense, the medium machine gun provides sustained direct fires that cover the most likely or most dangerous enemy dismounted avenues of approach. It also protects friendly units against the enemy's dismounted close assault. The PL positions the machine guns to concentrate fires in locations to inflict the most damage to the enemy. They are also placed where they can take advantage of grazing enfilade fires, stand off or maximum engagement range, and best observation of the target area.

10-23. Machine guns provide overlapping and interlocking fires with adjacent units, and cover tactical and protective obstacles with traversing or searching fires. When final protective fires are called for, machine guns (aided by M249 fires) place an effective barrier of fixed, direct fire across the platoon front. Leaders position machine guns to—
- Concentrate fires where they want to kill the enemy.
- Fire across the platoon front.
- Cover obstacles by direct fire.
- Tie in with adjacent units.

10-24. In the defense, the MK19 and M2 machine guns may be fired from the vehicle mount or dismounted from the vehicle and mounted on a tripod at a defensive fighting position designed for the weapon system. These guns provide sustained direct fires that cover the most likely enemy-mounted avenue of approach. Their maximum effective range enables them to engage enemy vehicles and equipment at far greater ranges than the platoon's other direct-fire weapons.

10-25. When mounted on the tripod, the M2 and MK19 are highly accurate to their maximum effective ranges. Predetermined fires can be planned for likely high payoff targets. The trade-off is that these weapon systems are heavy and slow to move. These guns are less accurate mounted on vehicles than when fired from the tripod-mounted system. However, they are more easily maneuvered to alternate firing locations, should the need arise.

Chapter 10

CONTROL OF MACHINE GUNS

10-26. Leaders use control measures, coordinating instructions, and fire commands to control the engagements of their machine guns. Rehearsals are key in a leader's ability to control machine guns. The noise and confusion of battle may limit the ability of a leader to control the machine guns. Therefore, a leader uses a combination of methods that accomplish the mission. The following are several successful methods for a leader to control fires:
- Oral.
- Hand and arm signals.
- Prearranged signals.
- Personal contact.
- Range cards.

10-27. A fire command is given to deliver effective fire on a target quickly and without confusion. It is essential that the commands delivered by the WSL are understood and echoed by the assistant gunner or gun team leader and the gunner. The elements of a fire command are:
- **Alert** lets the gun crew know that they are about to engage a target.
- **Direction** lets the gun team know where to engage.
- **Description** lets the gun team know what they are engaging.
- **Range** (if not already set on predestined target), the gun team can adjust the traverse and elevation (T&E) mechanism.
- **Method of fire** element includes manipulation and rate of fire. Manipulation dictates the class of fire with respect to the weapon. It is announced as, "FIXED," "TRAVERSE," "SEARCH," or "TRAVERSE AND SEARCH." Rate controls the volume of fire (sustained, rapid, and cyclic).
- **Command to open fire** initiates the firing of the weapon system.

AMMUNITION PLANNING

10-28. Leaders carefully plan for the rates of fire to be employed by machine guns as they relate to the mission and the amount of ammunition available. The WSL fully understands the mission, the amount of available ammunition, and the application of machine gun fire needed to support all key events of the mission fully. Careful planning helps ensure the guns do not run out of ammunition.

10-29. A mounted platoon might have access to enough machine gun ammunition to support the guns throughout any operation. A dismounted platoon with limited resupply capabilities has to plan for only the basic load to be available. In either case, leaders take into account key events the guns support during the mission. They plan for the rate of machine gun fire needed to support the key events, and the amount of ammunition needed for the scheduled rates of fire.

10-30. The leader estimates how much ammunition is needed to support all the machine guns, and adjusts the amount used for each event to ensure enough ammunition remains for all phases of the operation. Examples of planning rates of fire and ammunition requirements for a platoon's machine guns in the attack are listed below.

Machine Gun Employment

Weapons Squad Tactics, Techniques, and Procedures

A. Use a starter belt when moving (about 50 to 70 rounds).

B. Ensure ammunition and NVDs are in packs, such as an assault pack for mounted and city operations; rucksack for long sustainment missions; and are readily accessible.

C. Carry the traverse and elevation (T&E) mechanism and tripod together.

D. Mission dependent on when the tripod is taken (such as urban operations).

E. Use optics, lasers, NVDs. For example, in urban operations, think about using a reflexive sight because most of the engagements are 150 m or less. Also, zero the iron sights.

This page intentionally left blank.

Chapter 11

Urban Operations

Today's security environment demands more from leaders than ever before. Leaders not only lead Rangers but also influence other people. They are able to work with other members of the armed services and government agencies. They win the willing cooperation of multinational partners, military and civilian. Urban offensive operations pose great risks to Army forces and noncombatants. Yet, the military demands self-aware and adaptive leaders who can compel enemies to surrender in war and master the circumstances facing them in stability operations and peace. Victory and success depend on the effectiveness of these leaders' organizations. Developing effective organizations requires hard, realistic, and relevant training.

PLANNING

11-1. Urban operations include decisive action—the continuous, simultaneous combinations of offensive, defensive, and stability or defense support of civil authorities tasks that may be executed, either sequentially or (more likely) simultaneously during an urban operation. (For further study, see ADRP 3-0, ATTP 3-06.11, and ATP 3-21.8.

11-2. Urban areas are strategically important. Several factors attract armies to combat in urban areas, such as—
- Using the defensive advantages of the urban environment.
- Developing allegiance and support of populace.
- Adapting urban resources for operational or strategic purposes: infrastructure, capabilities, and other resources.
- Drawing the enemy in.
- Playing on area's symbolic importance.
- Using the area's geographical advantages:
  - Dominance of a region.
  - Avenue of approach.

TASK ORGANIZATION

11-3. Task-organizing subordinate units for urban operations depends largely on the nature of the operation. Some units are always part of the task organization to ensure the success of urban operations (UOs). Infantry, special operations, civil affairs, aviation, military police, military information support operations (MISO), military intelligence, and engineers are units required for decisive action in urban operations. Other types of forces such as Armor, artillery, and chemical units, have essential roles in specific types of urban operations, and may apply less to other operations.

11-4. Military forces conduct decisive action within urban areas. Commanders conduct decisive action abroad by executing offensive, defensive, and stability urban operations as part of a joint, interagency, and multinational effort. The situation mandates that one type of operation—offense, defense, stability, or defense support of civil authorities—dominates the urban operation. Commanders often find themselves executing offensive, defensive, and stability operations at the same time. In fact, waiting until all combat operations are concluded before beginning stability operations often results in lost, sometimes irretrievable, opportunities. The dominant type of operation varies between different urban areas, even in the same campaign.

Chapter 11

PREPARATION

11-5. Operating successfully in a complex urban environment requires a thorough understanding of the environment and rigorous, realistic urban operation (UO) training. Training should cover every aspect of decisive action, including appropriate tactics, techniques, and procedures related to offense, defense, and stability operations. Training should also replicate the following:
- The psychological impact of intense, close combat against a well-trained, relentless, and adaptive enemy.
- The effects of noncombatants, including governmental and nongovernmental organizations, and agencies in close proximity to Army forces. This necessitates—
    - An in-depth understanding of culture and its effects on perceptions.
    - An understanding of civil administration and governance.
    - The ability to mediate and negotiate with civilians, including the ability to communicate through an interpreter effectively.
    - The development and use of flexible, effective, and understandable ROE.
- A complex intelligence environment requiring lower echelon units to collect and forward essential information to higher echelons for rapid synthesis into timely and useable intelligence at all levels of command. The multifaceted urban environment requires a bottom-fed approach to developing intelligence.
- The communications challenges imposed by the environment, as well as the need to transmit large volumes of information and data.
- The medical and logistical problems associated with operations in an urban area, including constant threat interdiction against lines of communications and sustainment bases

11-6. In a complex urban environment, every Ranger, regardless of branch or military occupational specialty, is committed and prepared to close with and kill or capture threat forces. Every Ranger is prepared to effectively interact with the urban area's noncombatant population and assist in the unit's intelligence collection efforts.

11-7. In UO, every Ranger is likely to perform advanced rifle marksmanship, including advanced firing positions, short-range marksmanship, and night firing techniques (unassisted and with the use of optics). While not all inclusive and necessarily urban specific, other critical individual and collective UO tasks might include:
- Conduct troop-leading procedures.
- Operate unit's crew-served weapons.
- Conduct urban reconnaissance and combat patrolling.
- Enter and clear buildings and rooms as part of an urban attack or cordon and search operation.
- Sensitive site exploitation (SSE).
    - Utilization of metal detectors.
    - Utilization of military working dogs (MWDs).
    - Conduct tactical call out.
    - Work with local Army, police, or special operation forces.
- Defend an urban area.
- Act as a member of a mounted patrol (including specific driver training).
- Recover own vehicles.
- Control civil disturbances.
- Navigate in an urban area.
- Prepare for follow-on missions.
- Identify explosives, bombs, booby traps, materials used, and methods for making and clearing them.
- Linkup with battlespace owner.
- React to contact, ambush, snipers, indirect fire, and IEDs.
- Set up personnel or vehicle checkpoint, or blocking positions around target location.

Urban Operations

- Establish overwatch positions and support-by-fire positions, such as sniper positions.
- Simultaneous clearing of top and bottom floors of the building.
- Assign climbing and roof-clearing teams for overwatch or sniper support.
- Teach how to use long-range surveillance, scout, and sniper teams effectively.
- Secure a disabled vehicle or downed aircraft.
- Call for indirect fire and CAS.
- Create and employ explosive charges.
- Handle detainees and enemy prisoners of war. Know how to extract high value targets.
- Treat and evacuate casualties.
- Accurately report information.
- Understand the society and culture specific to the area of operations.
- Use basic commands and phrases in the region's dominant language.
- Conduct tactical questioning (TQ).
- Interact with the media.
- Conduct thorough after action reviews.

ANALYZING THE URBAN ENVIRONMENT

11-8. Urban operations often differ from one operation to the next. However, some fundamentals apply to UOs regardless of the mission, geographical location, or level of command. They are particularly relevant to the urban environment that is dominated by man-made structures and a dense noncombatant population. These fundamentals help to ensure every action taken by a commander conducting UO contributes to the desired end state.

11-9. Maintaining close combat is inherent in decisive action UO. Close combat in any UO is resource intensive, requires properly trained and equipped forces, and has the potential for high casualties. The ability to decisively close with and destroy enemy forces as a combined arms team remains essential. In stability UO, a lack of respect and fear of Army forces can hinder recovery as much as the ill-advised use of force. All BCT Soldiers should be properly equipped and trained to fight in an urban environment. This allows the BCT to deter aggression, compel compliance, morally and physically dominate an enemy and destroy their means to resist, and terminate or transition UO on the BCT commander's terms.

11-10. Previous Army doctrine inclined towards a systematic linear approach to urban combat. This attrition approach emphasized standoff weapons and firepower. It can result in significant collateral damage, a lengthy operation, and an inconsistency with the political situation and strategic objectives. Enemy forces that defend urban areas want Army forces to adopt this approach because of the likely costs in resources. BCT commanders should only consider this approach to urban combat as an exception and justified by unique circumstances. Instead, commanders should seek to achieve precise, intended effects against multiple decisive points that overwhelm an enemy's ability to react effectively.

CONTROL THE ESSENTIAL AND MINIMIZE COLLATERAL DAMAGE

11-11. Rangers need to analyze the urban environment carefully. Things to consider include:
- **Mission.** Know correct task organization to accomplish the mission (offense, defense, or stability and support operations).
- **Enemy:**
  - **Disposition.** Analyze the array of enemy forces in and around the objective, known and suspected, such as known or suspected locations of minefields, obstacles, and strong points.
  - **Composition and strength.** Analyze the enemy's task organization, troop's available, suspected strength, and amount of support from the local civilian population based on intelligence estimates. Is the enemy a conventional or unconventional force?

Chapter 11

- **Morale.** Analyze the enemy's current operational status based on friendly intelligence estimates. For example, is the enemy well supplied? Have they recently won against friendly forces or taken many casualties? What is the current weather?
- **Capabilities.** Determine what the enemy can employ against friendly forces; for example, what weapon systems do they have? Are there snipers? What about IED or chemical, biological, radiological, nuclear, and high yield explosives (CBRNE) threats? Are there artillery, engineer, or air defense assets? Do they have thermal or night vision device capabilities, CAS, or armor threats? Be able to discriminate between threats and nonthreats, such as suicide vests.
- **Probable course(s) of action.** Based on friendly intelligence estimates, determine how the enemy will fight within the AO (in and around the friendly AO). Know the enemy AO tactics, techniques, and procedures (TTPs) such as trip wires, pressure plate IEDs, or snipers. Analyze historical data from attacks: where, what, how, and time of day.

Terrain

11-12. Leaders conduct a detailed terrain analysis of each urban setting, considering the types and composition of existing structures. They use OAKOC; political, military, economic, social, infrastructure, information, physical environment, and time (PMESII-PT); and areas, structures, capabilities, organizations, people, and events (ASCOPE) when analyzing terrain in and around the AO. (Refer to Chapter 2 for more information.)

- **Observation and fields of fire.** Always be prepared to conduct UOs under limited visibility conditions.
- **Cover and concealment.** Thoroughly analyze areas inside and on the edge of urban areas. Identify routes to objectives that give assault forces the best possible cover and concealment. Take advantage of limited visibility, which allows forces to move undetected to their final assault or breaching positions. Use overwatch elements and secondary entry teams for security while initial entry or breaching teams move forward. When in the final assault position, forces should move as rapidly as tactically possible to access structures, which afford cover and concealment.
  - It is human nature to stick together and seek safety, but try to avoid bunching up at entry points, funnels, walls, or indoors. Maintain a safe but securable distance between teams and squads. This helps ensure that one grenade cannot take out the whole team at once.
  - Learn to properly use obscurants, and use "tactical patience" to take full advantage of these effects.
  - Practice noise and light discipline. Avoid unnecessary voice communications, learn the proper use of white light, and limit contact with surfaces that could draw the enemy's attention.

Obstacles, Key Terrain, and Avenues of Approach

11-13. Many man-made and natural obstacles exist on the periphery, as well as within the urban environment. Conduct a detailed reconnaissance of routes and objectives, including subterranean complexes, and consider route adjustments and special equipment needs. Ensure routes are clear (not blocked). Avoid roads that run along or through market places, as these roads can become easily blocked.

11-14. Analyze which buildings, intersections, bridges, landing zones (LZs) and pickup zones (PZs), airports, and elevated areas provide a tactical advantage to either side. The leader also identifies critical infrastructure within the area of operations, which would provide the enemy with a tactical advantage on the battlefield. These may include, but are not limited to, communication centers, medical facilities, governmental facilities, and facilities with psychological significance.

11-15. Consider roads, intersections, inland waterways, and subterranean constructions (subways, sewers, and basements). Leaders should classify areas as go, slow go, or no go based on the navigability of the approach. Always have alternate infiltration and exfiltration routes. Keep in mind that a wall can be breached as an emergency exfiltration route.

Urban Operations

TROOPS AND TIME

11-16. Analyze friendly forces using their disposition, composition, strength, morale, capabilities, and so on. Leaders also consider the type and size of the objective to plan effective use of the available troops.

11-17. Operations in an urban environment have a slower pace and tempo. Leaders consider the amount of time required to secure, clear, or seize the urban objective, along with the stress and fatigue Rangers will encounter. Additional time is also allowed for area analysis efforts. This may include, but is not limited to:
- Maps, urban plans, and aerial photographs. Collect historical data from other units and indigenous forces.
- Hydrological data analysis.
- Line of sight surveys.
- Long range surveillance and scout reconnaissance.
- Is artillery supporting you and someone else at the same time?
- How long does it take to shift a 155-mm Howitzer and prepare the gun?
- What is the priority level for getting Armor assets?
- How close is Armor to the target?
- Does their presence compromise the mission?
- How long will it take them to move to a location?
- If Armor assets are not previously coordinated, how long will it take to get them? • How much preparation, survey, and emplacement time of charges do the engineers need?

CIVILIANS

11-18. The National Command Authority establishes the ROE. Commanders at all levels may provide further guidance for dealing with civilians in the AO. Leaders remind subordinates daily of the latest ROE, and immediately inform them of any changes. Rangers have the discipline to identify the enemy from noncombatants, and ensure civilians understand and follow all directed commands.

Note: Civilians may not speak English, may be hiding (especially small children), or may be dazed from a breach. Do not give them the means to resist. Rehearse how clearing or search teams react to these variables. Never compromise the safety of the Rangers. Consider having the interpreter (TERP) use a marking system to separate military-aged males (MAMs) from women and children. Have designated dirty and clean rooms, and a tactical questioning area.

11-19. The complexity of the urban environment, particularly the human dimension, requires rapid information sharing at all levels, to include joint services, multinational partners, and participating governmental and nongovernmental agencies. The analysis of urban information necessary to refine and deepen a commander's understanding of the urban environment and its infrastructure of systems also demands collaboration among the various information sources and consumers.

CLOSE QUARTERS COMBAT

11-20. Due to the nature of a close quarters combat (CQC) encounter, engagements are very close (within 10 m) and very fast (targets are exposed for only a few seconds). Most of these engagements are won by the side that hits first and puts the enemy down. It is more important to knock an enemy down as soon as possible than it is to kill them. In order to win a close quarter's engagement, Rangers make quick, accurate shots by mere reflex. This is accomplished by reflexive fire training. Remember; always fire until the enemy goes down. All reflexive fire training is conducted with the eyes open.

Chapter 11

> *Note:* Research has determined that only three out of ten people actually fire their weapons when confronted by an enemy during room cleaning operations. Close quarters combat success for the Ranger begins with being psychologically prepared for the close quarter's battle. The foundation for this preparedness begins with proficiency in basic rifle marksmanship. Survival in the urban environment does not depend on advanced skills and technologies. Rangers are proficient in the basics.

REHEARSALS

11-21. Similar to the conduct of other military operations, leaders need to designate time for rehearsals. UOs require a variety of individual, collective, and special tasks that are not associated with operations on less complex terrain. These tasks require additional rehearsal time for clearing, breaching, obstacle reduction, CASEVAC, and support teams. Additionally, time is identified for rehearsals with combined arms elements.

11-22. In a **stance**, feet are shoulder-width apart, toes pointed straight to the front (direction of movement). The firing side foot is slightly staggered to the rear of the nonfiring foot. Knees are slightly bent and the upper body leans slightly forward. Shoulders are not rolled or slouched. Weapon is held with the butt stock in the pocket of the shoulder, maintaining firm rearward pressure into the shoulder. This allows for more accurate shot placement on multiple targets. The firing side elbow is kept in against the body and the hand should be forward on the weapon, not on the magazine well. This allows for better control of the weapon. The stance should be modified to ensure the Ranger maintains a comfortable boxer stance.

11-23. In a **low carry technique**, the butt stock of the weapon is placed in the pocket of the shoulder. The barrel is pointed down so the front sight post and day optic are just outside of the Ranger's field of vision. The head is always up, identifying targets. This technique is safest and is recommended for use by the clearing team once inside the room.

11-24. For the **high carry technique**, the butt stock of the weapon is held in the armpit. The barrel is pointed slightly up, with the front sight post in the peripheral vision of the individual. To assume the proper firing position, push out on the pistol grip, thrust the weapon forward, and pull the weapon straight back into the pocket of the shoulder. This technique is best suited for the line-up outside the door. Exercise caution with this technique, always maintaining situational awareness, particularly in a multifloored building.

> *Note:* Muzzle awareness is critical to the successful execution of close quarter's operations. Rangers never, at any time, point their weapons at or cross the bodies of their fellow Rangers. They always avoid exposing the muzzle of their weapons around corners; this is referred to as "flagging."

11-25. If a Ranger has a **malfunction** with the weapon during any close quarters combat (CQC) training, he takes a knee to conduct immediate action. Once the malfunction is cleared, there is no need to stand up to engage targets immediately. Rangers can save precious seconds by continuing to engage from one knee. Whenever other members of the team see a Ranger down, they automatically clear his sector of fire. Before rising to his feet, the Ranger warns team members of the movement and only rises after checking the rear to make sure no one is shooting over him, and after they acknowledge him. If a malfunction occurs after he has committed to a doorway, the Ranger enters the room far enough to allow those following him to enter, and then moves away from the door. This drill is continually practiced until it is second nature.

11-26. Special consideration is given to the **approach to a building or breach point**. One trademark of Ranger operations is the use of limited visibility conditions. Whenever possible, breaching and entry operations should be executed during hours and conditions of limited visibility. Rangers should always take advantage of all available cover and concealment when approaching breach and entry points. When natural or manmade cover and concealment is not available, Rangers should employ obscurants

Urban Operations

to conceal their approach. Obscurants can also enhance existing cover and concealment sites. Members of the breach and entry team should be numbered for identification, communication, and control purposes.
- Ranger #1 should always be the most experienced and mature member of the team, *other than the team leader. Ranger #1 is responsible for frontal and entry, and breach point security.
- Ranger #2 is directly behind Ranger #1 in the order of movement, and moves through the breach point in the opposite direction from Ranger #1.
- Ranger #3 simply goes in the opposite direction as Ranger #2 inside the room; at least one meter from the door.
- Ranger #4 moves in the opposite direction as Ranger #3, and is responsible for rear security (and is normally the last Ranger into the room). An additional duty of Ranger #4 is breaching.
- The team leader is responsible for initiating all voice and physical commands, and exercises situational awareness at all times with respect to the task, friendly force, and enemy activity. He is in a position to maintain control of the team.
- With the possibility of civilians in the building or rooms, Rangers may decide to only enter with precision weapons such as M4s (not M249s) to avoid civilian casualties.

Note: Consider how much firepower each Ranger delivers. Where do you put the SAW gunner in the order? Weigh firepower against quick, accurate shots. If Ranger #4 has breaching responsibilities, it should not be the SAW gunner, because this would reduce the firepower.

11-27. Consideration is given for **actions outside the point of entry.** Entry point position and individual weapon positions are important. The clearing team members should stand one-to-two feet from the entry point, ready to enter. They should orient their weapons so the team provides its own 360-degree security at all times. Team members signal to each other that they are ready at the point of entry. This is best accomplished by sending up a "squeeze," or rocking motion. If a tap method is used, an inadvertent bump may be misunderstood as a tap.

11-28. For **enter and clear a room**, see battle drill 07-4-D9509 in Chapter 8. Figures 11-1 and 11-2 on page 11-8, and figure 11-3 on page 11-9, depict how Rangers clear a room. When **locking down the room:**
- Control the situation within the room.
- Use clear, concise arm and hand signals. Voice commands should be kept to a minimum to reduce the amount of confusion. It prevents the enemy who might be in the next room from discerning what is going on. This enhances the opportunity for surprise and allows the assault force to detect any approaching force
- Physically and psychologically, dominate the room's inhabitants.
- Assess the situation. In a less hostile situation, it may be better to slow clear instead of dominating the room with brute force. This keeps noncombatants calm and more manageable.
- Establish security and report status.
- Do a cursory search of the room to include the ceiling (three-dimensional fight).
- Identify the dead using reflexive response techniques (eye thump method or kick to the groin for males).
- Search the room for PIR while considering your time available on target.
- Evacuate personnel.
- Mark the room as clear by using chemical lights, engineer tape, chalk, paint, VS-17 signal panels, and so on.

Chapter 11

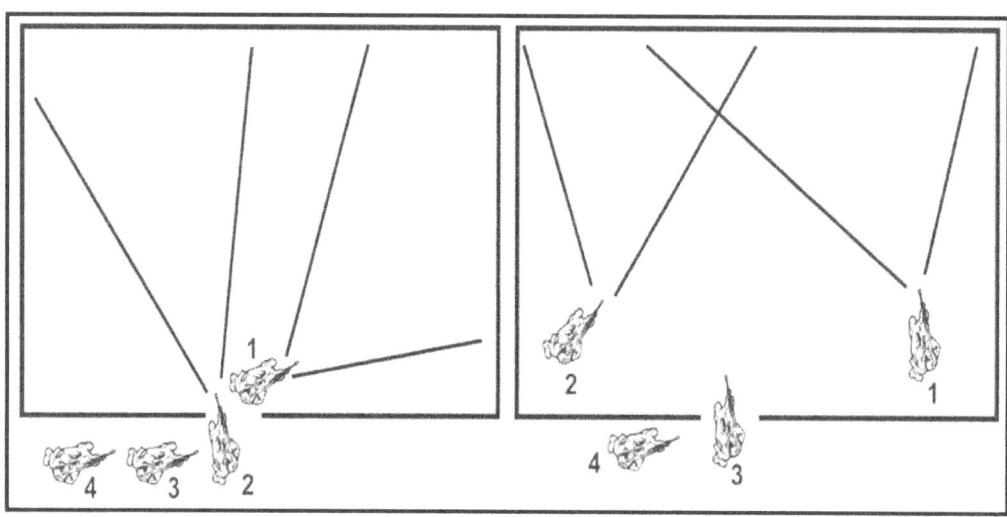

Figure 11-1. 07-4-D9509. Clear a room, first two Soldiers enter

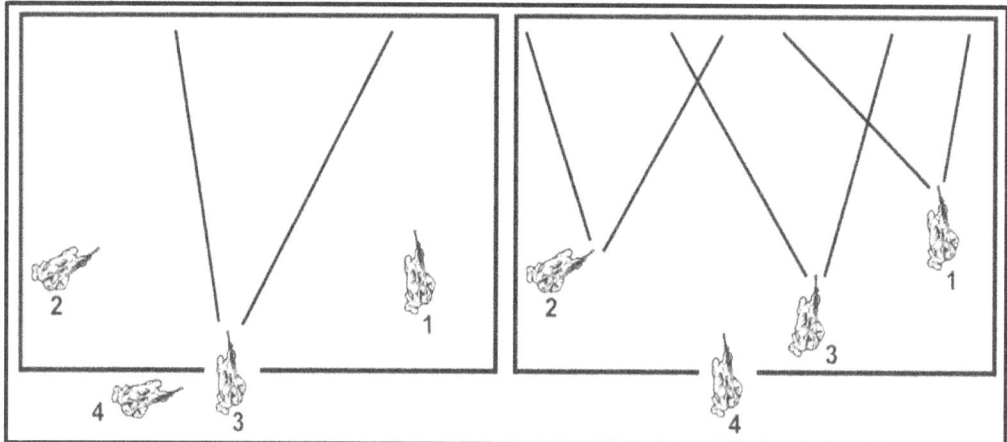

Figure 11-2. 07-4-D9509. Clear a room, third Soldier enters

Urban Operations

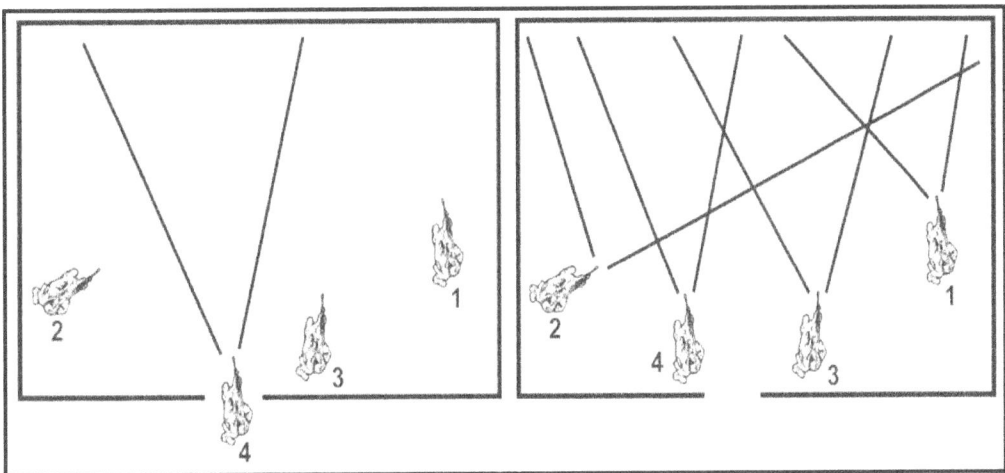

Figure 11-3. 07-4-D9509. Clear a room, third Soldier enters dominating his sector

Procedures for Marking Buildings and Rooms

11-29. Units have long identified a need to mark specific buildings and rooms during UOs. Sometimes, rooms need to be marked as having been cleared, or buildings need to be marked as containing friendly forces. Chalk is the most common marking material. It is light and easily obtained but less visible than other markings. Other techniques include spray paint and paintball guns. (Refer to ATTP 3-06.11 for more information.)

Note: Avoid permanently marking buildings and rooms, as this may cause collateral damage and is likely to deteriorate relationships built with local nationals.

11-30. Chemical lights (chemlights) and scrim-backed, pressure-sensitive tapes (100 miles per hour [mph] tapes) come in a variety of colors, can be seen easily from a distance, and can be removed when necessary. The colors of the chemlights and 100 mph tapes have different meanings:
- Red: casualty collection point.
- Green: room clear.
- Orange: unexploded ordnance.
- Blue: clean room.
- Infrared: breach point.

This page intentionally left blank

Chapter 12

Waterborne Operations

This chapter discusses rope bridges, poncho rafts, and watercraft. While conducting waterborne operations, all Rangers will be in the waterborne uniform. Equipment is worn in the following order:
1. Pant leg unbloused.
2. Top zipped up, collar fastened.
3. Cuffs fastened.
4. Swimmer safety line tied utilizing an around-the-waist bowline with an end of rope bowline at arm's length with carabiners attached to collar.
5. Field load carrier (FLC) is unzipped.

ROPE BRIDGE

Rope bridges are used when a battalion or smaller unit is required to conduct a covert gap crossing. A covert crossing is a planned crossing of an inland water obstacle or other gap that is intended to be undetected. A covert gap crossing can be used in a variety of situations to support various missions, but should be considered (as opposed to deliberate or hasty) only when there is a need or opportunity to cross a gap without being discovered.

12-1. The Ranger patrol seldom have ready-made bridges, so they must know how to employ covert gap crossing techniques. The personnel needed to make a rope bridge are:
- Ranger #1: lead safety swimmer and farside lifeguard.
- Ranger #2: swims water obstacle pulling 150-foot rope and ties off rope on farside anchor point.
- Ranger #3: nearside lifeguard is the last Ranger to cross water obstacle.
- Ranger #4: BTC is the most knowledgeable person on the team.
- Rangers #5 and #6: rope pullers.

12-2. For a wet crossing (or one-rope bridge), special equipment is needed. This includes:
- Two carabiners for each piece of heavy equipment.
- Three steel carabiners for each 150 feet of rope.
- One seven-foot utility rope for each person (swimmer safety line).
- Two carabiners for each person. (One clipped to swimmer safety line and one tied to top center frame of rucksack.)
- Two waterproof bags for each RTO.
- Two carbon dioxide (CO_2) inflatable life preservers (Scout swimmer vests).
- Three noninflatable life preservers (work vests).
- Two 150-foot nylon ropes.

Chapter 12

12-3. A gap crossing annex (see table 2-14 on page 2-27) is prepared with the unit's OPORD. Special organization is accomplished at this time. For a platoon-sized patrol, a squad is normally given the task of providing the rope bridge team. The squad leader designates the most technically proficient Ranger in the squad as the BTC. Rehearsals and inspections take place. (See figure 12-1) Emphasize—
- Security and actions on enemy contact.
- Actual construction of the rope bridge on dry land within the eight-minute time standard.
- Individual preparation.
- Order of crossing.
- All signals and control measures.
- Reorganization.

12-4. Conduct rehearsals as realistically as possible. Ensure personnel are proficient in the mechanics of a covert gap crossing operation. Inspect for equipment completeness; correct rigging and preparation; finalize weapon configurations, personnel knowledge, and understanding of the operation. During the preparation phase, Ranger #4 (BTC) rehearses the bridge team, accounts for all equipment in the bridge kit, and ensures the 150-foot rope is back-stacked and properly coiled.

Waterborne Operations

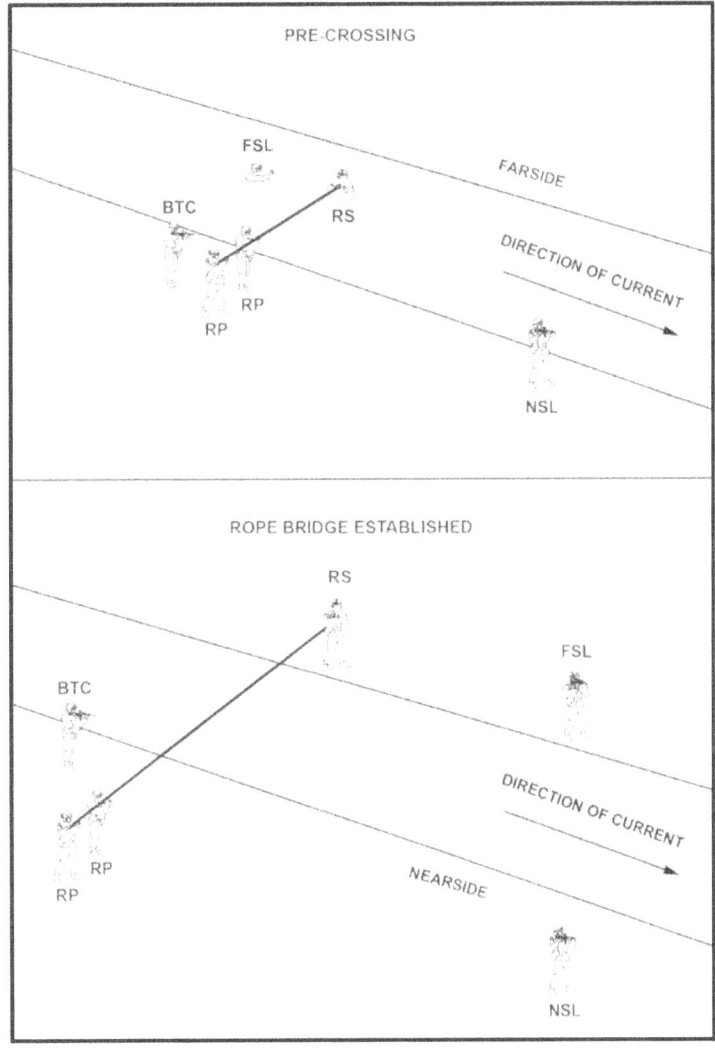

Figure 12-1. Position of bridge team personnel

LEGEND
BTC – bridge team commander; FSL –farside lifeguard; NSL – nearside lifeguard; RP – rope puller; RS – rope swimmer

Chapter 12

12-5. Once the execution phase is reached, several actions take place. This includes establishing and conducting a bridge for the gap crossing:
- Leader halts short of the river, establishes local security and reconnoiters the area for the presence of the enemy, and for crossing site suitability or necessity. The leader directs the BTC to construct the bridge.
- The BTC constructs a one-rope bridge and selects nearside and (visibility permitting) farside anchor points. To anchor himself to the bridge, the BTC ties a swimmer's safety line around his waist and secures it with an overhand knot, then ties the free running end of the bowline into an overhand knot, and attaches a carabiner to the loop in the knot. Ensure the bowline is just long enough to place the carabiner at arm's length. This allows the BTC to remain within reach of the rope bridge, should he lose his grip.
- The bridge team begins to establish the rope bridge while unit members begin individual preparation.
- Each Ranger puts a carabiner in his end of the bowline and in the front sight assembly of every M4, or M16. M240 gunners put a carabiner through the front sight assembly and rear swivel of their M240 machine gun. RTO, FO, and others with heavy rucksacks place an additional carabiner on the top center of their rucksack frames.
- Team establishes security upstream and downstream, while unit leader briefs the BTC on anchor points. The leader counts the Rangers across.
- The BTC enforces noise and light discipline, and maintains security.

12-6. The bridge team is responsible for constructing the rope bridge. Ranger #1 (lead safety swimmer and farside lifeguard) grounds his rucksack (with a carabiner through the top of frame) to the rear of the nearside anchor point. He carries a knotted hand line or safety line to assume duties of farside lifeguard. Wear equipment in the following order (from the body out):
- Waterborne uniform (top zipped up, neck collar fastened, and pants unbloused).
- Noninflatable life preserver.
- CO_2 inflatable life preserver.
- Field load carrier.
- Weapon (across the back).
- Swimmer safety line routed over all equipment and secured to the collar of the Army combat uniform (ACU) blouse.

12-7. Ranger #1 enters the water upstream from Ranger #2 and stays an arm's length away from Ranger #2. Ranger #1 identifies the farside anchor point upon exiting the water. Once Ranger #2 has exited the water, he moves to his farside lifeguard position downstream of the rope bridge, with knotted hand line in hand, FLC or weapon grounded, and noninflatable life preserver held in throwing hand. He continues to wear the CO_2 inflatable life preserver.

12-8. Ranger #2 (rope swimmer) in waterborne uniform (same as Ranger #1) grounds the rucksack with a carabiner through top of the frame to the rear of the nearside anchor point. His duties are to swim across the water obstacle pulling the rope and tie off the rope on the anchor point identified by Ranger #1 with a round turn and two half hitches with a quick release. The direction of the round turn is the same direction as the flow of the water current to facilitate exit off the rope bridge. Wear equipment in the following order (from the body out):
- Noninflatable life preserver.
- Field load carrier.
- Weapon (across the back).
- Swimmer safety line tied utilizing an around-the-waist bowline. The carabiner of the swimmer safety line is routed through an end-of-line bowline at arm's length and then secured by reattaching it to the swimmer safety line routed around the rope swimmer's waist, in vicinity of the small of the back.

12-9. Ranger #3 positions self on the downstream side of the nearside anchor point before Rangers #1 and #2 enter the water. Ranger #3 (near side lifeguard) wears the same type of waterborne uniform as the farside lifeguard. He grounds the rucksack with a carabiner through top of the frame on rear of the nearside anchor point. After the PSG crosses and verifies the headcount, Ranger

Waterborne Operations

#1 unties the quick release at the nearside anchor point. Ranger #3 is the last pulled across the water obstacle. Before crossing the water obstacle, Ranger #3 dons equipment in the following order:
- Noninflatable life preserver.
- CO_2 inflatable life preserver.
- Field load carrier.
- Weapon (across the back).
- Swimmer safety line routed over all equipment and secured to the two carabiners already secured in the figure-eight slip of the transport tightening system. (See figure 12-2.)

Figure 12-2. Transport-tightening system

12-10. Ranger #4 (BTC) wears the standard waterborne uniform with FLC and sling rope tied in the safety line around the waist bow line with end-of-line bow line no more than one arm's length. Ranger #4 is responsible for construction of the rope bridge and organization of bridge team, and is responsible for back feeding the rope and tying the transport tightening system. He designates the nearside anchor point, ties the figure-eight slip of the transport-tightening system, and hooks all personnel to the rope bridge. He ensures that the transport-tightening knot is on the upstream side of the rope bridge, and ensures that all individuals are in the waterborne uniform, hooked into the rope facing upstream. Ranger #4 ensures that the weapon is hooked onto the rope, and controls the flow of traffic on the bridge. He is responsible for crossing with the rucksack of Ranger #1, and is generally the next-to-last Ranger to cross (follows PSG, who is keeping a head count).

Chapter 12

12-11. Rangers #5 and #6 (rope pullers) wear the waterborne uniform with FLC and safety line. They tighten the transport-tightening knot. They also take the rucksacks of Rangers #2 and #3 across. Once they reach the far side, Rangers #5 and #6 pull the last Ranger (#3) across.

12-12. Rangers #4, #5, and #6 transport the rucksacks of Rangers #1, #2, and #3 across. To do so, they hook the rucksacks into the rope by running the carabiner through the top of the frames, then pulling the rucksacks across. They attach their own weapons between themselves and the rucksack they are pulling across the bridge.

12-13. BTC rehearses the bridge team during the planning sessions, and then directs the construction and emplacement. The unit leader selects the crossing site, which complements the tactical plan.
- Ranger #3 positions himself downstream of crossing site.
- Ranger #1 enters the water upstream of #2, staying one arm's length from Ranger #2 while being prepared to render any assistance to Ranger #2. They stay together to help compensate for the current. BTC feeds rope out of the rucksack positioned on the downstream side of the nearside anchor point.

12-14. Ranger #2 exits and identifies the farside anchor point (if BTC cannot identify it for Ranger #2). Ranger #2 exits on the upstream side of the farside anchor point. The rope is now routed to facilitate movement on and off the bridge. Radios and heavy equipment are double waterproofed and rigged. Rangers don waterborne uniforms and tie safety lines. PSG moves to anchor point and maintains accountability utilizing a head count.

12-15. Ranger #2 signals the BTC that the rope is temporarily attached to the farside anchor point. The BTC pulls out excess slack and ties the transport-tightening system using a figure-eight slipknot. The BTC signals Ranger #2 to pull the knot 12-to-15 feet from the nearside anchor point. After this, Ranger #2 ties round turns 18-to-24 inches off the water with the remaining rope, and secures the rope to itself with a carabiner. Ranger #2 signals the BTC and the pulling team (Rangers #4, #5, and #6) tighten the bridge, pulling the transport-tightening system as close as possible to the near-side anchor point.

12-16. Ranger #1 moves downstream and assumes the duties as the farside lifeguard. The BTC ties off the rope with a round turn and two half hitches around the nearside anchor point. The BTC places himself on the upstream side of the bridge (facing downstream) and starts hooking individuals into the rope and inspecting them for safety. Ranger #2 moves to the upstream side of the rope bridge, assists personnel off the rope on the far side, and keeps the head count going. Rangers #5 and #6 cross with the rucksacks of Rangers #1 and #2.

Note: *Any Ranger identified as a weak swimmer crosses alone so the nearside and farside lifeguards can watch him without distraction.*

12-17. The BTC maintains the flow of traffic, ensuring that no more than three Rangers are on the bridge at any one time (one hooking up, one near the center, and one being unhooked). Once the PSG has accounted for everyone on the near side, he withdraws left and right security and sends them across. PSG follows security across. Ranger #3 hooks the BTC (with #3's rucksack) onto the rope. Once the BTC crosses, Ranger #3 unhooks the nearside anchor point and the BTC unties the farside anchor point. Ranger #3 ties an Australian rappel seat with a carabiner to the front, hooks onto the carabiner that is in the end of the line bowline on the 150-foot rope, and then signals Rangers #4, #5, and #6 to take in the slack. Ranger #3 extends his arms in front of his head upstream) to fend off debris, and is pulled across by the #4, #5, and #6 Rangers. Except for Rangers #1, #2, and #3, everyone wears a rucksack. Rangers #4, #5, and #6 hook the rucksacks of Rangers #1, #2, and #3 onto the bridge by the carabiners. All the Rangers cross facing upstream.

12-18. The PSG and Ranger #5 verify weapons and equipment between themselves. After that, personnel reorganize and continue the mission. For Rangers with heavy equipment: all major groups are tied together with quarter-inch cord. An anchor line bowline runs through the rear swivel, down the left side of the gun. Tie a round turn through the trigger guard. Route the cord down the right side and tie off two half hitches around the forearm assembly with a round turn and two half hitches through the front sight

Waterborne Operations

posts. Tie off the rest of the working end with an end of the rope bow line about one foot from the front sight post large enough to place leading hand through. (See figure 13-3 on page 13-10.)

Note: More information is in Chapter 9.

M240 AND AN/PRC-119F

12-19. The M240 is secured to the bridge by carabiners on the front sight post and rear swivel. The M240 is pulled across by the trailing arm of the M240 gunner.

12-20. AN/PRC-119F are waterproofed before crossing a one-rope bridge. Once farside FM communications are set up, the nearside RTO breaks down and waterproofs the radio, and prepares to cross the bridge. A carabiner is placed in the top center of the rucksack frame (same as for Rangers #1, #2, and #3). The BTC hooks the rucksack to the rope.

Note: Using two carabiners binds the load on the rope. Adjust arm straps all the way out. The RTO pulls the radio across the rope bridge.

PONCHO RAFT

12-21. Normally, a poncho raft is constructed to cross rivers and streams when the current is not swift. A poncho raft is especially useful when the unit is still dry, and when the PL wants to keep their equipment dry. There are several things to consider when constructing and using a poncho raft:
- **Equipment requirements:**
  - Two serviceable ponchos.
  - Two weapons (poles can be used in lieu of weapons).
  - Two rucksacks for each team.
  - Ten feet of utility cord for each team.
  - One sling rope for each team.
- **Conditions.** Poncho rafts are used to cross water obstacles when at least one of the following conditions is found:
  - The water obstacle is too wide for a 150 foot-long section of rope.
  - No sufficient near-shore or far-shore anchor points are available to allow rope bridge construction.
  - Under no circumstances are poncho rafts used to cross a water obstacle if the current is unusually swift.
- **Choosing a crossing site.** Before a crossing site is used, a thorough reconnaissance of the immediate area is made. Analyzing the situation using METT TC, the patrol leader chooses a crossing site that offers as much cover and concealment as possible, and has entrance and exit points that are as shallow as possible. For speed of movement, it is best to choose a crossing site that has near-shore and far-shore banks that are easily traversed by an individual Ranger.
- **Execution phase.** Construct a poncho raft:
  - Pair-off the unit or patrol in order to have the necessary equipment.
  - Tie-off the hood of one poncho and lay it out on the ground with the hood up.
  - Place weapons in the center of the poncho, about 18 inches apart, muzzle to butt.
  - Place rucksacks and FLC between the weapons, with the two people placing their rucksacks as far apart as possible.
  - Start to undress, bottom to top, boots first. Take laces completely out for subsequent use as tie-downs, if necessary.
  - Place the boots over the muzzle or butt of weapon, toe in.
  - Continue to undress, folding each item neatly, and placing on top of boots.

Chapter 12

- Once all of the equipment is placed between the two weapons or poles, snap the poncho together. Lift the snapped portion of the poncho into the air and tightly roll it down to the equipment. Start at the center and work out to the end of the raft, creating pigtails at the end. This is faster and easier with two Rangers working together. Fold the pigtailed ends inward and tie them off with a single bootlace.
- Layout the other poncho on the ground with the hood up. In the center of this poncho, place the other poncho with equipment. Snap, roll, and tie the whole package up as before. Tie the third and fourth bootlaces (or utility cord) around the raft about one foot from each end for added security. The poncho raft is now complete.

Note: The patrol leader analyzes the situation using METT TC and makes a decision on the uniform to be worn for crossing the water obstacle; such as whether to place weapons inside the poncho raft or slung across the back, and whether to remain dressed or strip down with clothes placed inside the raft.

WATERCRAFT

12-22. Use of inland and coastal waterways may add flexibility, surprise, and speed to tactical operations. Use of these waterways also increases the load-carrying capacity of normal dismounted units. Watercraft are employed in reconnaissance and assault operations.

12-23. A waterborne insertion annex (see table 2-13 on pages 2-25 and 2-26) is prepared with the unit's OPORD and special organization is accomplished at this time. The PL designates the most technically proficient Rangers as coxswains.

12-24. The combat rubber raiding craft (CRRC) is a lightweight, inflatable watercraft that can be used on inland and coastal waterways. There are four separate valves inside the buoyancy tubes and eight separate airtight compartments. To pump air into the boat, turn all valves to the "orange" or "inflate" section of the valve. Once the assault boat is filled with air, turn all valves to the "green" or "navigation" section. This sections the assault boat into eight separate compartments. The characteristics of the watercraft are:

- Maximum payload: 2760 pounds (including engine, personnel, equipment, fuel, and deck).
- Crew: 11.
- Overall length: 15 feet, 5 inches.
- Overall width: 6 feet, 6 inches.
- Weight: (roll up floor, aluminum) 304 pounds.
- Weight: (roll up floor, composite) 274 pounds.
- Weight: (hard deck) 285 pounds.
- Weight: (without deck) 183 pounds.

Note: Characteristics may vary depending on the model.

Waterborne Operations

PREPARATION, PERSONNEL, AND EQUIPMENT

12-25. Crew-served weapons, radios, ammunition, and other bulky equipment are lashed securely to the boat to prevent loss in the event the boat is overturned. Machine guns with hot barrels are cooled prior to being lashed inside the boat. Refer to figure 13-4 on page 13-11 for equipment tie down, and figure 13-5 on page 13-12 for boat rigging. There are specific preparations, personnel, equipment, and procedures associated with watercraft. These are:

- **The rubber boat:**
  - Each rubber boat will have a 12-foot bowline secured to the front starboard D-ring. This rope is tied with an anchor line bowline, and the knot is covered with 100-mph tape.
  - Each rubber boat will have a 15-foot centerline tied to the rear floor D-ring. The same procedure for securing the bowline is used for the centerline.
  - Each rubber boat is filled to 240 millibars of air, and checked to ensure that all valve caps are tight and set in the NAVIGATE position.
  - Each rubber boat has one foot pump, which is placed in the boat's front pouch or, if no pouches are present, the foot pump is placed on the floor.
  - Each rubber boat is inspected using the maintenance chart.
- **Personnel and equipment:**
  - All personnel wear work vest or kapok (or another suitable positive flotation device).
  - FLC is worn over the work vest, unbuckled at the waist.
  - Individual weapon is slung across the back, muzzle pointed down and facing toward the inside of the boat.
  - Crew-served weapons, radios, ammunition, and other bulky equipment is lashed securely to the boat to prevent loss if the boat should overturn. Machine guns with hot barrels are cooled prior to being lashed inside the boats.
  - Radios and batteries are waterproofed.
  - Pointed objects are padded to prevent puncturing the boat.

12-26. When rigging weapons to be lashed to the boat, two carabiners are attached to the front and rear of the M240 with a middle-of-the-line bowline connecting the two carabiners. Make sure to connect a third carabiner to the center of the rope. The M249 machine gun also has two carabiners attached to the front and rear with parachute cord attached to both carabiners. (See figure 12-3 on page 12-10.) Rucksacks have one carabiner attached to the top of each pack. (See figure 12-4 on page 12-11.)

12-27. Rucksacks are placed in the boat with fames facing inward and tied down through the carabiners attached to the top of the packs. The end of the centerline is tied near the bow to the left or right D-rings on the bouncy tube with a round turn and two quick releases. (See figure 12-5 on page 12-12.) Attach the M240 machine gun to the D-ring at the bow and the M249 machine gun to the centerline by the carabiners, making sure the weapon is on top of the rucksacks.

Chapter 12

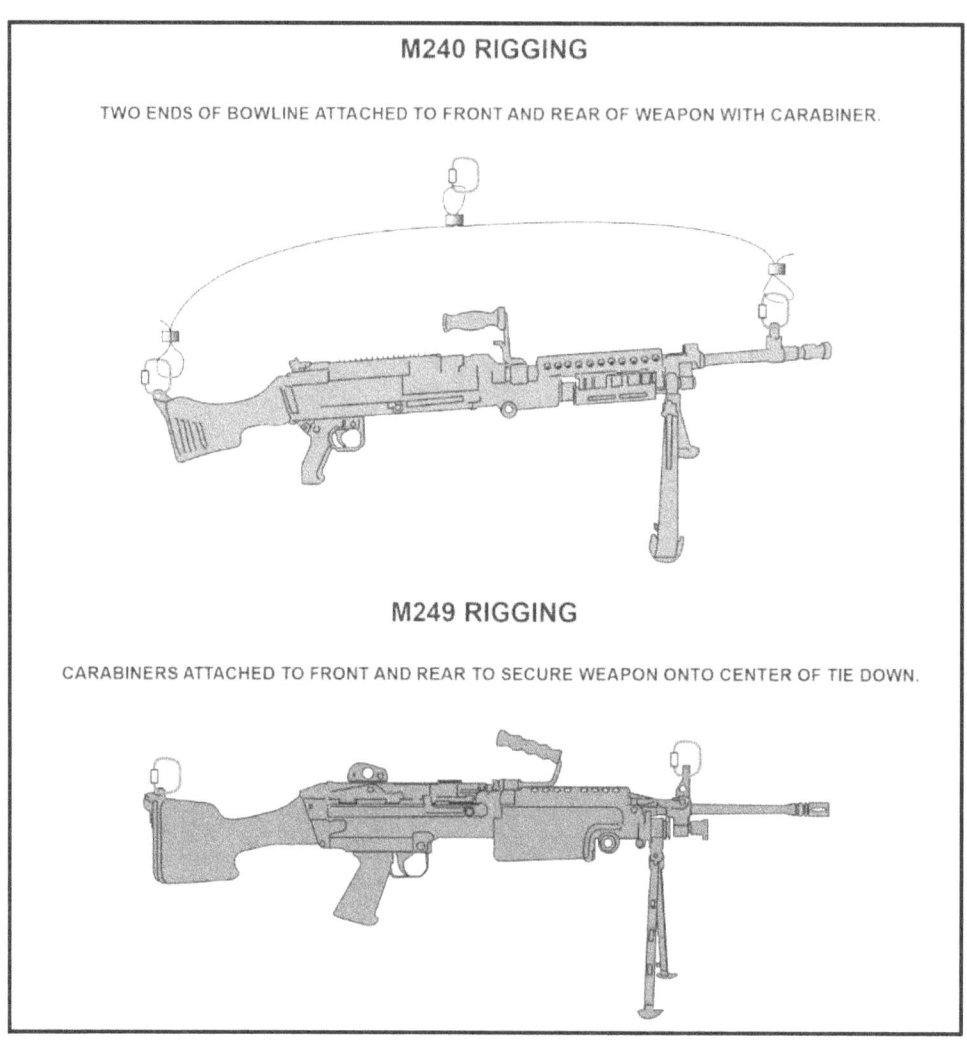

Figure 12-3. Weapon rigging

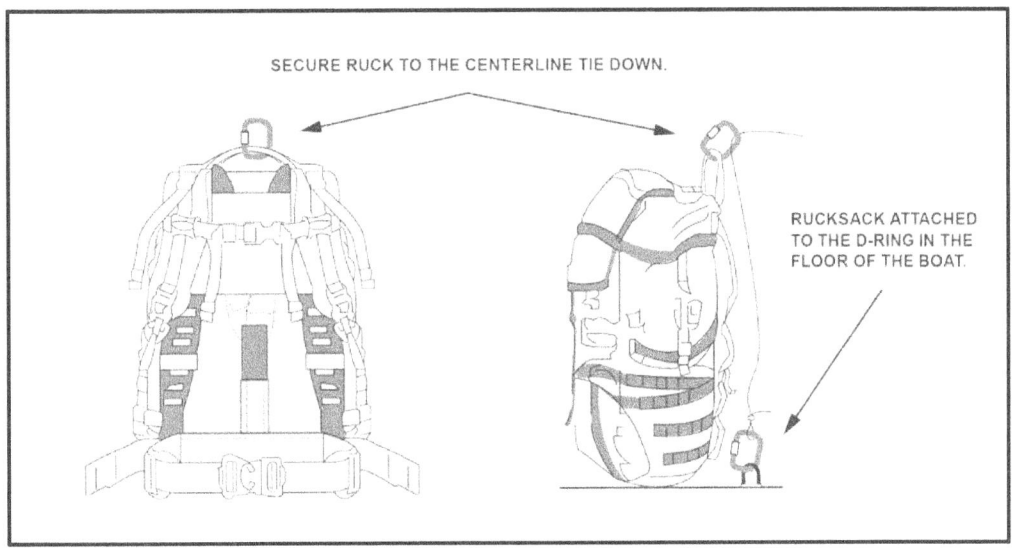

Figure 12-4. Rucksack rigging

Chapter 12

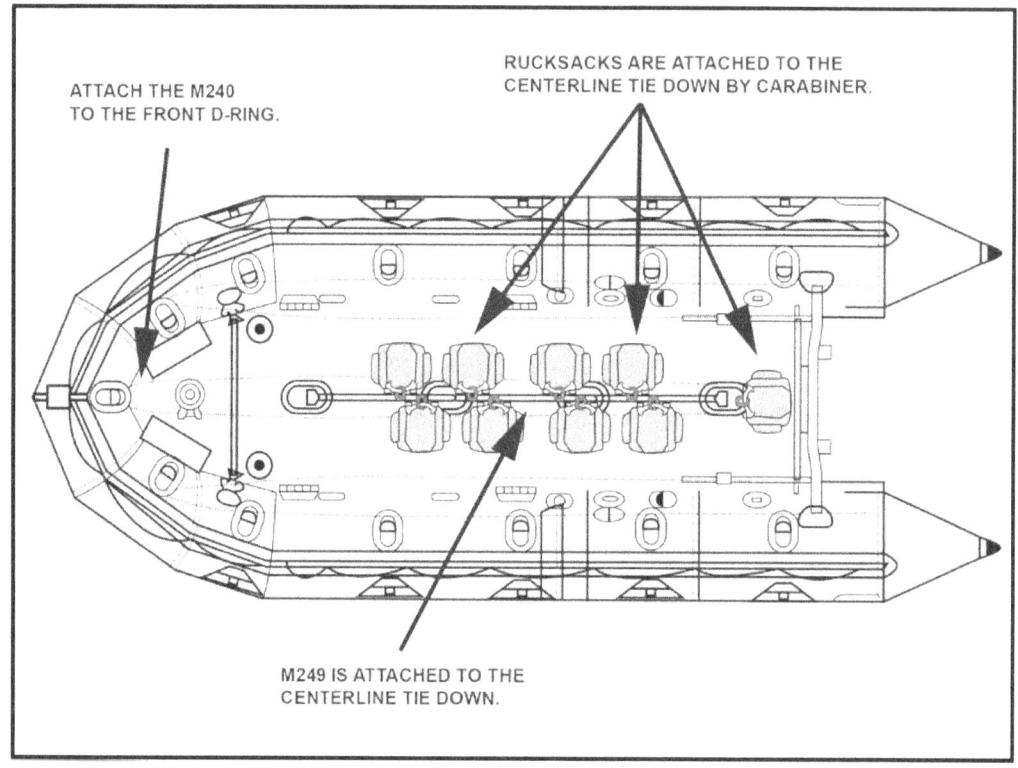

Figure 12-5. Equipment rigging

12-28. Each Ranger is assigned a specific boat position (see figure 13-6) and all have various duties, as well as embarkation and debarkation procedures. This includes:

  a. **Duties:**
  - Designate a commander for each boat, normally the coxswain.
  - Designate a navigator, normally a leader within the platoon, and observer team, as necessary.
  - Position crew as shown in figure 12-6.
  - Duties of the coxswain:
    - Responsible for control of the boat and actions of the crew.
    - Supervises the loading, lashing, and distribution of equipment.
    - Maintains the course and speed of the boat.
    - Gives all commands.
  - Paddler #2 (long count) is responsible for setting the pace.
  - Paddler #1 is the observer, stowing and using the bowline unless another observer is assigned.

Waterborne Operations

b. **Embarkation and debarkation procedures:**
- When launching, the crew maintains a firm grip on the boat until they are inside it. When beaching or debarking, the crew hold onto the boat until it is completely out of the water. Loading and unloading is done using the bow as the entrance and exit point.
- Keep a low center of mass when entering and exiting the boat to avoid capsizing. Maintain three points of contact at all times.
- The long count is a method of loading and unloading by which the boat crew embarks or debarks individually over the bow of the boat. It is used at riverbanks, on loading ramps, and when deep water prohibits the use of the short count method.
- The short count is a method of loading or unloading by which the boat crew embarks or debarks in pairs over the sides of boat while the boat is in the water. It is used in shallow water, allowing the boat to be quickly carried out of the water. The short count method of organization is primarily used during surf operations.
- Beaching the boat is a method of debarking the entire crew at once into shallow water and quickly carrying the boat out of the water.

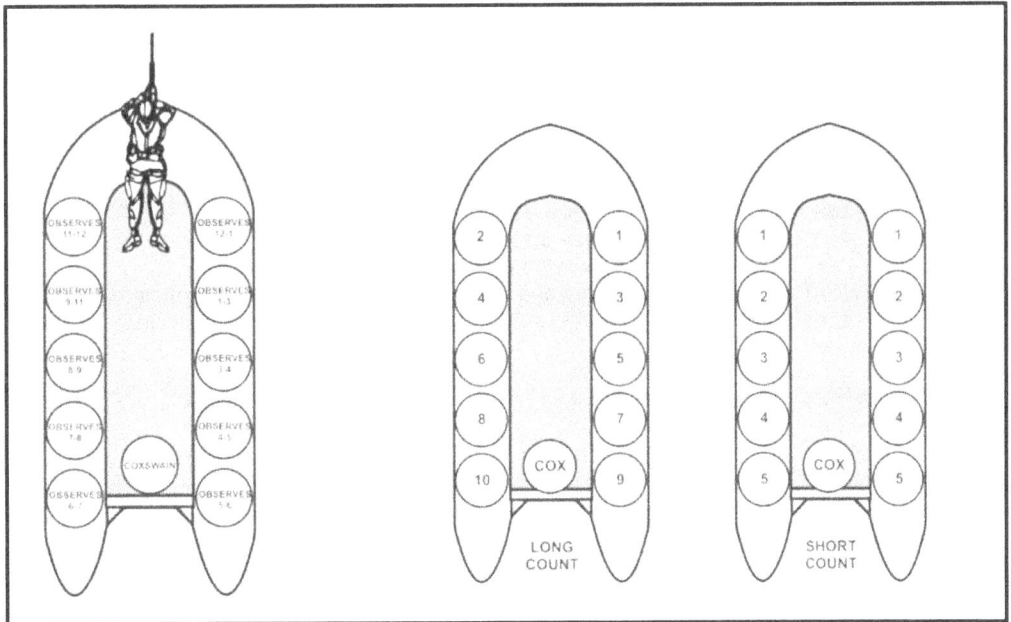

Figure 12-6. Crew positions, long count and short count

LEGEND
COX - coxswain

Chapter 12

COMMANDS

12-29. Commands are issued by the coxswain to ensure the boat is transported over land and controlled in the water. All crewmembers learn and react immediately to all commands issued by the coxswain. The various commands are:
- "SHORT COUNT. COUNT OFF." Crew counts off their position by pairs, such as one, two, three, four, five (passenger #1, #2, if applicable), coxswain.
- "LONG COUNT. COUNT OFF." Crew counts off the position by individual, such as one, two, three, four, five, six, seven, eight, nine, ten, (passenger #1, #2, if applicable), coxswain.
- "BOAT STATIONS." Crew takes positions alongside the boat.
- "HIGH CARRY. MOVE." (Used for long distance move over land.)
    - . On the preparatory command of "HIGH CARRY," the crew faces the rear of the boat and squats down, grasping carrying handles with the inboard hand.
    - . On the command "MOVE," the crew swivels around, lifting the boat to their shoulders so that the crew is standing and facing to the front with the boat on their inboard shoulders.
    - Coxswain guides the crew during movement.
- "LOW CARRY. MOVE." (Used for short distance moves over land.)
    - On preparatory command of "LOW CARRY," the crew faces the front of the boat, bent at the waist, and grasps the carrying handles with their inboard hands.
    - On the command of "MOVE," the crew stands up straight raising the boat about six-to-eight inches off the ground.
    - Coxswain guides the crew during movement.
- "LOWER THE BOAT. MOVE." Crew lowers the boat gently to the ground using the carrying handles.
- . "GIVE WAY TOGETHER." Crew paddles to front with #2 Ranger setting the pace.
- "HOLD." Entire crew keeps paddles straight downward and motionless in the water, stopping the boat.
- "LEFT SIDE HOLD." Left crew holds, right continues with previous command.
- "BACK PADDLE." Entire crew paddles backward, propelling the boat to the rear.
- "BACK PADDLE LEFT." Left crew back paddles causing the boat to turn left, right crew continues with previous command.
- "REST PADDLES." Crewmembers place paddles on their laps with blades outboard. This command may be given in pairs such as, "#1s, REST PADDLES."

CONDUCT CAPSIZE PROCEDURES

12-30. The capsize drill prepares Rangers to safeguard lives and equipment in the event that the boat overturns.

TASK: Conduct Capsize Procedures.

CUE: The boat team will need to capsize the boat intentionally. This may be necessary when the boat is full of water due to rough seas or heavy rainfall.

CONDITIONS: The platoon is conducting a waterborne insertion, using CRRC. The platoon has organized into nine-man boat teams. The platoon has all safety equipment required; all platoon organic equipment is tied down according to the unit SOP. The unit has communications with higher and adjacent units. Some iterations of this task should be conducted during hours of limited visibility.

STANDARDS: Each boat team will properly rig and lash their boat. Each boat team will intentionally capsize their boat, and then right their boat. Each team will recover all personnel and equipment back into the boat and continue their mission.

Waterborne Operations

SAFETY NOTE: During training, the unit should adhere to the water submersion chart during cold weather months. The unit should provide a safety boat for each boat team conducting this task during training. All personnel will be in the waterborne uniform (page 13-1) and wearing a serviceable noninflatable life vest.

Note: Boats must be rigged for capsizing. The coxswain ensures that all equipment is secured. This may require additional tie downs.

1. The coxswain gives the command, "LONG COUNT, COUNT OFF." After the long count, the coxswain gives the command "PASS PADDLES." All paddles are passed to the rear of the boat. This is done by each crewmember raising their paddle over their heads (except the #7 and #8 Rangers) to allow the crewmember behind them to take it. The #7 and #8 Rangers (or last two crewmembers) hold the paddles until the boat is righted.

2. The coxswain designates three crewmembers (#2, #4, and #6) to remain in the boat. (They will capsize the boat after the others are in the water.) The coxswain then orders the other members out of the boat by commanding, "ONE OUT, THREE OUT, FIVE OUT," until only three Rangers remain in the boat. Once out of the boat, the Rangers move about three meters away from the boat.

3. The coxswain designates the #1 Ranger who is in the water, to hold onto the boat in order to be pulled over onto the boat once capsized (this is done by holding onto two carrying handles). The three crewmembers in the boat each grasp a capsize line that are attached to three D-rings. They stand-up and lean backwards until the boat is capsized. This pulls the #1 Ranger onto the boat when it is capsized.

4. The coxswain designates a Ranger to pull the quick release that attaches the centerline tie down to the D-ring on the bow of the boat. (If the three crewmembers can right the boat without disconnecting the centerline quick release, then this step may be omitted.)

5. The #1 Ranger assists the #2 and #4 Rangers onto the boat to help in righting it. The #6 Ranger (who is in the water), holds onto the boat in order to be pulled over onto the boat once righted. This is done by holding onto two carrying handles. The #1, #2, and #4 Rangers each grasp a capsize line, stand-up, and lean backward until the boat is righted. This pulls the #6 Ranger onto the boat when it is righted.

6. Once the boat is righted, all crewmembers move to and hold onto the boat. The #6 Ranger assists other crewmembers back into the boat. The #7 and #8 Rangers pass the paddles to the other crewmembers and are assisted onto the boat.

7. If the quick release on the centerline tie down was released, then crewmembers will recover attached equipment and re-tie the centerline tie down.

8. Once all equipment and crewmembers are in position, the coxswain again has everyone count-off using the long count method. The Rangers also check to see that their equipment is accounted for. The coxswain then gives the crew the appropriate orders and continues the mission.

RIVER MOVEMENT, NAVIGATION, AND FORMATIONS

12-31. It is very important that Rangers understand the characteristics of the river and how to navigate the water using various formations. Before embarking, it is vital to know the local conditions of the river and its movement. Common knowledge and terminology used in water navigation includes:
- A *bend* is a turn in the river course.
- A *reach* is a straight portion of river between two curves.

Chapter 12

- A *slough* (pronounced "sloo") is a dead-end branch from a river. They are normally quite deep and can be distinguished from the true river by their lack of current.
- *Dead water* is a part of the river, due to erosion and changes in the river course that has no current. Dead water is characterized by excessive snags and debris.
- An *island* is usually a pear-shaped mass of land in the main current of the river. Upstream portions of islands usually catch debris and are avoided.
- The current in a narrow part of a reach is normally greater than in the wide portion. The current is greatest on the outside of a curve; sandbars and shallow water are found on the inside of the curve.
- *Sandbars* are located at those points where a tributary feeds into the main body of a river or stream.

12-32. Because Rangers #1 and #2 are sitting on the front left and right sides of the boat, they observe for obstacles as the boat moves downriver. If either notices an obstacle on either side of the boat, the coxswain is notified. The coxswain then adjusts steering to avoid the obstacle.

12-33. The patrol leader is responsible for navigation. Rangers have three acceptable methods of river navigation, They are:
- **Checkpoint and general route.** These two methods are used when the drop site is marked by a well-defined checkpoint and the waterway is not confused with a lot of branches and tributaries. They are best used during daylight hours and for short distances.
- **Navigator observer method.** This is the most accurate means of river navigation and is used effectively in all light conditions. Navigation equipment needed includes—
    - Compass.
    - Global Positioning System (GPS).
    - Photo map (first choice).
    - Topographical map (second choice).
    - Poncho (for night use).
    - Pencil or grease pencil.
    - Flashlight (for night use).

12-34. The navigator is positioned in the center of the boat and does not paddle. During hours of darkness, the flashlight is used under the poncho to check the map. The observer (or Ranger #1) is at the front of the boat. Working together:
- The navigator keeps the map and compass oriented at all times.
- The navigator keeps the observer informed of the configuration of the river by announcing bends, sloughs, reaches, and stream junctions, as shown on the map.
- The observer compares this information with the bends, sloughs, reaches, and stream junctions actually seen. When these are confirmed, the navigator confirms the boat's location on the map.
- The navigator also keeps the observer informed of the general azimuths of reaches as shown on the map, and the observer confirms these with actual compass readings of the river.
- The navigator announces only one configuration at a time to the observer and does not announce another until it is confirmed and completed.
- A strip map drawn on clear acetate backed by luminous tape may be used. The drawing is to scale or a schematic. It should show all curves, the azimuth, and the distance of all reaches. It may also show terrain features, stream junctions, and sloughs.

Waterborne Operations

12-35. Various boat formations are used (day and night) for control, speed, and security. The choice of which formation is used depends on the tactical situation and the discretion of the patrol leader. Hand and arm signals should be used to control the assault boats. The formations are:
- Wedge.
- Line.
- File.
- Echelon.
- Vee.

SECURE THE LANDING SITE

12-36. If the patrol is going into an unsecured landing site, a security boat can land, reconnoiter the site, and then signal the remaining boats to land. This is the best way. If the landing site cannot be secured prior to the waterborne force landing, some form of early warning, such as scout swimmers, should be considered. These Rangers swim to shore from the assault boats and signal the boats to land. All signals and actions are rehearsed prior to the actual operation.

12-37. The landing site can be secured by force with all the assault boats landing simultaneously in a line formation. While this is the least desirable method of securing a landing site, it is rehearsed in the event that the tactical situation requires its use. Arrival at the debarkation point involves several steps:
- 1. Unit members disembark according to leaders order.
- 2. Local security is established.
- 3. Leaders account for personnel and equipment.
- 4. Unit continues movement:
  - Rangers pull security, initially with work vest on.
  - Coxswain and two Rangers unlash and de-rig rucksacks.
  - Rangers return in buddy system or teams to secure rucksack and drop off work vest.
  - Boats are camouflaged and cached prior to movement, if necessary.

QUARTERING PARTY PROCEDURES

12-38. A quartering party (QP) is a patrol that departs ahead of a main body (MB). The purpose is to secure, reconnoiter, and organize an area for the MBs arrival and occupation. During waterborne operations, the QP leaves early in order to inspect and prepare small boats (such as the CRRC) for rigging and lashing. This saves time and facilitates an expedient and tactical occupation and departure from the beach landing site. Procedures include but are not limited to:
- QP departs ahead of MB. QP consists of a senior leader, RTO, security element, all coxswains.
- QP issues contingency plan.
- QP is counted out. Communication is maintained with MB. Perimeter security is readjusted.
- QP arrives at the beach landing site and establishes local security.
- Senior leader of QP conducts a partisan linkup in order to coordinate for small boats.
- Once boats are identified, coxswains inspect boats for serviceability and equipment.
- Coxswains use proper commands and lifts to move boats into position to the actual launch point.
- Coxswains ready equipment (paddles, work vests, centerline rope) in preparation for the MBs arrival.
- MB arrives and conducts linkup with the QP and security is readjusted. Information is disseminate among leaders.
- Coxswains begin supervising and directing boat crews to line up rucksacks and secure work vests.
- With the assistance of a crewmember, coxswains begin rigging and lashing rucksacks and other heavy equipment.

Chapter 12

> *Note: Coxswains must ensure that the boat remains afloat during loading. Equipment and personnel will cause the boat to rest on the bottom while in shallow water, which could result in damaging the boat.*

- Once all rucksacks and heavy equipment are secured, coxswains begin directing the loading of boat crews.
- Remaining security elements are pulled from the perimeter. Security is continued while on boats to ensure there is no security gaps.
- Accountability is given to PL and PSG.
- Platoon is postured for boat movement.

Chapter 13

Mounted Patrol Operations

This chapter outlines a technique for conducting vehicle mounted patrol operations. Mounted patrol operations present a challenge to the Ranger leader. Trucks and other combat vehicles produce a large signature on the battlefield and increase the unit's value as a target. Vehicle movement is restricted to roads and terrain that can be traversed. (Refer to ATP 4-01.45 and ATP 3-21.8 for more information.)

PLANNING

13-1. When conducting a mounted patrol as part of the operation, it is important to incorporate the mounted patrol as a leader uses the eight steps of the troop leading procedures. The following information should be included when conducting a mission analysis using METT-TC.
- **Mission.** The PL extracts the following information from the company OPORD:
  - Vehicle support (number and type of vehicles, and the allowable combat load).
  - Weather: road conditions.
  - Vehicle pick up and drop off location and markings.
  - Vehicle movement timeline (pick up time, movement time, and other information).
  - Vehicle routes (primary and alternate checkpoints).
- **Enemy.**
  - Known or suspected enemy locations in the AO or along planned routes.
  - Potential locations for enemy ambush or improvised explosive device (IED) emplacement.
  - Recent enemy activities or reactions to mounted patrol operations.
- **Terrain.**
  - Identify potential pick up and drop off locations.
  - Evaluate routes, and pick up and drop off locations using OAKOC.
  - Consider weather and road conditions.
- **Troops.**
  - Number of passengers for each vehicle.
  - Chalks and chalk leaders identified.
  - Tactical cross load.
  - Linkup and marking teams identified.
  - Pick up and drop off security plan.
- **Time.** Backwards planning sequence:
  - Ground tactical plan.
  - Unload plan.
  - Ground movement plan.
  - Loading plan.
  - Staging plan.
- **Civilians.**
  - Rules of engagement (ROE) actions with civilians during movement.
  - ROE actions with civilian vehicles during movement.

Chapter 13

> *Note: Allocate time for movement, reconnaissance, and establishment of security.*

13-2. There are five phases of mounted patrol. Each phase supports the ground tactical plan, which specifies actions in the objective area to accomplish the commander's intent for the assigned mission, whether it is a raid, ambush, reconnaissance, or other follow on missions.

13-3. The five phases are the staging plan, loading plan, ground movement plan, unloading plan, and ground tactical plan. This involves:

- **Staging plan:**
    - Establish security.
    - Employ markings and recognition signals for day and night.
    - Linkup.
    - Conduct final friendly unit coordination.
    - Disseminate information and any changes to subordinate leaders.
- **Loading plan: task organization and tactical cross loading.** Each Ranger is assigned to a vehicle, ensuring tactical cross-load of weapon systems and key personnel:
    - Vehicle number, key leader, key weapon systems, additional personnel, and communications.
    - Location of PL.
    - Location of PSG and medic.
    - Location of WSL.
    - Location of communication (FO, RTO, or both).
- **Ground movement plan:**
    - Troops awake and alert pulling active security during movement.
    - Platoon leader and vehicle commanders tracking route progress.
    - Compromise and contingency plan.
    - React to IED.
    - React to ambush.
    - Vehicle breakdown.
- **Unloading plan:**
    - Dismount vehicles according to SOP and the reverse load plan.
    - Establish security.
    - PSG accounts for personnel and clears all vehicles for departure.
    - Establish security at the halt or perimeter.
    - Adjust perimeter as vehicles depart area.
- **Ground tactical plan:**
    - Prepare to continue movement.
    - Conduct follow-on mission

13-4. The warning order brings together the vehicle movement. It contains basic information on the situation, mission, task organization, any special instructions, initial time organization, and the uniform and equipment common to all. (See table 13-1 on pages 13-3 through 13-5.)

Table 13-1. Mounted tactical movement brief

ADMINISTRATIVE. PERSONNEL (ROLL CALL):
  a. Responsibilities Driver/NAV
     VCS drivers (primary/alternate):
     CSW operator:
     Counter-assault element leader:
     Designated marksman:
     Medics/combat lifesaver:
     Guide/Interpreter:
     Higher HQ rep :
  b. Sectors of fire (by priority, weapon system, vehicle, and phase).
  c. Task organization: (internal organization for mounted patrol—manifest).
1. SITUATION:
  a. Enemy forces: discuss enemy.
     Identification of enemy (if known).
     Composition/capabilities/strength/equipment.
     Location (danger areas highlighted on map).
     Most likely/most dangerous COA (defend, reinforce, attack, withdraw, and delay [DRAW-D]).
  b. Weather: general forecast.
  c. Light data (EENT, percent illumination, MR, MS, BMNT):
  d. Friendly forces:
     Units along the route.
     Operational support provided by higher HQ.
     Aviation support:
        ASOC   Call sign_____Frequency _____
        DASC   Call sign_____Frequency _____
        JSTARS Call sign_____Frequency _____
     Mobile security forces/Quick reaction forces (QRFs)
       EOD. SOF.
       Fire support elements.
          Element_____Location_____Frequency/Call sign _____
     Attachments (from outside the organization)
2. MISSION. (WHO, WHAT, WHEN, WHERE, WHY):
  Example: Unit X conducts tactical mounted patrol to FOB YY and returns to FOB XX NLT 231000ZDECO3 in order to provide resupply of CL V (ammo).

Chapter 13

Table 13-1. Mounted tactical movement brief (continued)

3. **EXECUTION:**
   a. Concept of operations: mounted patrol execution and task(s) of elements, teams, and individuals at the objective(s). (Broad general description beginning to end.)
   b. Tasks to subordinate units. (Include attached or OPCON elements.)
   c. Coordinating instructions: (Instructions for all units.)
   SAFETY. (See Appendix E, Risk Management.)
   Overall risk to force: Low_____ Medium_____ High_____

   Overall risk to mission accomplishment: Low_Medium_____High _____

   Fratricide reduction measures.
   (1) Order of march (spacing of serials/location of support elements):
   (2) Routes (ensure strip map is attached):
   (3) Additional movement issues (speed, intervals, lane, parking, accidents, and other potential issues).
   (4) Mounted patrol execution
   (5) Timeline:
       (a) Vehicle/personal gear preparation, and preventive maintenance checks and services (PMCS) completed.
       (b) Briefing.
       (c) Put on equipment.
       (d) Load vehicle.
       (e) Rehearsals/test fires.
       (f) Back brief/confirmation brief from key leaders.
       (g) Start point (SP)/departure.
       (h) Return to base (RTB).
       (i) Debrief.
       (j) Recovery: maintain vehicles/personal gear.
   (6) Sectors of fire: cover assigned sectors while mounted/dismounted. Cover up/down bridges, rooftops, balconies, storefronts, multistory structures, and cross streets.
   (7) Scanning: scan crowds, vehicles, and roadsides for attack indicators. Note communicate indicators throughout the mounted patrol.
       (a) Beware of motorcycles, vans with side doors, and dump trucks.
       (b) Beware of objects in the road (cars, potholes, fresh asphalt/concrete, and trash).
   (8) Mounted patrol speed:_____Min/ Max: _____
       (a) Speed is dictated by either the rear vehicle's ability to keep up or placing slower vehicles in the lead.
       (b) Highways/open roads example: 50+ mph.
       (c) Urban/channeled areas. As fast as traffic will allow. (Brief evasive maneuvers, bumping and blocking technique, and use of ramming techniques to allow for continuous movement of the mounted patrol.

Mounted Patrol Operations

Table 13-1. Mounted tactical movement brief (continued)

(9) Vehicle interval.
    (a) Highways/open roads/cloverleaf's/bridges/ramps: open spacing, but do not allow vehicles to enter the mounted patrol.
    (b) Urban/channeled areas: close interval, but must have visual of tires on vehicle in front of your vehicle. Drive on wrong side, if necessary.
(10) Headlight status (on/off, blackout, use of night observation).
(11) ROE for mounted patrol operations (theater specific).
(12) Battle drills will be rehearsed (no need to cover in brief).
4. **ADMINISTRATION AND LOGISTCS:** (Equipment.)
  a. Individual equipment (precombat inspections [PCIs], see checklist).
  b. Vehicles (see PCI checklist).
5. **MISSION COMMAND:**
  a. Chain of command (positioning of mounted patrol).
  b. Mounted patrol call sign(s): _____
  c. Area of operations communications/MEDEVAC and CASEVAC plan.
  d. Mounted patrol primary/alternate/contingency/emergency (PACE) communications.
  e. Vehicle internal (back to:_____).
  f. Hand and arm/visual signals (according to the unit SOP).

LEGEND ammo - ammunition; ASOC – air support operations center; BMNT – begin morning nautical twilight; CASEVAC – casualty evacuation; CL V – Class V—ammunition; COA – course of action; CSW – crew-served weapon; DASC – direct air support center; EENT - end evening nautical twilight ; EOD – explosive ordnance disposal; FOB – forward operating base; FSE – fire support element; HQ – headquarters; MEDEVAC – medical evacuation; mm – millimeter; MP – military police; mph – miles per hour; MR - moonrise; MS - moonset; NAV - navigator; NLT – not later than; OPCON – operation control; QRF – quick reaction force; rep – representative; ROE – rules of engagement; SOF – Special Operations Forces; SOP – standard operating procedure; VCS – Vehicle Control System

POTENTIAL SITUATIONS

13-5. Whenever there is a mounted patrol, especially in hostile environments, there is the possibility of an ambush, forced stop, or other potentially hazardous situation. Rangers are well trained in maneuvers to protect themselves and their fellow Soldiers in these circumstances. Figure 13-1 on page 13-6, and figure 13-2 on page 13-7, detail various methods used in mounted patrols.

Chapter 13

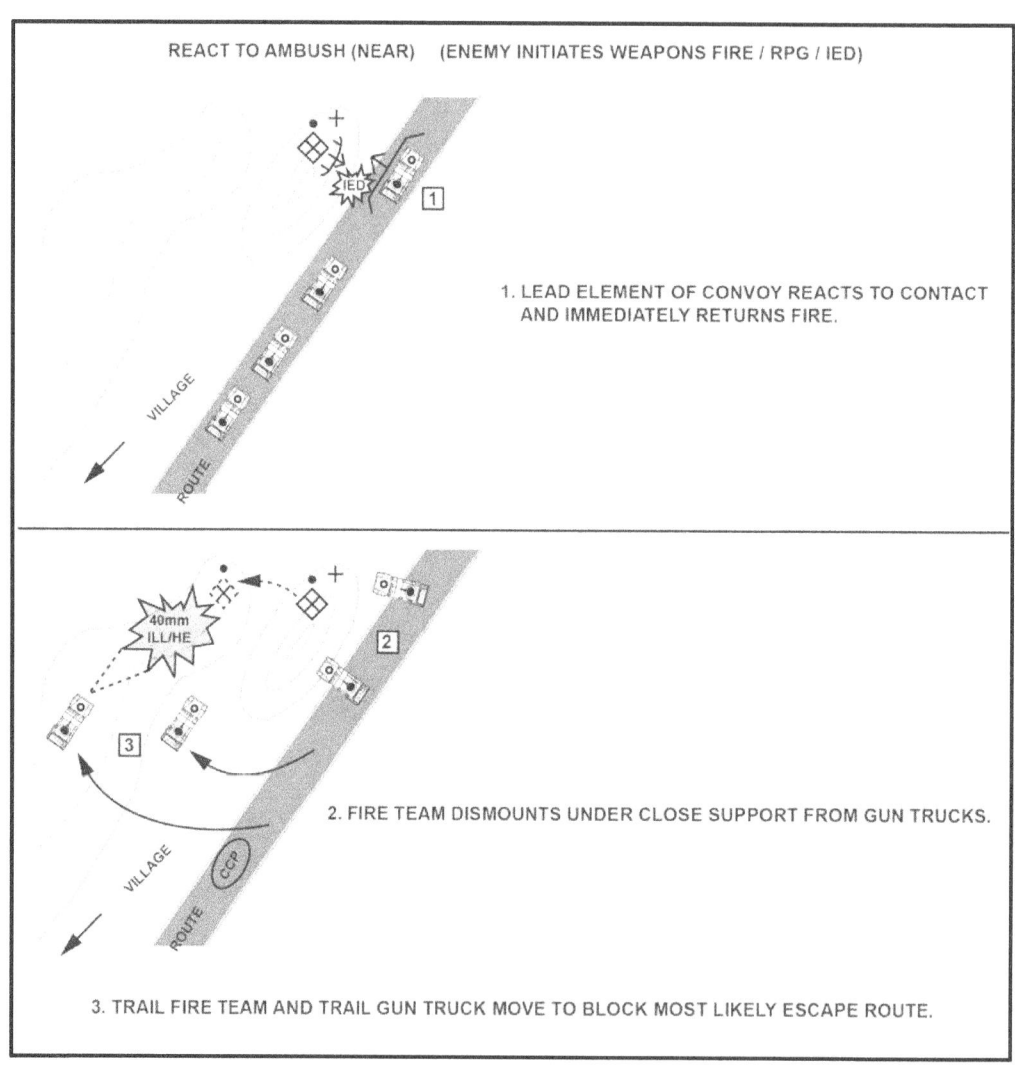

Figure 13-1. React to ambush (near)

LEGEND
CCP – casualty collection point; HE – high explosive; IED – improvised explosive device; ILL - illumination; mm – millimeter; RPG – rocket propelled grenade

Mounted Patrol Operations

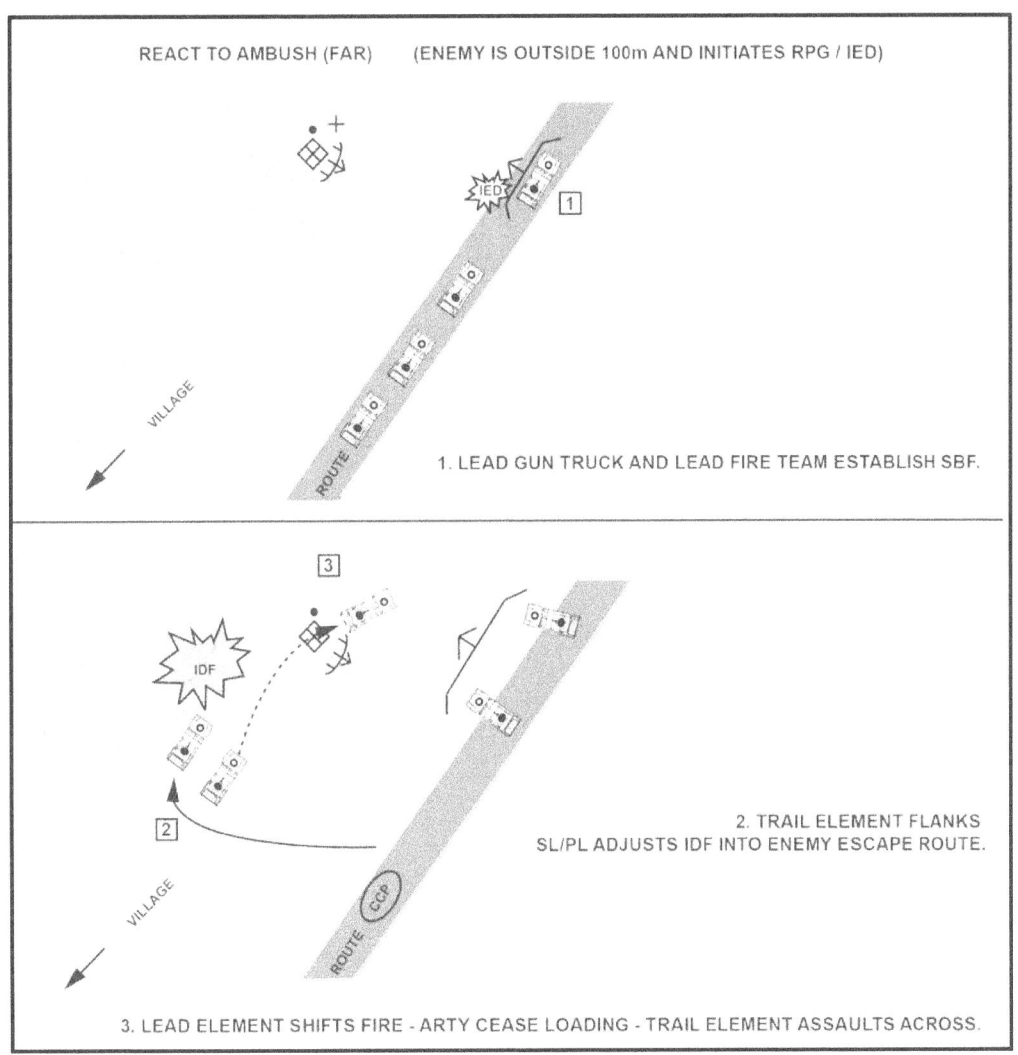

Figure 13-2. React to ambush (far)

LEGEND
CCP – casualty collection point; IED – improvised explosive device; IDF – indirect fire; RPG – rocket propelled grenade; m – meter; PL – platoon leader; SBF – support by fire; SL – squad leader

Chapter 13

FORCED STOPS

13-6. When vehicle(s) are forced to stop due to weapons fire, RPGs, IEDs, or indirect fire, activate the turn signal to indicate the direction of contact. If the vehicle(s) are not in direct contact, report using internal communication the identity of the vehicle, type of contact, clock direction, and grid coordinates (if available).

13-7. Personnel on vehicle(s) that are forced to stop dismount on the noncontact side, assume covered positions, and provide initial base of fire. The entire patrol halts, personnel dismount on the noncontact side, and provide additional fire. Vehicles not in contact reposition and provide supporting fire.

METHOD ONE

13-8. The PL assesses the situation and maneuver, in order to suppress the enemy and gain fire superiority. Once the PL determines the threat is eliminated, recovery and CASEVAC operations can begin.

13-9. If the PL determines the patrol cannot gain fire superiority to eliminate the threat, the patrol executes break contact procedures. Figure 13-3 details method one to use when mounted patrols are forced to stop.

Mounted Patrol Operations

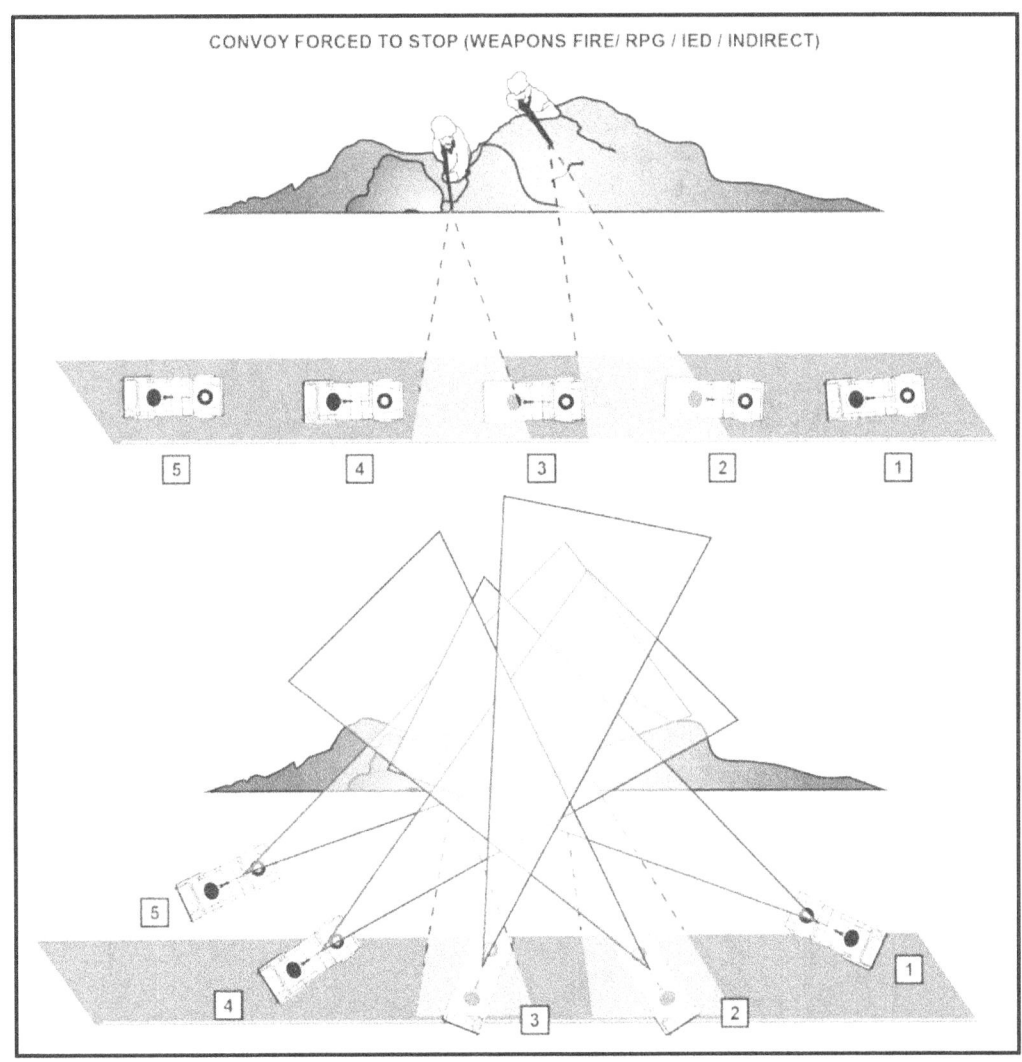

Figure 13-3. Mounted patrol forced to stop, method one

LEGEND
IED – improvised explosive device; RPG – rocket propelled grenade

Chapter 13

METHOD TWO

13-10. When vehicle(s) are forced to stop due to weapons fire, RPGs, IEDs, or indirect fire, activate the turn signal to indicate the direction of contact. All personnel STAY IN VEHICLES.

13-11. Drive the vehicle(s) out of the kill zone. The vehicle(s) directly behind disabled vehicle(s) push the disabled vehicle(s) out of the kill zone. The vehicle(s) not disabled establish a base of fire toward the suspected or known enemy.

13-12. If fire superiority can be gained, the PL uses the minimum amount of force necessary to destroy the enemy. If the PL determines the patrol cannot gain fire superiority, the leader breaks contact. Figure 13-4 details method two to use when mounted patrols are forced to stop.

Mounted Patrol Operations

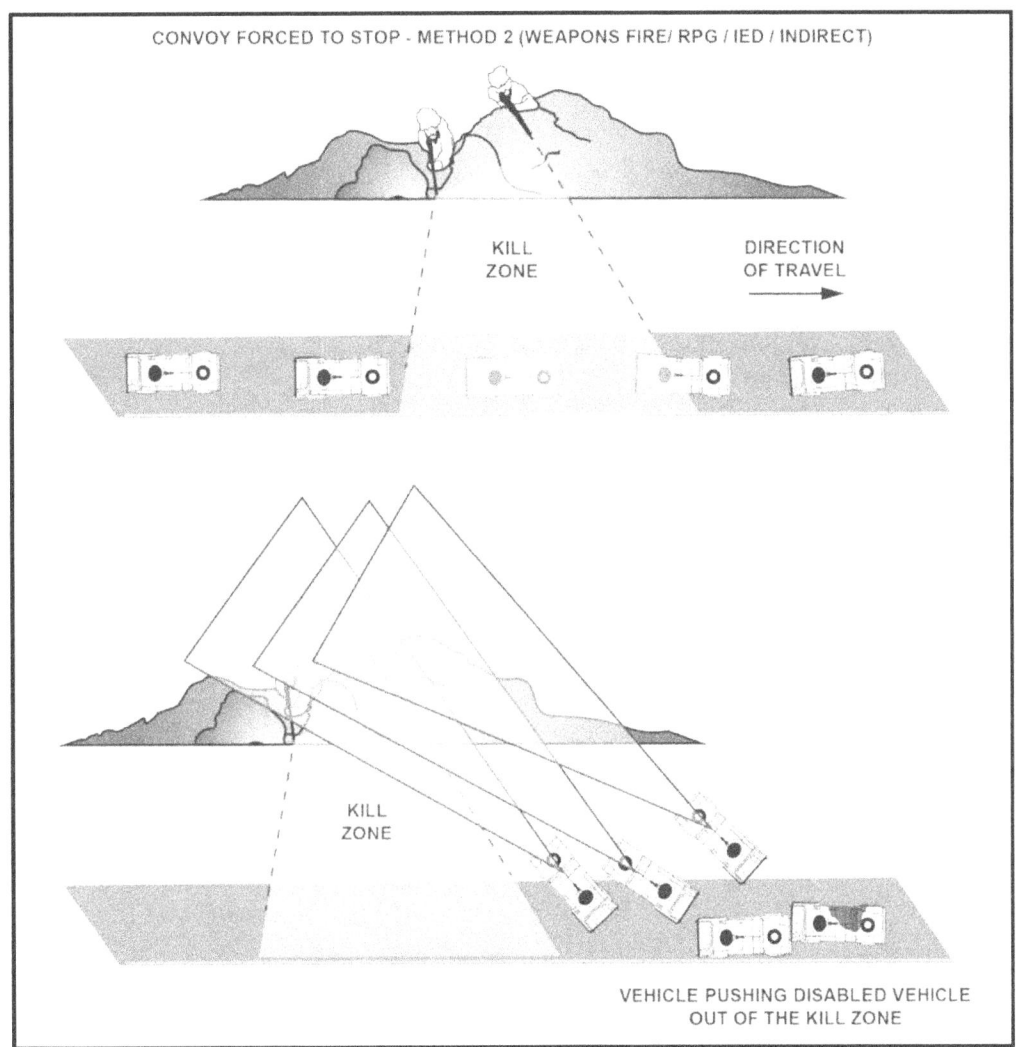

Figure 13-4. Mounted patrol forced to stop, method two

LEGEND
IED – improvised explosive device; RPG – rocket propelled grenade

Chapter 13

BREAK CONTACT

13-13. Always try to close with the enemy first, so they cannot come back later to attack the patrol. If the PL determines the patrol cannot gain fire superiority and decides to break contact, the PL determines a rally point (RP) to the front or rear (or both). Communications and pyrotechnic signals are used to break contact and occupy the rally point(s). The patrol deploys obscuration measures, if available.

13-14. Using cover and concealment, the aid and litter team(s) evacuate all casualties under fire. The patrol maintains position and fire suppression in the contact zone, and assists the aid and litter team(s) as necessary.

13-15. Disabled vehicles are towed or destroyed, as directed by leaders. Vehicles displace forward or backward under the control of leaders. The most forward vehicle in the contact zone moves first, followed by the next most forward vehicle. Vehicles continue to displace under supporting fires until contact is broken.

13-16. If break contact occurs with vehicles on both sides of the contact zone, displacement of vehicles occurs using an alternating technique. Upon occupation of the ORP, leaders immediately position vehicles to establish 360-degree security, consolidate, and reorganize. Figure 13-5 details how to break contact.

Mounted Patrol Operations

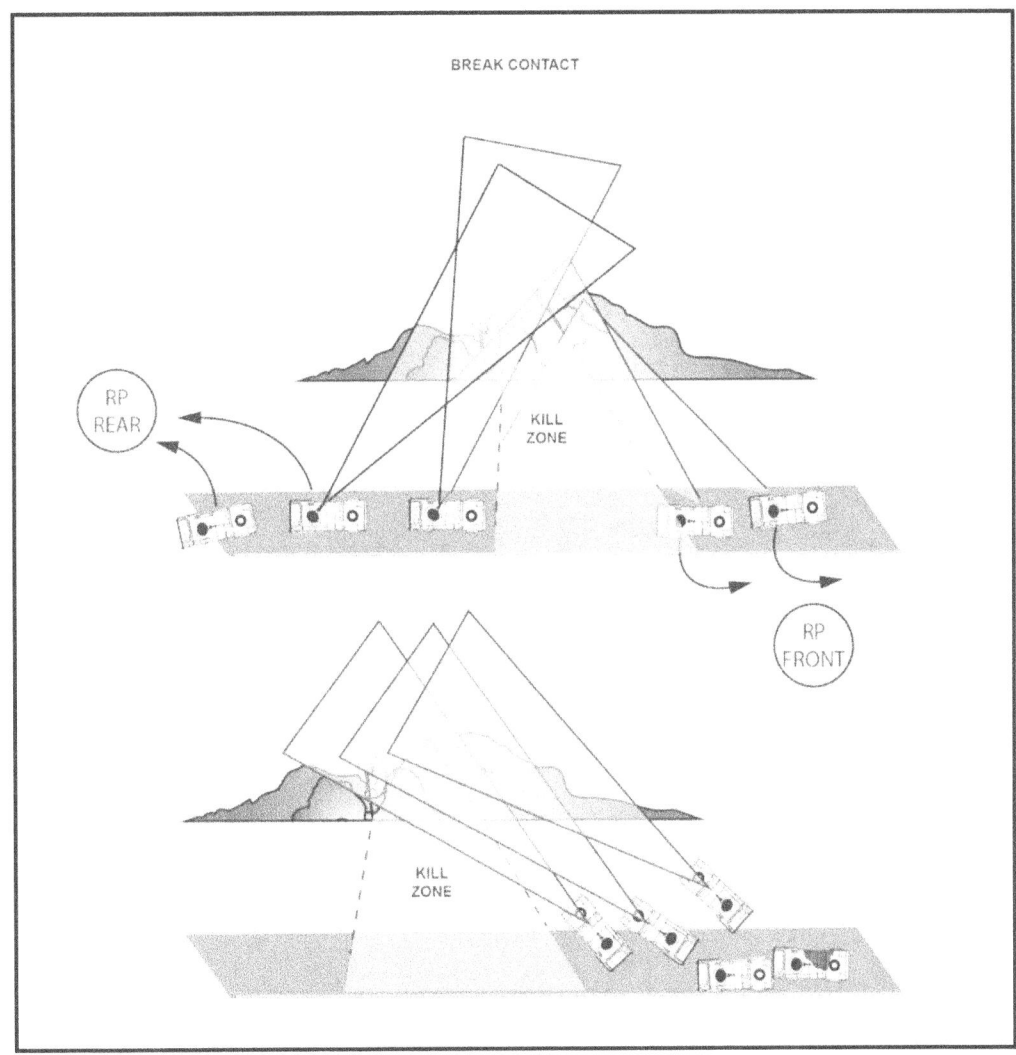

Figure 13-5. Break contact

LEGEND
RP – rally point

Chapter 13

CASEVAC AND RECOVERY OPERATIONS

13-17. Once the leader assesses the enemy threat is destroyed, neutralized, defeated, and the area is secure, CASEVAC and recovery operations begin. This helps keep Soldiers focused on defeating and destroying the threat.

13-18. During CASEVAC operations, the aid and litter team(s) position themselves on the safe side, extracting casualties and personnel. Casualties are treated after they are safely removed from the contact area.

13-19. During vehicle recovery procedures, the recovery team position themselves on the safe side of the disabled vehicle. The truck commander (TC) dismounts and assesses the disabled vehicle. If the TC determines the vehicle can be safely recovered, the TC guides the recovery vehicle into position and conducts a hasty hookup. If necessary, the TC can operate the disabled vehicle.

13-20. Upon exiting the contact area, complete and correct hookup procedures occur. If it is assessed that outside support is necessary for recovery, the leader contacts higher HQ for guidance. Disabled vehicles may be abandoned or destroyed by leaders. Once recovery operations are complete, the team displaces and conducts linkup at the rally point. Figure 13-6 details CASEVAC and recovery operations.

Mounted Patrol Operations

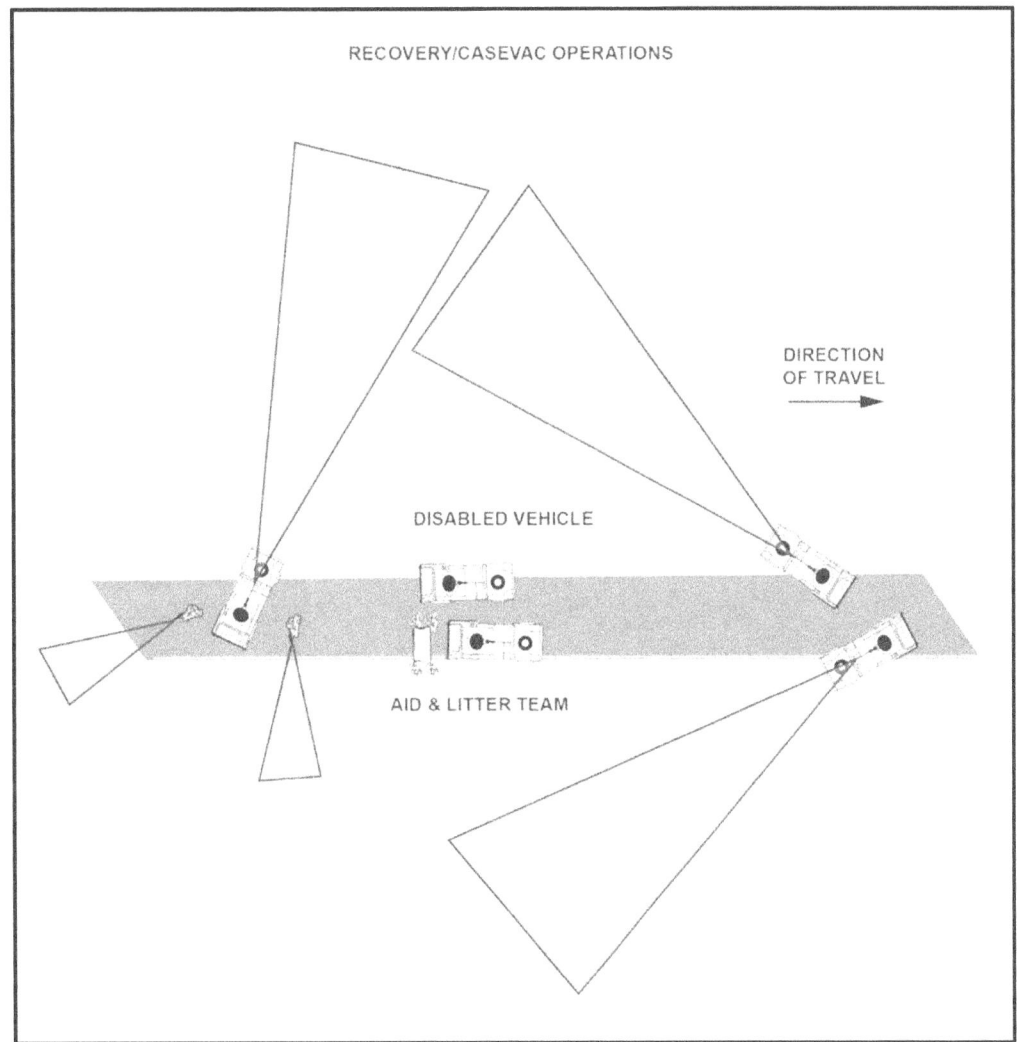

Figure 13-6. Recovery/CASEVAC operations

LEGEND
CASEVAC – casualty evacuation

This page intentionally left blank.

Chapter 14

Aviation

Army aviation and Infantry units can be fully integrated with other members of the combined arms team to form powerful and flexible air assault task forces. These forces can project combat power throughout the depth and width of the modern battlefield with little regard for terrain barriers, making these combat operations deliberate, precisely planned, and vigorously executed. They strike the enemy when and where they are most vulnerable.

REVERSE PLANNING SEQUENCE

14-1. Successful air assault execution is based on a careful analysis of METT-TC and detailed, precise, reverse planning. Five basic plans that comprise the reverse planning sequence are developed for each air assault operation. The battalion is the lowest level that has sufficient personnel to plan, coordinate, and control air assault operations. When company-size or lower operations are conducted, most of the planning occurs at the battalion or higher HQ. The five plans are:
- Ground tactical plan.
- Landing plan.
- Air movement plan.
- Loading plan.
- Staging plan.

14-2. The commander's ground tactical plan forms the foundation of a successful air assault operation. All additional plans support this plan. It specifies actions in the objective area to accomplish the mission, and addresses subsequent operations.

14-3. The landing plan supports the ground tactical plan. This plan outlines a sequence of events that allows elements to move into the area of operations, ensures that units arrive at designated locations at prescribed times, and that as soon as they arrive, they are prepared to execute the ground tactical plan.

14-4. The air movement plan is based on the ground tactical and landing plans. It specifies the schedule and provides instructions for air movement of troops, equipment, and supplies from PZs to LZs.

14-5. The loading plan is based on the air movement plan. It ensures that troops, equipment, and supplies are loaded on the correct aircraft. Unit integrity is maintained when aircraft loads are planned. Cross loading may be necessary to ensure survivability of mission command assets, and that the mix of weapons arriving at the LZ is ready to fight.

14-6. The staging plan is based on the loading plan. It prescribes the arrival time of ground units (troops, equipment, and supplies) at the PZ in the order of movement.

SELECTION AND MARKING OF PICKUP AND LANDING ZONES

14-7. Small unit leaders should consider the size, surface conditions, ground slope, obstacles, and the approach and departure when selecting a PZ or LZ. A minimal circular landing point separation from other aircraft and obstacles is needed. The sizes for the aircraft are:
- OH-6A = 25 m.
- AH-1 = 35 m.
- UH-60L, AH 64 = 50 m.

Chapter 14

- Cargo helicopters = 80 m.
- Any helicopter with a sling load = 100 m.

14-8. Surface conditions should avoid potential hazards such as sand, blowing dust, snow, tree stumps, or large rocks. Ground slope is another concern that affects landing. The degree of slope for landing is:
- Zero-to-six percent = land upslope.
- Seven-to-fifteen percent = land side slope.
- Over 15 percent = no touchdown (aircraft may hover).

14-9. When planning the approach and departure of the PZ and LZ, an obstacle clearance ratio of ten-to-one is used. For example, a tree that is 10 feet tall requires 100 feet of horizontal distance for approach or departure. Mark obstacles with a red chemical light (chemlight) at night, or red panels in the daytime. Avoid using markings if the enemy would see them.

14-10. Approach and depart facing into the wind and along the long axis of the PZ or LZ. The greater the load, the larger the PZ or LZ is in order to accommodate the insertion or extraction. The PZ and LZ are marked in different ways depending on if it is day or night. For example:
- **Day:** A ground guide marks the PZ or LZ for the lead aircraft by holding an M4 rifle over his head, by displaying a folded VS-17 signal panel chest high, or by other coordinated and identifiable means.
- **Night:** The code letter "Y" (inverted "Y") is used to mark the landing point of the lead aircraft at night. (See figure 14-1.) Chemlights or beanbag lights are used to maintain light discipline. A swinging chemical light (chemlight) may also be used to mark the landing point.

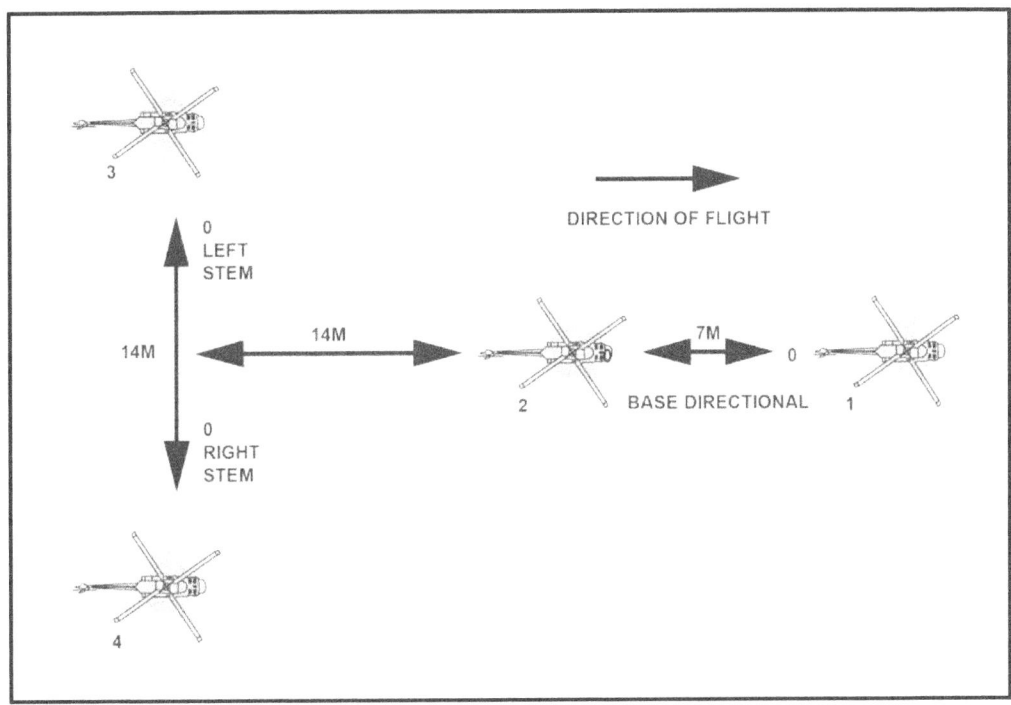

Figure 14-1. Inverted "Y"

LEGEND
M - meter

Chapter 14

AIR ASSAULT FORMATIONS

14-11. Aircraft supporting an operation may use any of the following PZ and LZ configurations. (See table 14-1.) These are prescribed by the air assault task force (AATF) commander working with the air mission commander (AMC).

Table 14-1. Air assault formations

FORMATION	PROS	CONS
Heavy left or heavy right	Provides firepower to front and flank.	Requires a relatively long, wide landing area. Presents difficulty in prepositioning loads. Restricts suppressive fire by inboard gunners.
Diamond	Allows rapid deployment for all-round security. Requires only a small landing area.	Presents some difficulty in prepositioning loads. Restricts suppressive fire of inboard gunners.
Vee	Requires a relatively small landing area. Allows rapid deployment of forces to the front.	Presents some difficulty in prepositioning loads.
Echelon left or echelon right	Allows rapid deployment of forces to the flank. Allows unrestricted suppressive fire by gunners.	Presents some difficulty in prepositioning loads. Requires a relatively long, wide landing area.
Trail	Requires a relatively small landing area. Allows rapid deployment of forces to the flank. Simplifies prepositioning of loads. Allows unrestricted suppressive fire by gunners.	Requires a long landing area.
Staggered trail left or right	Simplifies prepositioning of loads. Allows rapid deployment for all-around security.	Requires a relatively long, wide landing area. Somewhat restricts gunners' suppressive fire.

Aviation

14-12. The positive aspect of a heavy left or heavy right formation (see figure 14-2) is that it provides firepower to the front and flank. However, this formation requires a relatively long and wide landing area, presents difficulty in prepositioning loads, and restricts suppressive fire by inboard gunners.

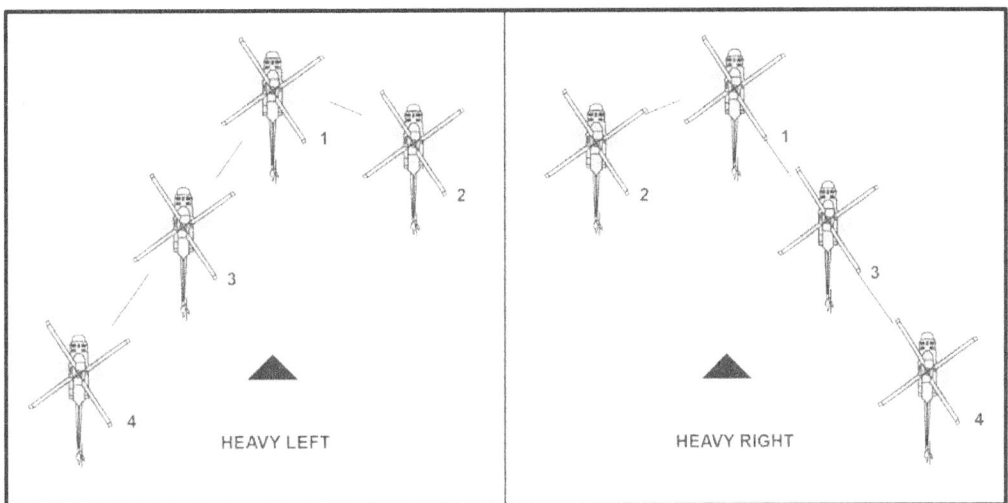

Figure 14-2. Heavy left or heavy right formation

Chapter 14

14-13. The diamond formation (see figure 14-3) allows rapid deployment for all-around security. Although it requires a small landing area, the diamond formation presents some difficulty in prepositioning loads. It also restricts suppressive fire by inboard gunners.

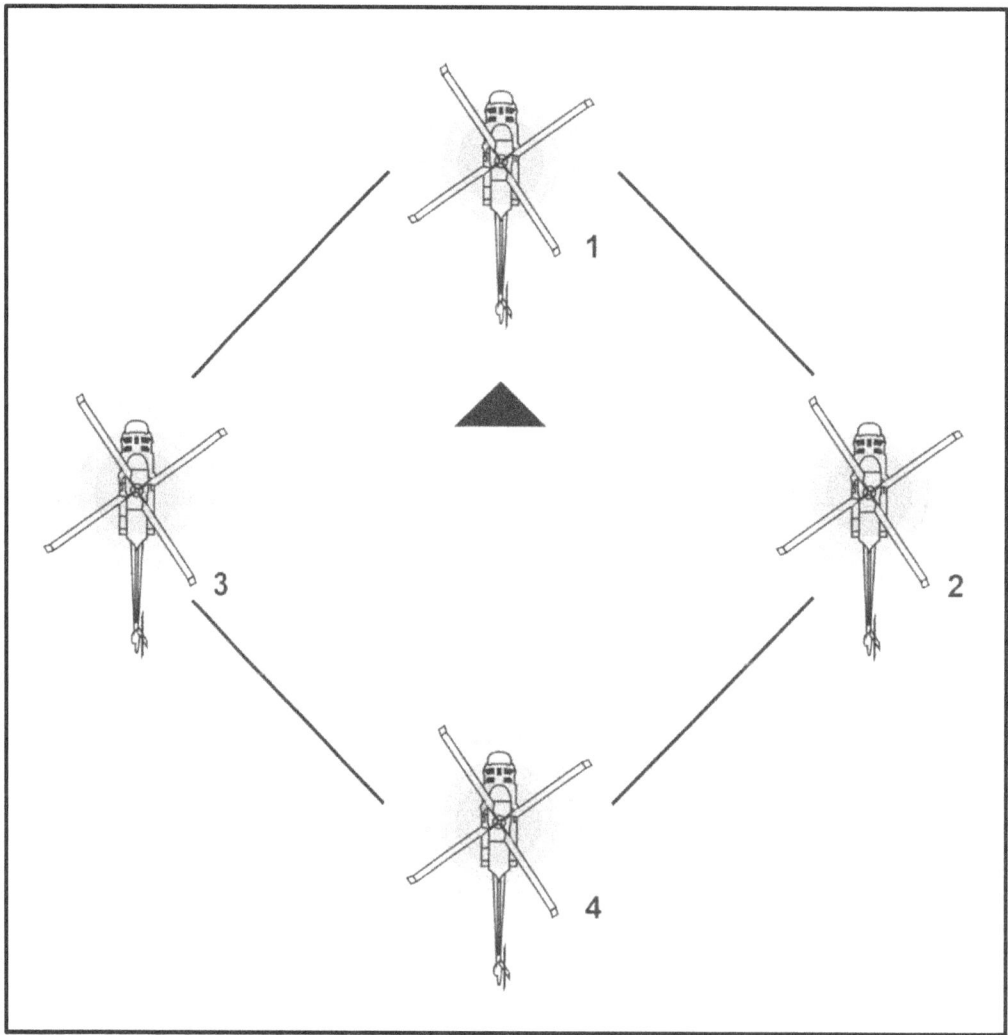

Figure 14-3. Diamond formation

Aviation

14-14. The vee formation (see figure 14-4) also requires a relatively small landing area and allows the rapid deployment of forces to the front. It restricts suppressive fire of inboard gunners and presents some difficulty in prepositioning loads.

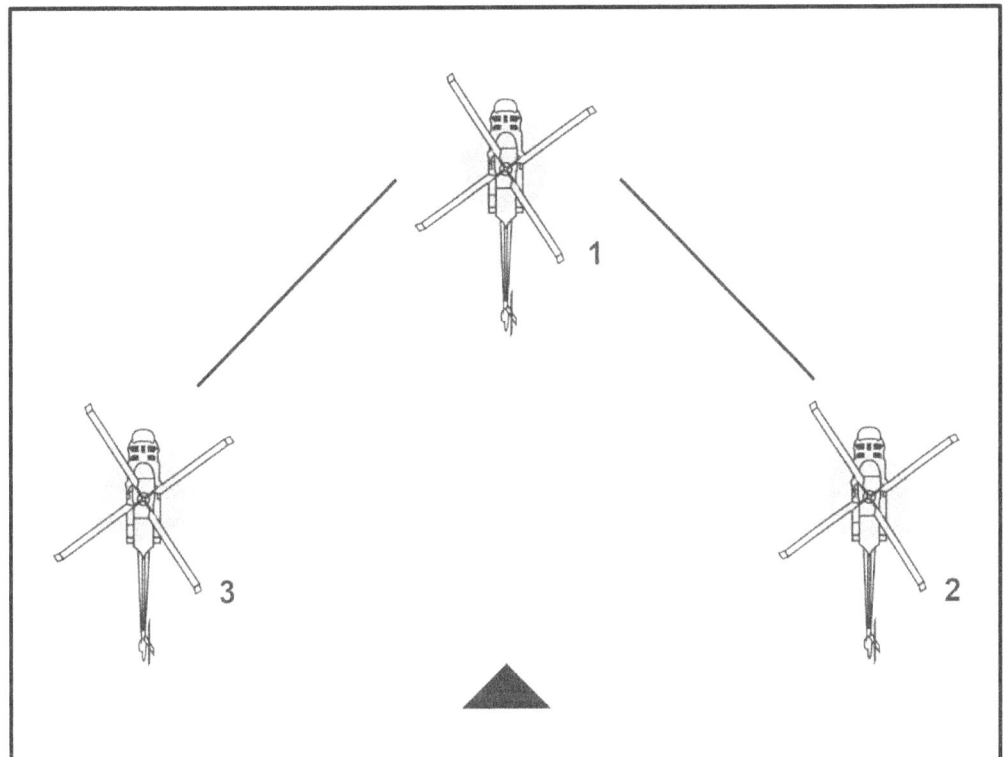

Figure 14-4. Vee formation

14-15. The echelon left or right formation (see figure 14-5) allows rapid deployment of forces to the flank and unrestricted suppressive fire by gunners. It also requires a relatively long and wide landing area, and presents some difficulty in prepositioning loads.

Chapter 14

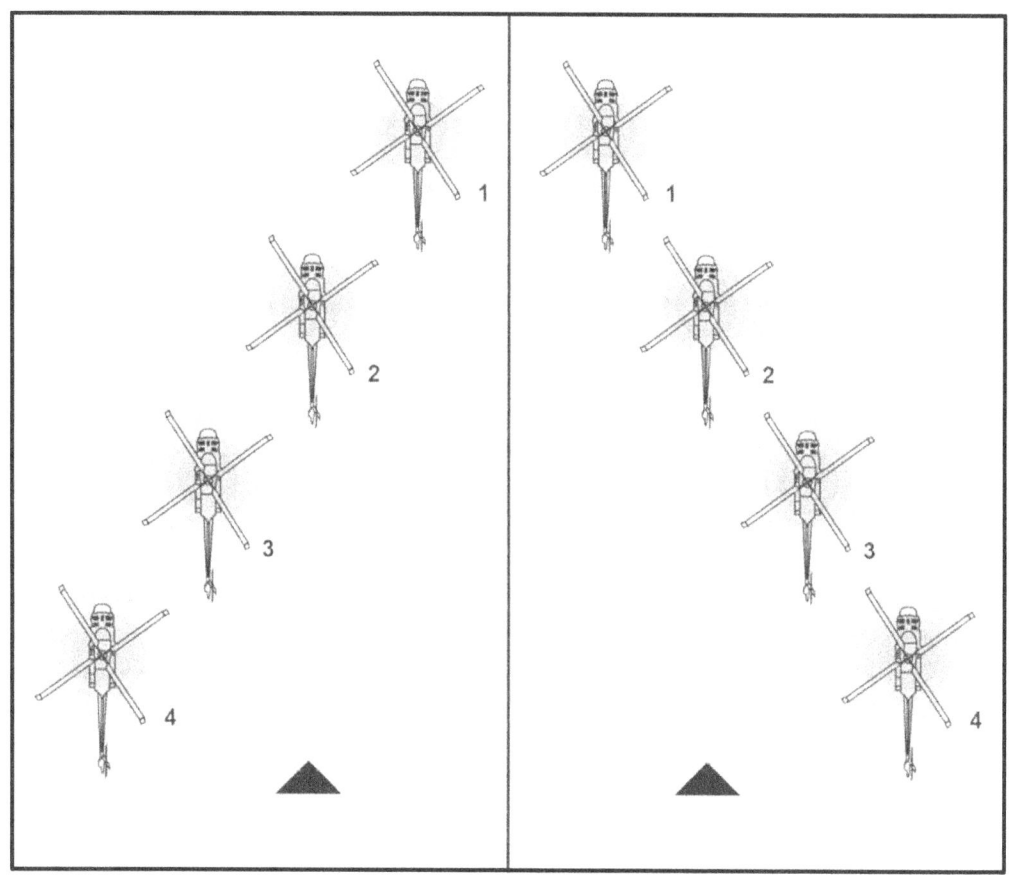

Figure 14-5. Echelon left or right formation

14-16. The trail formation (see figure 14-6) requires a relatively small landing area and allows rapid deployment of forces to the flank. It also simplifies prepositioning loads and allows unrestricted suppressive fire by gunners.

Aviation

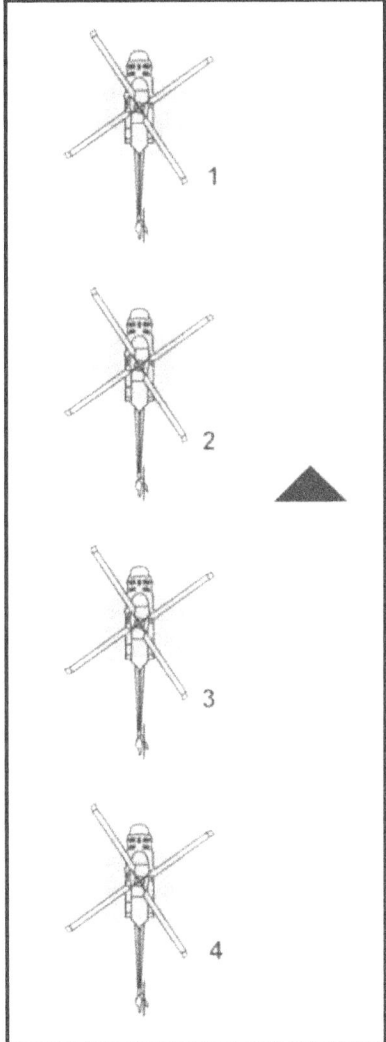

Figure 14-6. Trail formation

14-17. A staggered trail left or right formation (see figure 14-7) requires a relatively long, wide landing area, and the gunners' suppressive fire is somewhat restricted. It simplifies prepositioning loads and allows rapid deployment for all-around security.

Chapter 14

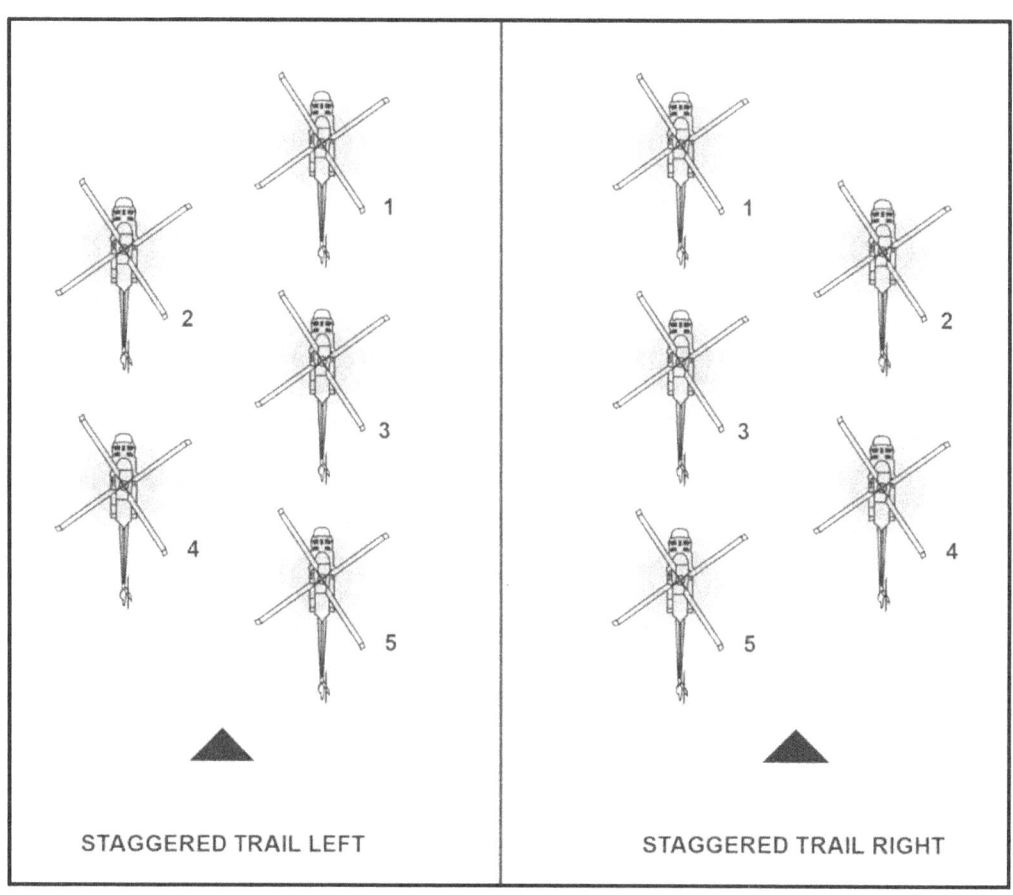

Figure 14-7. Staggered trail left or right formation

PICKUP ZONE OPERATIONS

14-18. Prior to the arrival of aircraft, the PZ is secured. The PZ control party is positioned, and the troops and equipment are positioned in the platoon and squad assembly areas. In occupying a patrol and squad assembly area, the patrol or squad leader maintains all-around security of the assembly area, maintains communications, organizes personnel and equipment into chalks and loads, and conducts safety briefings and equipment checks of the troops.

14-19. Figure 14-8 shows an example of a large, one-sided PZ. Figures 14-9 through 14-12 on pages 14-12 through 14-15, demonstrate loading and unloading procedures and techniques.

Aviation

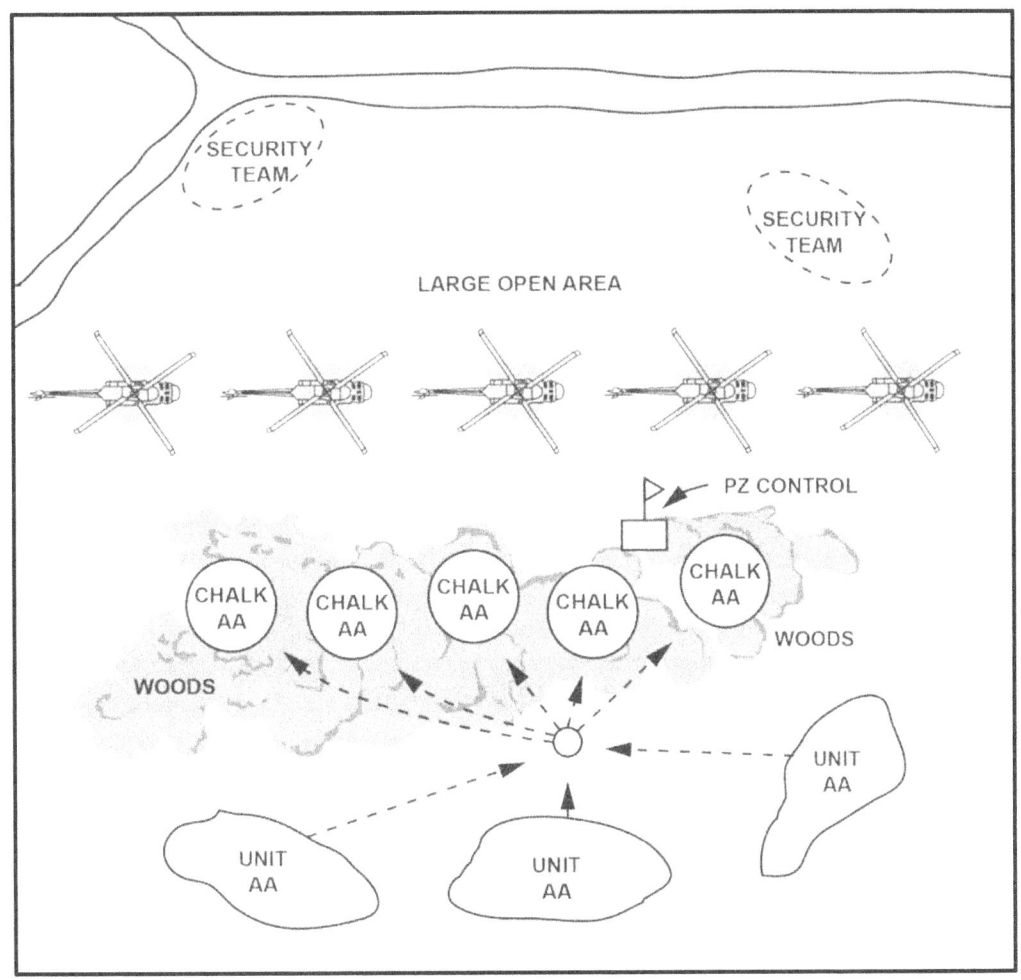

Figure 14-8. Large, one-sided pickup zone

LEGEND
AA – assembly area; PZ – pickup zone

Chapter 14

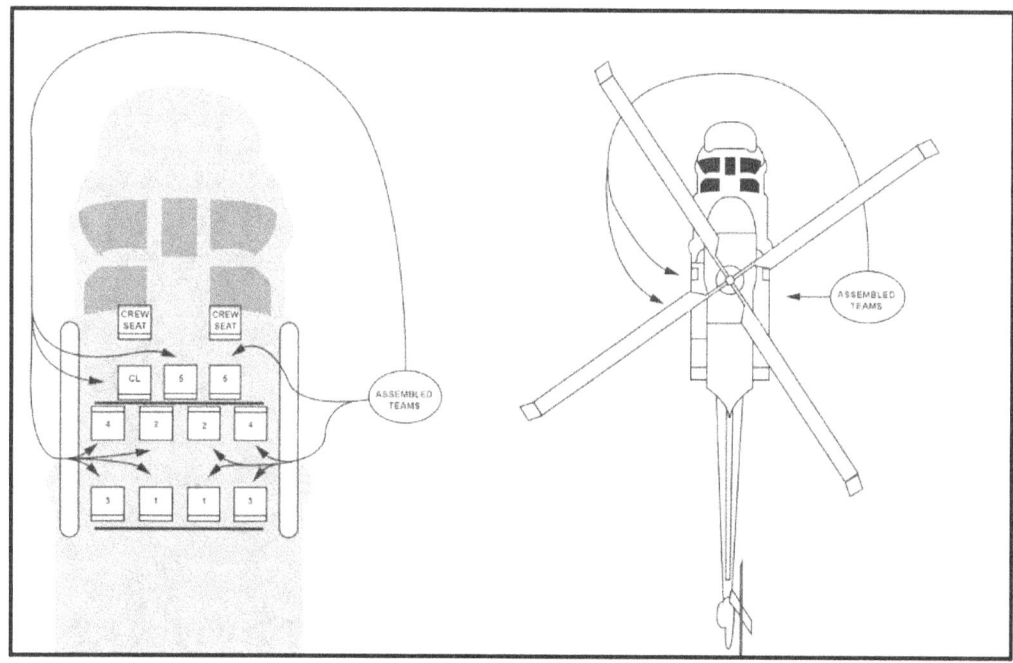

Figure 14-9. UH-60L loading sequence

LEGEND
CL – chalk leader

Aviation

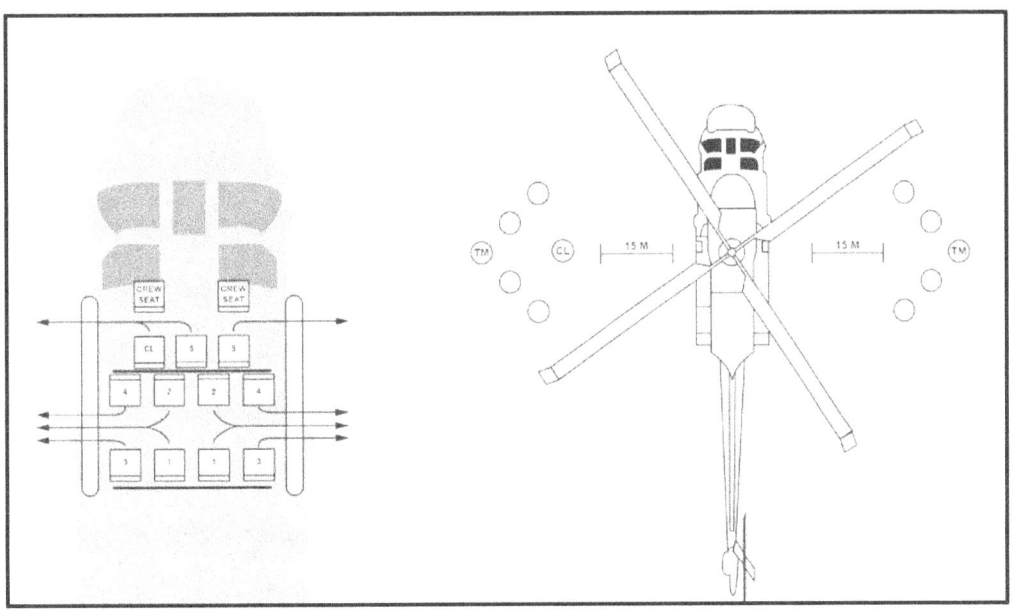

Figure 14-10. UH-60L unloading sequence

LEGEND
CL – chalk leader; M – meter; TM - team

Chapter 14

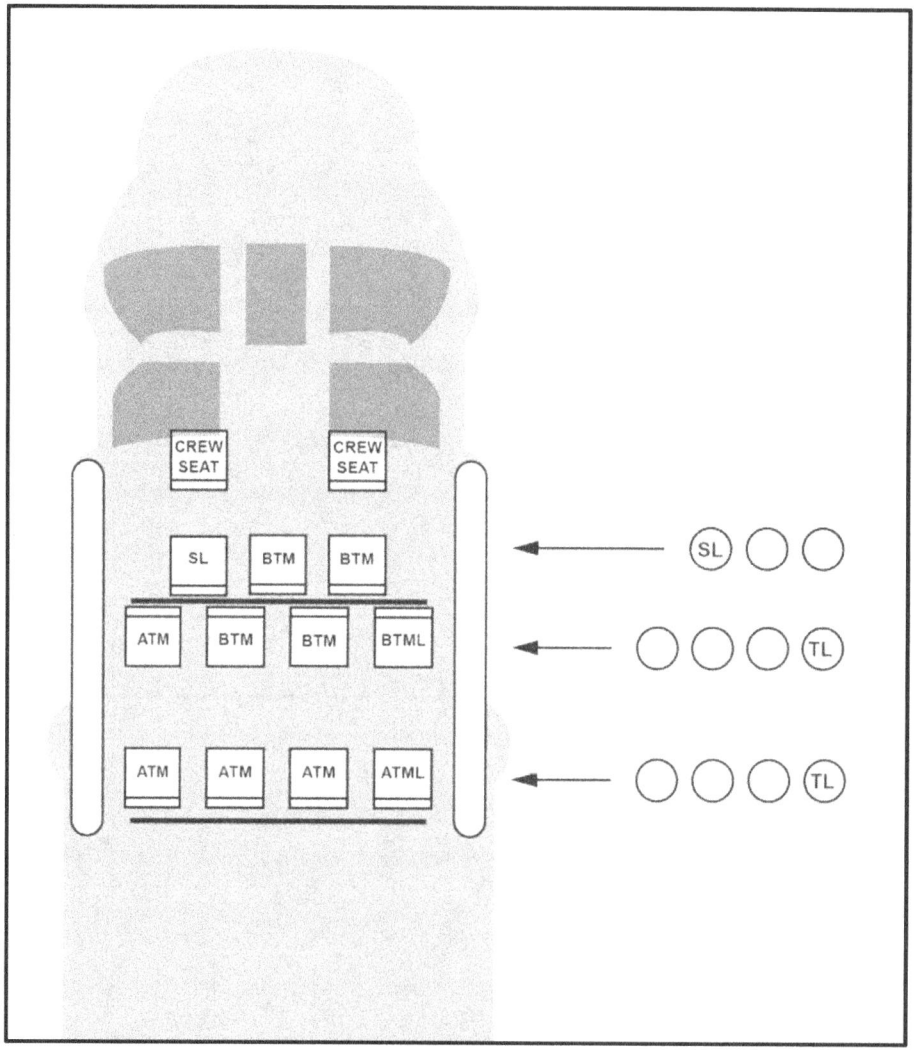

Figure 14-11. Tactical loading sequence

LEGEND
ATM – A team; ATML – A team leader; BTM – B team; BTML – B team leader; SL – squad leader; TL – team leader

Aviation

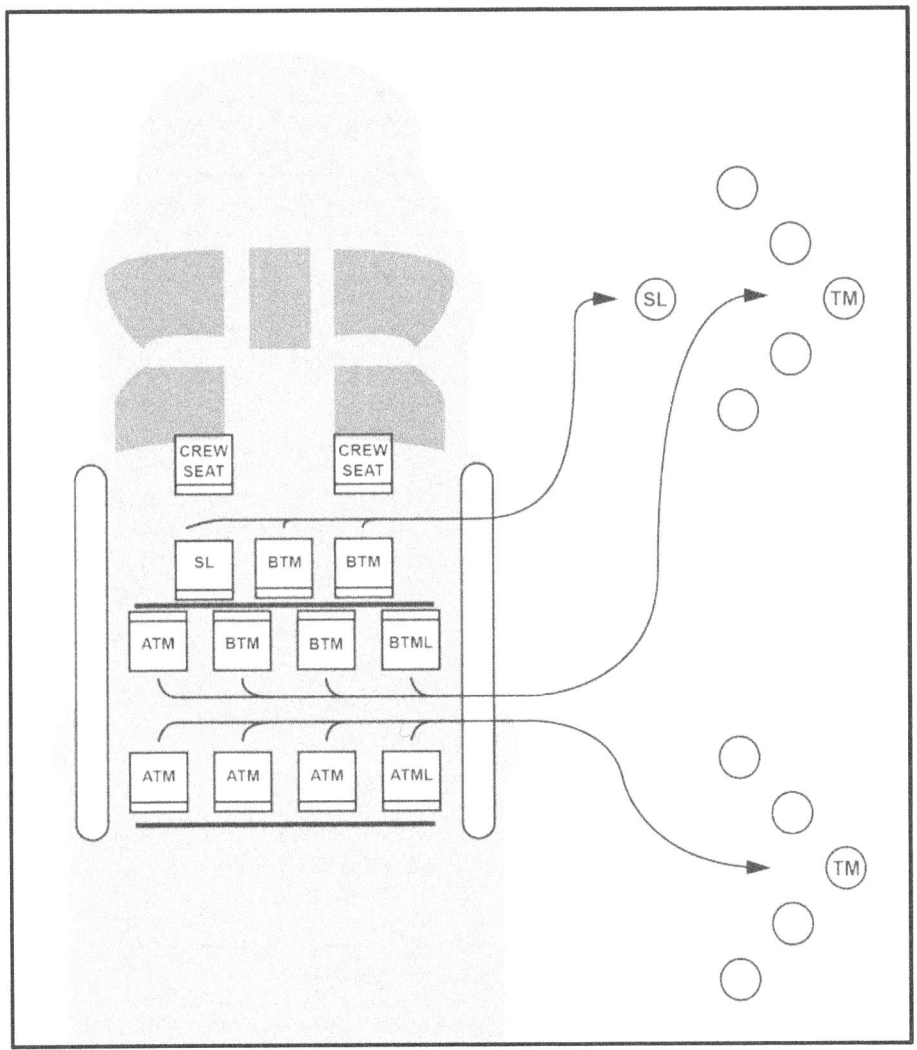

Figure 14-12. Tactical unloading sequence using door nearest cover, concealment
LEGEND
ATM – A team; ATML – A team leader; BTM – B team; BTML – B team leader; SL – squad leader; TL – team leader; TM -team

Chapter 14

CH-47/CV-22 Rear Ramp Off-load

14-20. In the rear ramp off-load method, Soldiers exit from the rear ramp of a CH-47 or other rear-exiting aircraft. Soldiers move out from the aircraft and drop to a prone fighting position, establishing 360-degree security until the aircraft lifts to depart the LZ. (See figure 14-13.) Once the aircraft departs the LZ, the unit may execute a one- or two-side LZ rush according to the landing plan or SOPs.

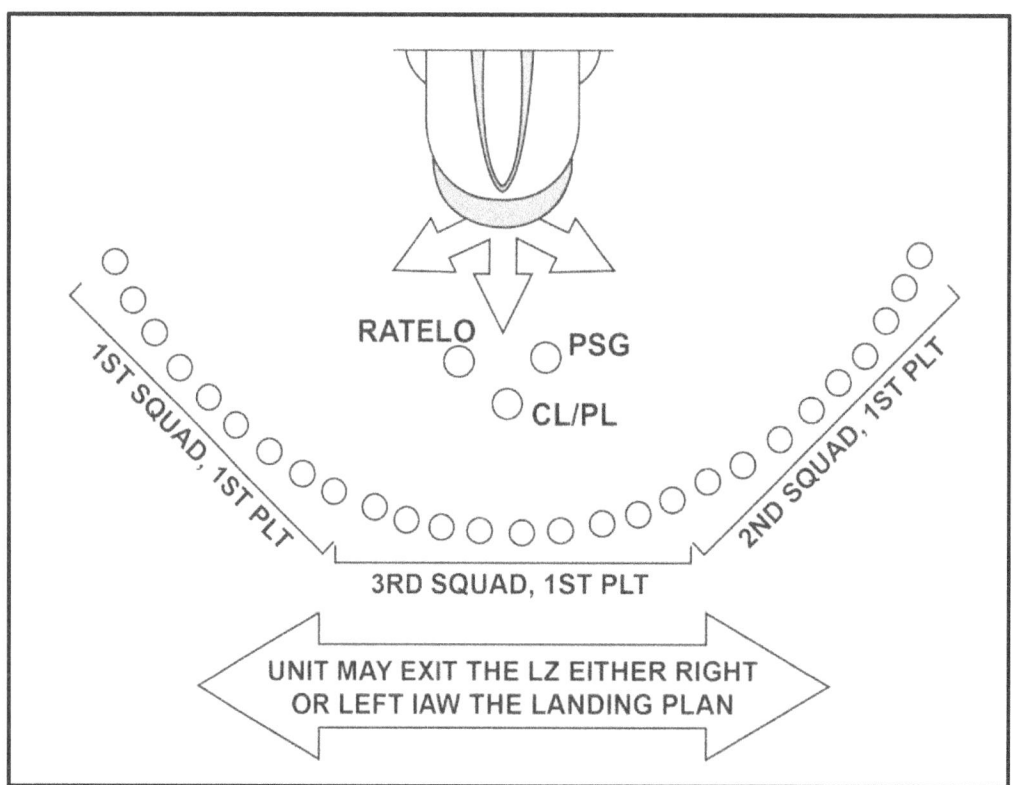

Figure 14-13. CH 47/CV-22 rear ramp off-load

LEGEND
1ST – first; 2ND – second; 3RD – third; CL/PL – crew leader/platoon leader; IAW – in accordance with; LZ – landing zone; PLT – platoon; PSG – platoon sergeant; RATELO – radiotelephone operator

14-21. In this method, Soldiers exit from the rear ramp of a CH-47 or other rear exiting aircraft. Soldiers move out from the aircraft and drop to a prone fighting position, establishing 360-degree security until the aircraft lifts to depart the LZ. Once the aircraft departs the LZ, the unit may execute a one- or two-side LZ rush according to the landing plan or SOPs.

Aviation

14-22. Safety is the primary concern of all leaders when operating in and around aircraft. The inclusion of aircraft into Ranger operations brings high risks. Consider the following, UH-60L:
- Approach the aircraft from 45-to-90 degrees off the nose.
- Point the muzzles of weapons upward when loaded with blank firing adapters.
- Point the muzzles of weapons downward when loaded with live ammunition.
- Wear the ballistic helmet.
- When possible, conduct an aircrew safety brief with all personnel.
- At a minimum, cover loading and off-loading, emergency, and egress procedures.
- Leaders carry a manifest and turn in a copy to higher HQ.

REQUIREMENTS

14-23. Minimum landing space requirements and minimum distance between helicopters on the ground depend on many factors. If the aviation unit SOP does not spell out these requirements, the aviation unit commander works with the Pathfinder leader. The final decision about minimum landing requirements rests with the aviation unit commander. In selecting helicopter landing sites from maps, aerial photographs, and actual ground or aerial reconnaissance, consider the following factors:
- **Number of helicopters.** To land a large number of helicopters at the same time, the commander might have to provide another landing site(s) nearby. Alternatively, land the helicopters at the same site, but in successive lifts.
- **Landing formations.** Helicopter pilots should try to match the landing formation to the flight formation. Pilots should have to modify their formations no more than necessary to accommodate the restrictions of a landing site. However, in order to land in a restrictive area, they might have to modify their formation.
- **Surface conditions.** Rangers choose landing sites that have firm surfaces. This prevents helicopters from bogging down, creating excessive dust, or blowing snow. Rotor wash stirs up any loose dirt, sand, or snow. This can obscure the ground, especially at night. Rangers remove these and any other debris from landing points, since airborne debris could damage the rotor blades or turbine engine(s).
- **Ground slope.** Rangers choose landing sites with relatively level ground. For the helicopter to land safely, the slope should not exceed seven degrees. Whenever possible, pilots should land upslope rather than downslope. All helicopters can land where ground slope measures seven degrees or less.
  - **Day operation signals.** For daylight operations, use different smoke colors for each landing site. The same color can be used more than once, just spread them out. Use smoke only when necessary, because the enemy can see it, too. Try to use it only when the pilot asks for help locating the helicopter site.
  - **Night operation signals.** For night operations, use pyrotechnics or other visual signals in lieu of smoke. As in daylight, red signals mean, "DO NOT LAND," but they can also be used to indicate other emergency conditions. Everyone plans and knows emergency codes. Each flight lands at the assigned site according to messages and the visual aids displayed. Arm and hand signals can be used to help control the landing, hovering, and parking of helicopters.

Planning Considerations

14-24. To ensure success of the ground mission, leaders plan their own missions in detail. The more time they have to make plans, the more detailed plans they can make. As soon as the senior leader receives word of a pending operation, a mission alert is issued, immediately followed with a warning order. Just enough information is issued to allow the subordinate leaders to start preparing for the operation. This includes:
- Roll call.
- Enemy and friendly situations (in brief).
- Mission.
- Chain of command and task organization.

Chapter 14

- Individual uniform and equipment (if not discussed in the SOP).
- Required equipment.
- Work priorities (who does what, when, and where).
- Specific instructions.
- Attached personnel.
- Coordination times.

14-25. On receiving the alert or warning order, leaders inspect and augment personnel and equipment, as needed. Leaders prepare equipment in the following order, from the most to the least important:

- 1. Radios.
- 2. Navigation aids (electronic and visual).
- 3. Weapons.
- 4. Essential individual equipment.
- 5. Assembly aids.
- 6. Other items as needed.

14-26. To succeed, an operation has security. Each person receives only the information necessary to complete each phase of the operation. For example, the commander isolates any Soldiers who know the details of the operation. The situation dictates the extent of security requirements.

ROTARY WING AIRCRAFT SPECIFICATIONS

14-27. Rotary wing aircraft are vital for the success of certain missions. The Army relies on different types: the UH-60L Blackhawk (see table 14-2), CV-22 Osprey (see table 14-3 on page 14-20), and the CH-47D Chinook (see table 14-4 on page 14-21).

14-28. When fitted with a sling load, the Chinook technical data package (TDP) is #5 (100-m diameter). Without the sling load, it is #4 (80-m diameter). Specifications for all three helicopters are in the following tables.

Table 14-2. Specifications for the UH-60L Blackhawk

OPTICS	Pilots use AN/AVS-6 to fly the aircraft at night.
NAVIGATION EQUIPMENT	Doppler navigation set or Global Positioning System (GPS).
FLIGHT CHARACTERISTICS	Maximum speed (level): 156 knots (kts). Normal cruise speed: 120 to 145 kts. Maximum speed (with external sling loads): 90 kts.
ADDITIONAL CAPABILITIES	External Stores Support System (ESSS) allows for extended operations without refueling (more than five hours) with two 230-gallon fuel tanks. ESSS also allows configuration for ferry and self-deployment flights with four 230-gallon fuel tanks. Enhanced mission command. Console provides the maneuver commander with an airborne platform that can support six secure very high frequency (VHF) radios, one high frequency (HF) radio, two VHF radios, and two ultrahigh frequency (UHF) radios. UH-60L is capable of inserting and extracting troops with Fast Rope Infiltration, Exfiltration System (FRIES) and Special Purpose Infiltration, Exfiltration System (SPIES). Sling load lift rating of 9000 pounds (lbs.).

Chapter 14

Table 14-3. Specifications for the CV-22 Osprey

WEAPONS SYSTEM AND RANGES	M2 (.50-caliber machine gun): 1800 meters.	
COMMUNICATION EQUIPMENT	Internal.	AN/AIC-30.
	External.	ARC-210 radio.
NAVIGATION EQUIPMENT	Navigational aid.	ARN-147.
FLIGHT CHARACTERISTICS	Cruise airspeed: 240 knots (kts).	Payload: 24 troops (seated), 32 troops (floor loaded), or 10,000 pounds cargo.
	Max airspeed: 250 kts (at sea level) and 305 kts (at 15,000 ft).	Endurance: 500 nautical miles with troops.
AIRCRAFT SURVIVABILITY EQUIPMENT	Radar warning receiver (RWR).	AN/APR-39A(V)2.
	Laser warning.	AN/AVR-2A Laser Detection System.
	Missile warning.	AN/AAR-47.
	Electronic countermeasures.	ALE-47 Countermeasures Dispensing System.
Fuel Capacity	2025 gallons.	
Other capabilities	Self-deployable, vertical, or short takeoff and landing.	

Table 14-4. Specifications for the CH-47D Chinook

OPTICS	
NAVIGATION EQUIPMENT	Pilots use AN/AVS-6 to fly the aircraft at night.
FLIGHT CHARACTERISTICS	Doppler navigation set or Global Positioning System (GPS).
ADDITIONAL CAPABILITIES	Max speed (level): 170 knots (kts). Normal cruise speed: 130 kts.
LIMITATIONS	Can be configured with additional fuel for mobile forward-arming refueling equipment system or for ferry and self-deployment missions. Has an internal load winch to ease loading of properly configured cargo. Can sling load virtually any piece of equipment in the light Infantry, Airborne, or air assault divisions.
CAPACITY	Carries 33 to 55 troops, 24 litters and three attendants, or 28,000 pounds (lbs.) of cargo.

Chapter 15
First Aid

Patrolling, more than some other types of missions, puts Rangers in harm's way. CASEVAC planning is vital. Trained medical personnel might be unavailable at the initial point of injury, so all Rangers know how to diagnose and treat injuries, wounds, and common illnesses. The unit should also have a plan for handling KIAs.

LIFESAVING STEPS AND CARE UNDER FIRE

15-1. Whatever the injury, (1) stop life-threatening bleeding; (2) open the airway and restore breathing; (3) stop the bleeding and protect the wound; (4) recheck, treat, and monitor for shock; and (5) MEDEVAC the casualty. The 9-line MEDEVAC Request and the Casualty Feeder Card can be found in Appendix B.

15-2. When still under fire, (1) maintain situational awareness; (2) return fire; determine if the casualty is dead or alive; have the casualty render self-aide (3) protect the casualty; (4) move the casualty to cover; and (5) identify and control severe bleeding with a bandage or tourniquet. Make sure any sensitive equipment is secured. While conducting the primary survey, use table 15-1 as a step-by-step process.

Table 15-1. ABCs of first aid

Airway	Open airway by patient position or with airway adjuncts. Switch A and C.
Breathing	Identify and seal open chest wounds with occlusive dressing.
Circulation	Identify uncontrolled bleeding and control with pressure or tourniquet. Start intravenous (IV) therapy, if needed.
Disability	Determine level of consciousness.
Exposure	Fully expose patient. (Environment dependent.)

AIRWAY MANAGEMENT

15-3. The airway is usually obstructed (blocked) at the base of the tongue. If this happens, open the airway using the chin lift method for nontraumatic injuries to the face or skull. (See figure 15-1 on page 15-2.) For traumatic injuries, keep the airway open by using the jaw thrust method.

Chapter 15

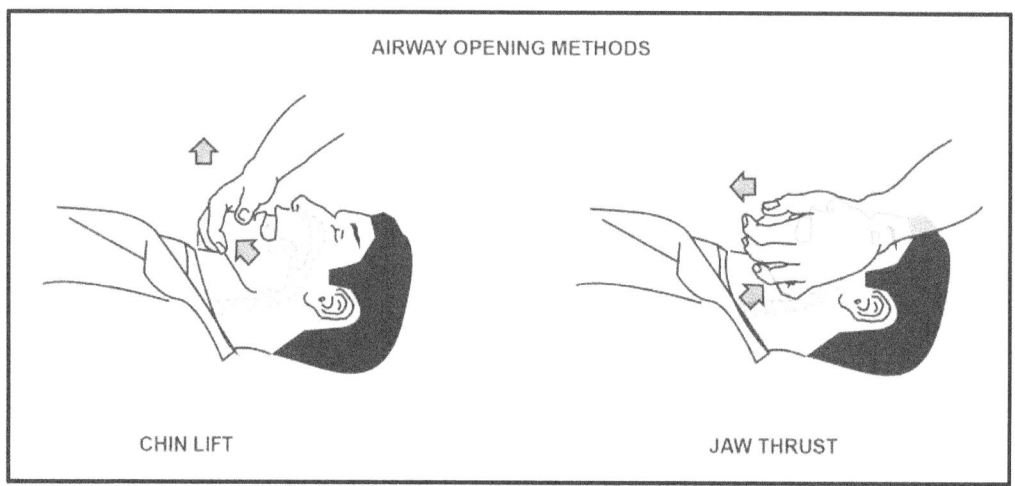

Figure 15-1. Chin lift and jaw thrust methods

15-4. Remove debris (teeth, blood clots, bone) from the oral cavity (use suction if available), and place airway adjuncts to allow the victim to breathe through their nose (see figure 15-2) or mouth (see figure 15-3 on page 15-4).

First Aid

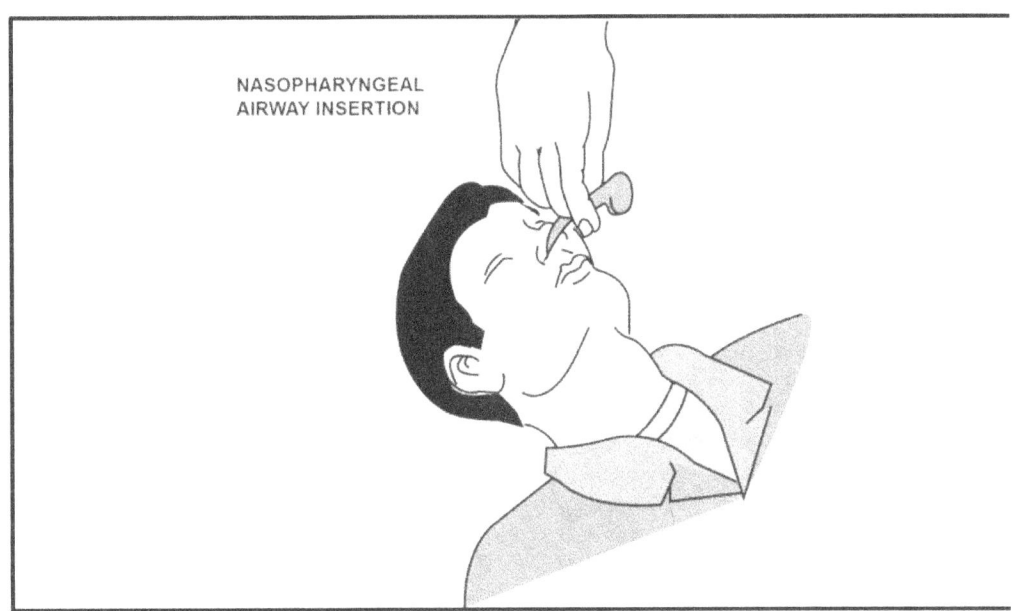

Figure 15-2. Nasal airway insertion

Chapter 15

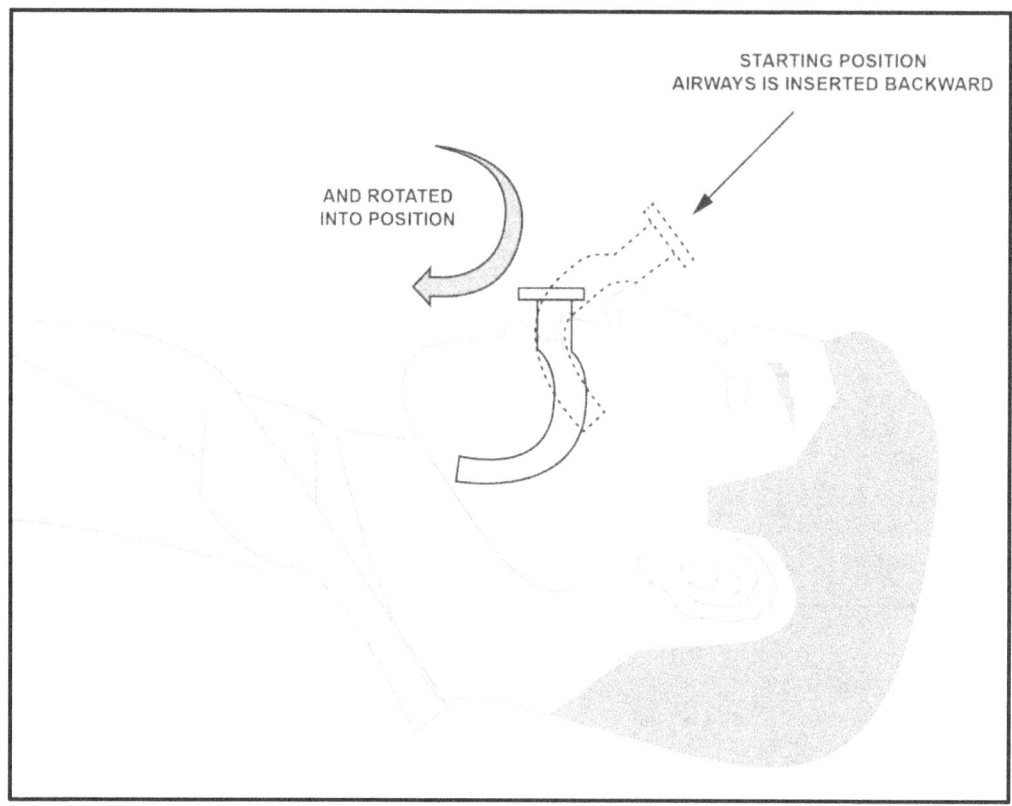

Figure 15-3. Mouth airway insertion

STABILIZING, BREATHING, BLEEDING, AND SHOCK

15-5. If the patient is having trouble breathing, expose the chest and identify open chest injuries (check for entrance and exit wounds). Apply an occlusive dressing to seal open entry and exit chest wounds. Place the patient on the injured side, or position him where he can breathe most comfortably.

15-6. Quickly identify and control bleeding. Apply a tourniquet to arterial bleeding of the extremities two-to-three inches above the elbow or knee. If this does not control the bleeding, apply a second tourniquet above the first and apply a pressure dressing. Control all other bleeding with a pressure dressing. Check dressings often to ensure bleeding is under control.

15-7. Shock is caused by an inadequate flow of oxygen to body tissues. The most common form of shock is hemorrhagic (due to uncontrolled bleeding). Signs and symptoms of shock include sweaty but cool, clammy skin; pale skin; restlessness;

First Aid

nervousness; agitation; unusual thirst; altered mental status; rapid breathing; blotchy, bluish skin around the mouth; and nausea. Basic treatment for shock is:
- Control bleeding.
- Open airway.
- Restore breathing.
- Position the casualty.
- Monitor condition.
- Evacuate the casualty.

INJURIES AND BURNS

15-8. For extremity injuries, identify and control the bleeding. If a fracture is suspected, splint the bone as it lies. Do not reposition the injured extremity. Check the distal pulse to make sure there is still adequate blood flow after splinting. If there is no pulse, redo the splint and reassess.

15-9. Identify and control the bleeding of abdominal injuries and then treat for shock. If internal organs are exposed, cover them with dry, sterile dressing. Do not place them back in the abdominal cavity. Place the patient in a comfortable position. Flex knees to relax the abdomen. Do not give anything by mouth to the patient.

15-10. For burn patients, remove them from the source. Remove all clothing and jewelry from the areas of the body with burns. Cover burns with dry, sterile dressings. Ensure fingers and toes have dressings between them before covering the entire area. Immediately evacuate any casualties with burns of the face, neck, hands, genitalia, or over 20 percent (one fifth) of the body surface. (See figure 15-4 on page 15-6.)

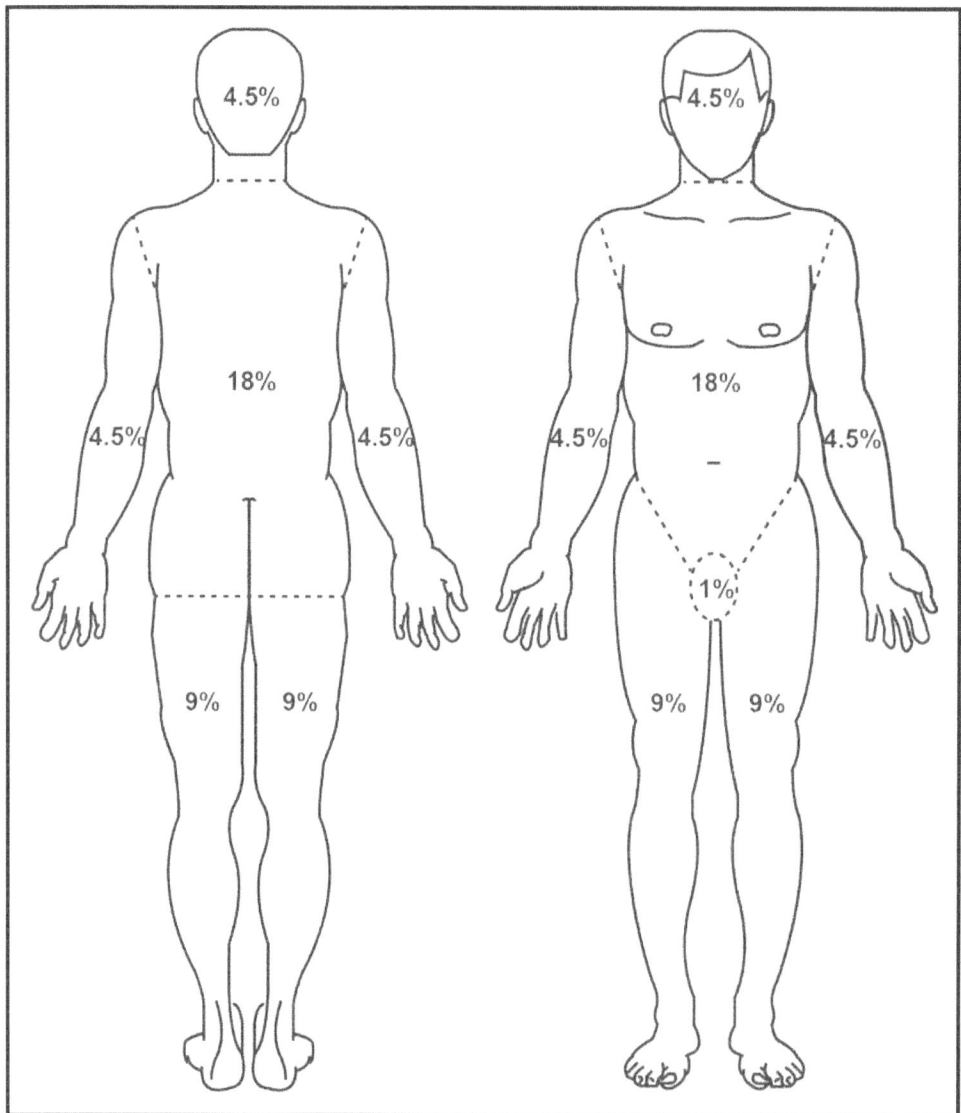

Figure 15-4. Percentage of body area

First Aid

TREATING INJURIES

15-11. Rangers pay close attention to weather temperatures and the environment to avoid injuries. Hot and cold weather injuries can be mild to life threatening. Environmental injuries range from bites and stings, to accidental exposure to poisonous plants.

15-12. Knowing the signs and symptoms, and the proper treatment is crucial. Tables 15-2, 15-3 on page 15-8, and 15-4 on page 15-9, detail the hot, cold, and environmental injuries that can afflict a Ranger, and the first aid steps for recovery.

Table 15-2. Heat injuries

INJURY	SIGNS OR SYMPTOMS	FIRST AID
Heat Cramp	Casualty experiences muscle cramps in arms, legs, or stomach. May also have wet skin and extreme thirst.	1. Move the casualty to a shaded area and loosen clothing. 2. Allow casualty to drink one quart of cool water slowly every hour. 3. Monitor casualty and provide water as needed. 4. Seek medical attention if cramps persist.
Heat Exhaustion	Casualty experiences loss of appetite, headache, excessive sweating, weakness or faintness, dizziness, nausea, or muscle cramps. The skin is moist, pale, and clammy.	1. Move the casualty to a cool, shaded area and loosen clothing. 2. Pour water on casualty and fan to increase cooling effect of evaporation. 3. Provide at least one quart of water to replace lost fluids. 4. Elevate legs. 5. Seek medical aid.
Heat Stroke (Sunstroke)	Casualty stops sweating (hot, dry skin), may experience headache, dizziness, nausea, vomiting, rapid pulse and respiration, seizures, mental confusion. Casualty may suddenly collapse and lose consciousness.	1. Move casualty to a cool, shaded area, loosen clothing, and remove outer clothing if the situation permits. 2. Immerse in cool water. If cool bath is not available, pour cool water on the head and body. Fan casualty to increase the cooling effect of evaporation. 3. If conscious, slowly consume one quart of water.
	DANGER Sunstroke is a medical emergency! Seek medical aid and evacuate as soon as possible. Perform any lifesaving measures.	

Chapter 15

Table 15-3. Cold injuries

INJURY	SIGNS OR SYMPTOMS	FIRST AID
Chilblain	Red, swollen, hot, tender, itchy skin. Continued exposure may lead to infected (bleeding, ulcerated) skin lesions.	1. Area usually responds to locally applied warming (body heat). 2. Do NOT rub or massage area. 3. Seek medical treatment.
Immersion (Trench) Foot	Affected parts are cold and numb. As body parts warm, they may become hot, with burning and shooting pains. **Advanced stage:** Skin is pale with bluish cast; pulse decreases; blistering, and swelling occur. Swelling, heat hemorrhages, and gangrene may follow.	1. Gradual warming by exposure to warm air. 2. Do NOT massage or moisten skin. 3. Protect affected parts from trauma. 4. Dry feet thoroughly and avoid walking. 5. Seek medical treatment.
Frostbite	**Superficial:** Redness, blisters in 24-to-36 hours followed by peeling skin. **Deep:** Preceded by superficial frostbite: skin is painless, pale-yellowish, waxy, "wooden" or solid to the touch, blisters form in 12-to-36 hours.	**Superficial:** 1. Keep casualty warm and gently warm affected parts. 2. Decrease constricting clothing, increase exercise and insulation. **Deep:** 1. Protect the part from additional injury. 2. Seek medical treatment as fast as possible.
Snow Blindness	Red, scratchy, or watery eyes; headache; increased pain in eyes with exposure to light.	1. Cover the eyes with a dark cloth. 2. Seek medical treatment.
Dehydration	Similar to heat exhaustion.	1. Keep warm and loosen clothes. 2. Replace lost fluids, rest, and seek additional medical treatment.
Hypothermia	Casualty is cold, shivers uncontrollably until shivering stops. A core (rectal) temperature below 95 degrees Fahrenheit can affect consciousness. Uncoordinated movements, shock, and coma may occur as body temperature drops.	**Mild hypothermia:** 1. Warm body evenly and without delay. (Provide a heat source.) 2. Keep dry, protect from elements. 3. Warm liquids may be given only to a conscious casualty. 4. Be prepared to start CPR. 5. Seek medical treatment immediately. **Severe hypothermia:** 1. Quickly stabilize body temperature. 2. Attempt to prevent further heat loss. 3. Handle the casualty gently. 4. Evacuate to nearest medical treatment facility as soon as possible.
LEGEND CPR – cardiopulmonary resuscitation		

First Aid

Table 15-4. Environmental injuries

TYPE	FIRST AID
Snake bite	1. Get the casualty away from the snake. 2. Remove all rings and bracelets from the affected extremity. 3. Reassure the casualty and keep them quiet. 4. Apply constricting band(s) two-to-three inches above the bite. 5. Immobilize the affected limb below the level of the heart. 6. Treat for shock and monitor. 7. **Kill the snake (without damaging its head or endangering yourself, if possible) and send it with the casualty.** 8. Evacuate and seek medical treatment immediately.
Brown Recluse or Black Widow spider bite	1. Keep the casualty calm. 2. Wash the area. 3. Apply ice or a freeze pack, if available. 4. Seek medical treatment.
Tarantula bite, scorpion sting, or ant bite	1. Wash the area. 2. If site of bite(s) or sting(s) is on the face, neck (possible airway blockage), or genital area, or if the reaction is severe (or it was a dangerous southwestern scorpion sting), keep the casualty as quiet as possible, administer an antidote, if needed, and seek immediate medical aid.
Wasp or bee sting	1. If the stinger is present, remove by scraping with a knife or fingernail. Do NOT squeeze venom sack on stinger, more venom may be injected. 2. Wash the area. 3. Apply ice or freeze pack, if available. 4. If allergic signs or symptoms appear, be prepared to administer an antidote and seek medical assistance.
Human or animal bites	1. Cleanse the wound thoroughly with soap or detergent solution. 2. Flush bite well with water. 3. Cover bite with a sterile dressing. 4. Immobilize injured extremity. 5. Transport casualty to a medical treatment facility. 6. For human bites, try to extract some of the attacker's saliva from the wound and send that in a sealed, identified container with the casualty. 7. For animal bites, kill the animal without endangering yourself or damaging the animal's head, and send its head with the casualty.
Poison ivy, oak, or sumac	1. Gently clean affected area two-to-three times daily. Wash clothing. 2. Apply topical anti-itch lotion or ointment as needed, and cover. 3. Avoid scratching the area. 4. Observe for signs of infection (increasing redness, tenderness, warm to the touch). 5. Seek medical attention, if needed.

Chapter 15

POISONOUS PLANT IDENTIFICATION

15-13. Poison plants include, among others, poison ivy, oak, and sumac, and stinging nettles, which is not discussed here. (See figure 15-5.) Poison ivy grows as a vine or shrub. The compound leaves of poison ivy have three pointed leaflets. The middle one has a much longer mini-stalk than the two side ones. The leaflet edges can be smooth or toothed but are rarely lobed (lobed leaves look something like a hand with fingers). The leaves vary greatly in size, from one-third inch to just over two inches long. In spring, the leaves appear reddish. They turn green in the summer, and then red, orange, and yellow in the fall. Small greenish flowers grow in bunches right where the leaf joins the main stem. The flowers are later replaced by clusters of poisonous white, waxy, plump, droopy fruit.

15-14. Poison oak is a widespread deciduous shrub throughout mountains and valleys of North America, generally below 5000 feet elevation. It commonly grows as a climbing vine with airy roots that cling to the trunks of oaks and sycamores. Poison oak can also form dense thickets. Leaves typically have three leaflets (sometimes five), with the terminal one on a slender mini-stalk, as opposed to Eastern poison ivy, whose terminal leaf is often on a longer mini-stalk, with leaves that tend to be less ragged and serrated (less "oak like"). Like many members of the sumac family (Anacardiaceae), new foliage and autumn leaves often turn brilliant shades of pink and red.

15-15. Poison sumac is a woody perennial shrub or small tree. It grows from 5-to-25 feet tall, and favors swampy areas. To identify it, look for the fruit that grows between the leaf and the branch. Look for red stems that stay red all year. Leaves grow adjacent to each other and grow in odd numbers totaling five to thirteen on each stem. They have a glossy, waxy look and turn bright red and orange during the fall.

15-16. Throughout the phases of Ranger School, students will encounter these poisonous plants. The rash is caused by contact with a sticky oil called urushiol found in poison ivy, oak, or sumac. You can get the rash by touching or brushing against any part of these plants, including the leaves, stems, flowers, berries, and roots, even if the plant is dead. The rash is only spread through the oils. You cannot catch a rash from someone else by touching the blister fluid.

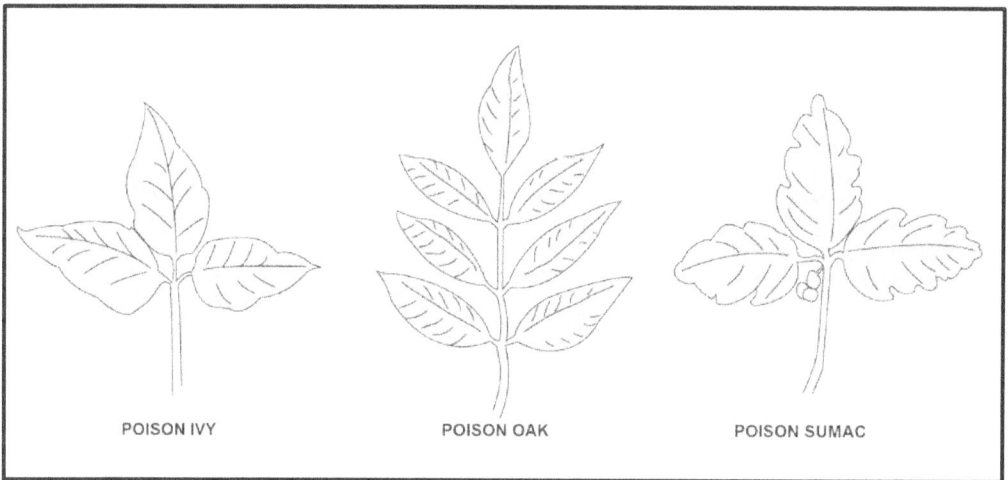

Figure 15-5. Poisonous plants

First Aid

Note: Knowing the proper procedures and methods to employ a litter can save a fellow Ranger's life. See Chapter 9 for more information.

FOOT CARE, HYDRATION, AND ACCLIMATIZATION

15-17. Use moleskin to prevent blisters prior to a movement or foot march. Keep feet as clean and dry as possible. Use foot powder and change socks. Let feet air dry, as mission permits. With blisters, seek medical help if needed or if infected.

15-18. Minimizing dehydration and increasing acclimatization is crucial for maintaining good health. There are various practices to help the Ranger improve in these aspects. Table 15-5 shows some strategies.

Table 15-5. Hydration management and acclimatization

STRATEGY	SUGGESTIONS FOR IMPLEMENTATION
Start Early	1. Start at least one month prior to school. 2. Be flexible and patient: performance benefits take longer than physiological benefits.
Mimic the Training Environment Climate	1. In warm climates, acclimatize in the heat of day. 2. In temperate climates, work out in a warm room wearing sweats.
Ensure Adequate Heat Stress	1. Induce sweating. 2. Work up to 100 minutes of continuous physical exercise in the heat. Be patient. The first few days, it may not be possible to go the full 100 minutes without resting. 3. Once exercising for 100 minutes in the heat is comfortable, continue doing so for seven days. Work up to at least fourteen days, and increase the exercise intensity each day (loads, or training runs).
Learn to Drink and Eat	1. The thirst mechanism improves as the body becomes acclimatized to the heat. Do not wait until feeling thirst to drink water, as this can actually cause dehydration. 2. Acclimatizing to heat increases personal water requirements. 3. Dehydration offsets most benefits of physical fitness and heat acclimatization. 4. More electrolytes leave the body through sweating during the first week of heat acclimatization, so add salt to food, or drink electrolyte solutions. 5. A convenient way to learn how much water is needed to replace lost fluids is weighing yourself before and after the 100 minutes of exercise in the heat. For each pound lost, drink about one-half of a quart of fluid; for example, if the weight loss is eight pounds, eight half-quarts equals four quarts (or one gallon) of fluid. 6. Do not skip meals, as this is when most of the water and salt losses are naturally replaced.

Chapter 15

WORK, REST, AND WATER CONSUMPTION

15-19. It is very important for Rangers to adhere to a proper work, rest, and water consumption schedule whenever possible. Table 15-6 provides a work, rest, and water consumption guide. This guidance applies to an average-sized, heat-acclimated Ranger wearing the Army combat uniform (ACU).

15-20. The work and rest times and fluid replacement volumes shown help the Ranger sustain work performance and hydration for at least four hours in the specified heat category. Fluid needs can vary based on individual differences (give or take one quart every hour).

15-21. In table 15-6, "NL" means that there is no limit to work time every hour. "Rest" means minimal physical activity such as sitting or standing, preferably in the shade. Consume no more than 1.5 quarts of fluid every hour, and no more than 12 quarts every day. If wearing body armor in a humid climate, add five degrees Fahrenheit (F) to the wet bulb globe temperature (WBGT). If wearing mission-oriented protective posture (MOPP) 4 clothing, add 10 degrees Fahrenheit to the WBGT. Work categories include easy, moderate, and hard:
- **Easy work** includes maintaining weapons, walking on hard surfaces at 2.5 mph with a load of no more than 30 pounds, participating in marksmanship training, and participating in drills or ceremonies.
- **Moderate work** includes walking in loose sand at 2.5 mph (maximum) or with no load, walking on a hard surface at 3.5 mph (maximum) with a load weighing no more than 40 pounds, performing calisthenics, patrolling, or conducting individual movement techniques such as the low or high crawl.
- **Hard work** includes walking on a hard surface at 3.5 mph with a load weighing 40 or more pounds, walking in loose sand at 2.5 mph while carrying a load, and conducting field assaults.

Table 15-6. Work, rest, and water consumption guidelines

HEAT CATEGORY	WBGT INDEX IN DEGREES F	EASY WORK		MODERATE WORK		HARD WORK	
		Work/Rest	Water Intake (Qt/H)	Work/Rest	Water Intake (Qt/H)	Work/Rest	Water Intake (Qt/H)
1	78 to 81.9	NL	0.50	NL	0.75	40/20	0.75
2 (Green)	82 to 84.9	NL	0.50	50/10	0.75	30/30	1.00
3 (Yellow)	95 to 87.9	NL	0.75	40/20	0.75	30/30	1.00
4 (Red)	88 to 89.9	NL	0.75	30/30	0.75	20/40	1.00
5 (Black)	90 or more	50/10 min	1.00	20/40	1.00	10/50	1.00

LEGEND
F- Fahrenheit; H – hour; HEAT - high-explosive antitank; NL – no limit; Qt – quart; WBGT – wet bulb globe temperature

First Aid

REQUESTING MEDICAL EVACUATION

15-22. Due to the nature of their work, at some point Rangers will request medical evacuation (MEDEVAC). This is done by following a well-rehearsed task and completing a 9-line MEDEVAC request. Additional information is required to ensure the best possible treatment of patients using a report based on the mechanism of injury, injuries sustained, signs and symptoms, and treatment given (MIST). This information is sent as soon as possible after the 9-line MEDEVAC request has been sent MEDEVAC missions should not be delayed while waiting for the MIST information.

15-23. An explanation of the filled-out North Atlantic Treaty Organization (NATO) 9-Line Request with MIST Report is in Appendix B on page B-6. An example of a Tactical Combat Casualty Care (TCCC) card is in Appendix B on pages B-10 and B-11. The information on the TCCC is used for the MIST report.

This page intentionally left blank

Appendix A
Resources

This appendix is a quick reference for some of the necessary techniques used by Rangers. Additional information for some methods described in previous chapters is discussed in depth, along with an outline on site exploitation that is not part of the main text.

TRAINING BOARDS

A-1. Figures A-1 through A-11 on pages A-2 through A-12, figure A-12 on page A-14, and figure A-13 on page A-19, depict training boards of critical actions Rangers routinely use. Table A-1 on page A-13, and tables A-2 through A-5 on pages A-15 through A-18 contain information vital for a successful mission.

A-2. Following the figures and tables, various techniques and formations used for clearing buildings are discussed in depth. Site exploitation is also outlined as a handy guide.

Appendix A

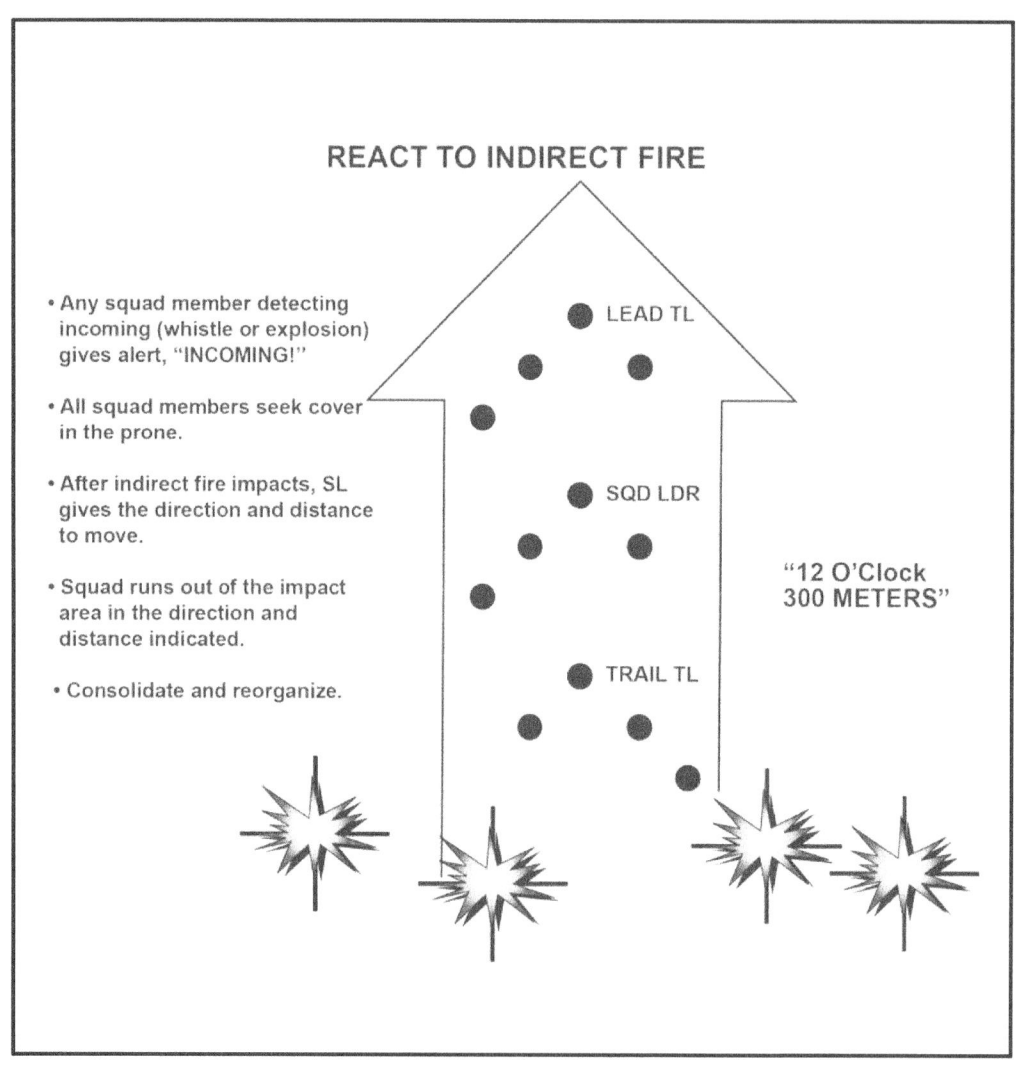

Figure A-1. React to indirect fire

LEGEND
LDR – leader; SL – squad leader; SQD – squad; TL – team leader

Resources

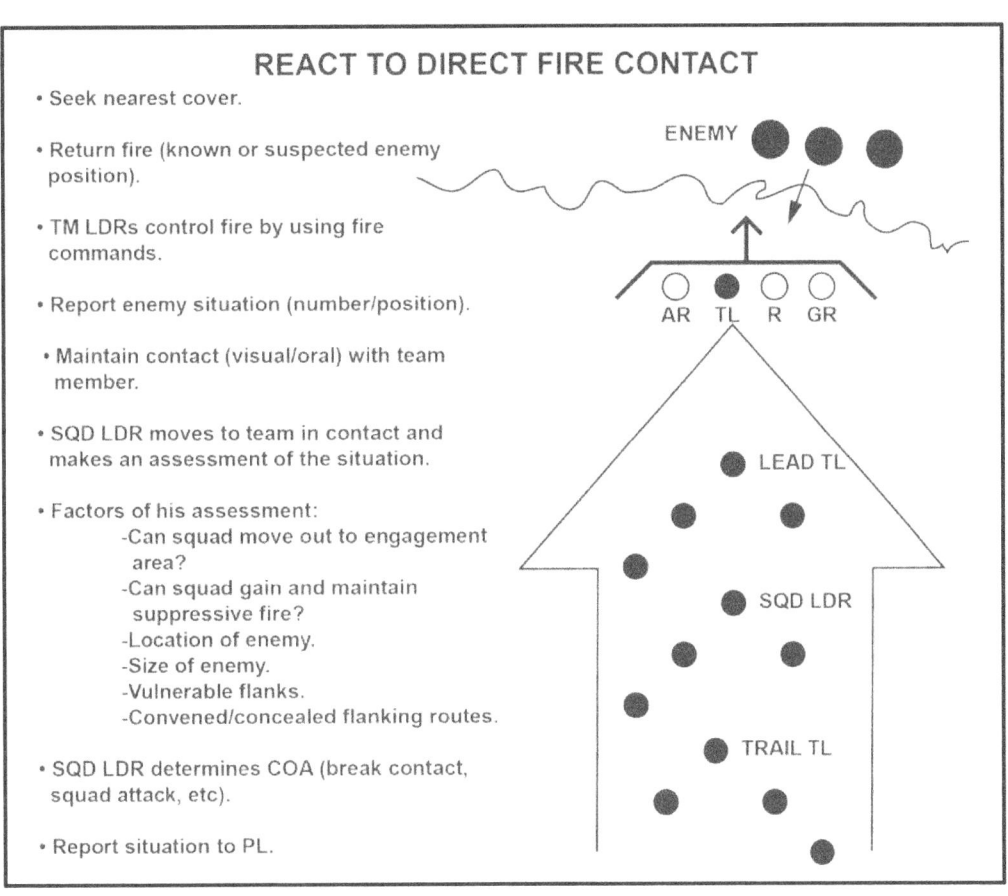

Figure A-2. React to direct fire contact

LEGEND
AR – automatic rifleman; COA – course of action; GR – grenadier; LDR – leader; PL – platoon leader; R - rifleman; SQD – squad; TL – team leader; TM – team

Appendix A

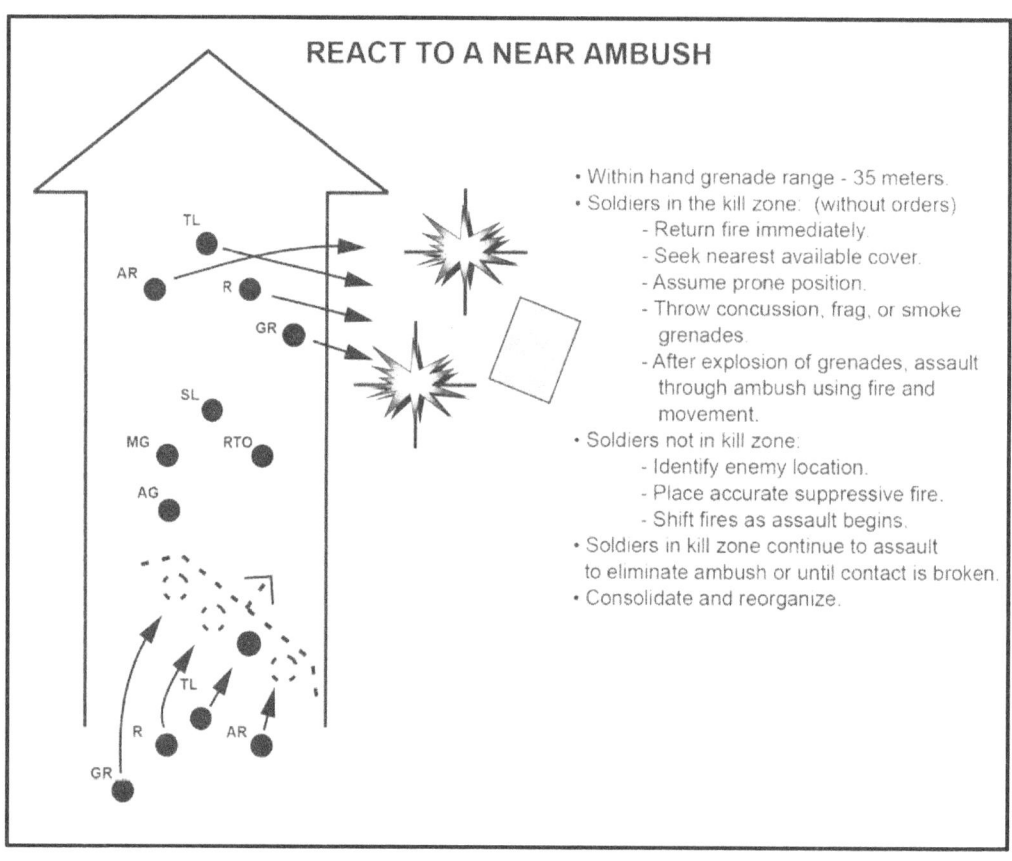

Figure A-3. React to a near ambush

LEGEND
AG – assistant gunner; AR – automatic rifleman; frag – fragmentary; GR – grenadier; MG – machine gunner; R - rifleman; RTO – radiotelephone operator; SL – squad leader; TL – team leader

Resources

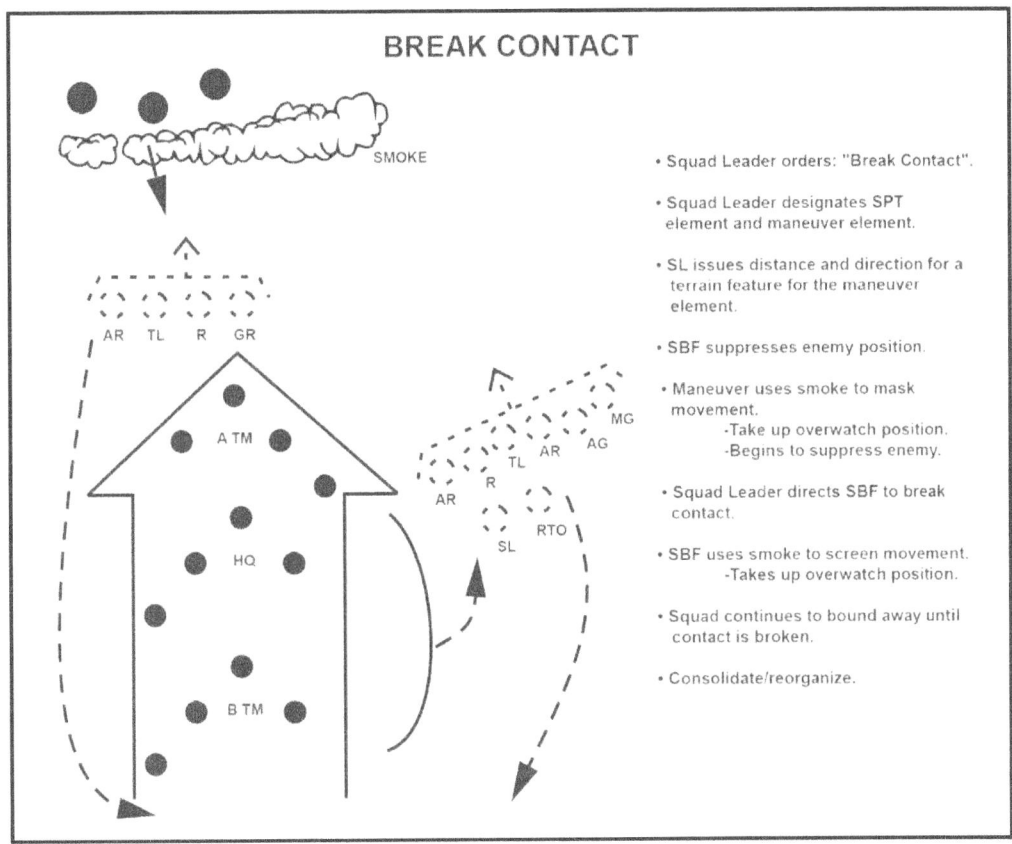

Figure A-4. Break contact

LEGEND
AG – assistant gunner; AR – automatic rifleman; GR – grenadier; HQ – headquarters; MG – machine gunner; R – rifleman; RTO – radiotelephone operator; SBF – support by fire; SL – squad leader; SPT – support; TL – team leader; TM - team

Appendix A

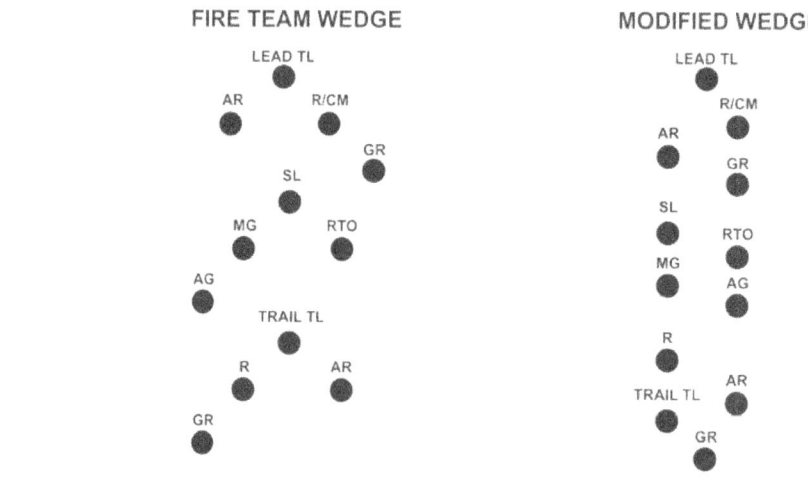

FORMATIONS AND ORDER OF MOVEMENT

I. Movement Formation: Fire Team Wedge: MG Team attached.
II. Three Movement Techniques used:
    A. Traveling technique used behind FFL when contact is not likely.
    B. Traveling Overwatch forward of the FFL when enemy contact is possible.
    C. Bounding Overwatch used forward of the FFL when enemy contact is expected.
III. Distances are based on but not dictated by visibility, terrain, and vegetation.
IV. Actions at Night: Modified Wedge
V. Actions at the Halt: Short and Long Halt (GV/LV)
VI. Leader Location: Fixed/Unfixed

Figure A-5. Formations and order of movement

LEGEND
AG – assistant gunner; AR – automatic rifleman; FFL – forward friendly line; GR – grenadier; GV/LV – good visibility/limited visibility; MG – machine gunner; R/CM – rifleman/compass man; R – rifleman; RTO – radiotelephone operator; SL – squad leader; TL – team leader

Resources

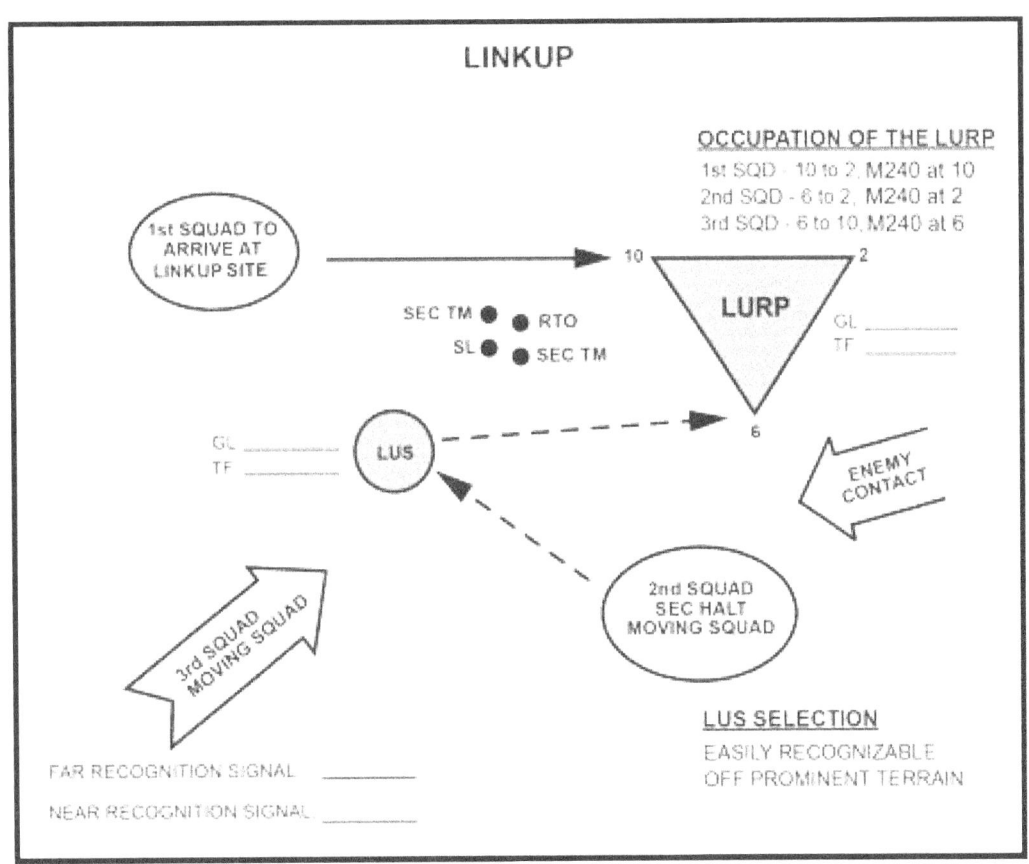

Figure A-6. Linkup

LEGEND
GL – grid location; LURP – linkup rally point; LUS – linkup site; RTO – radiotelephone operator; SL – squad leader; SEC – section; SQD – squad; TF – terrain feature; TM – team

Appendix A

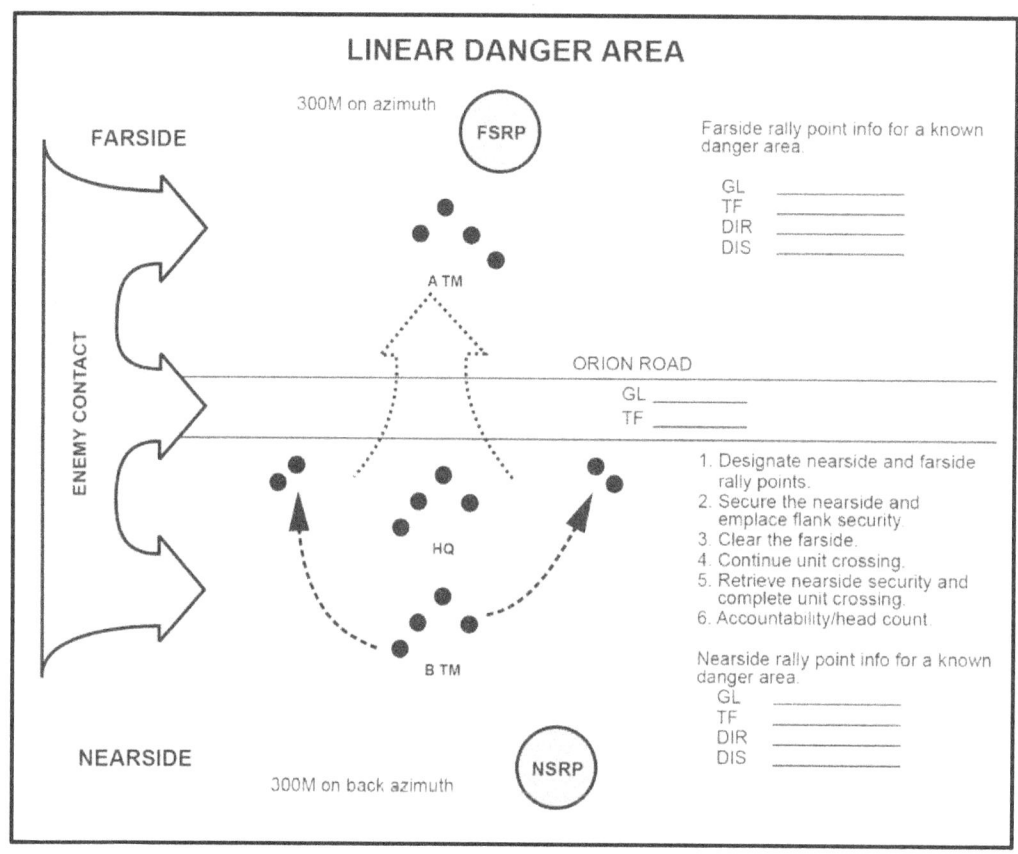

Figure A-7. Linear danger area

LEGEND
DIR – direction; DIS – distance; FSRP – farside rally point; GL – grid location; HQ – headquarters; info – information; M – meter; NSRP – nearside rally point; TF – terrain feature; TM - team

Resources

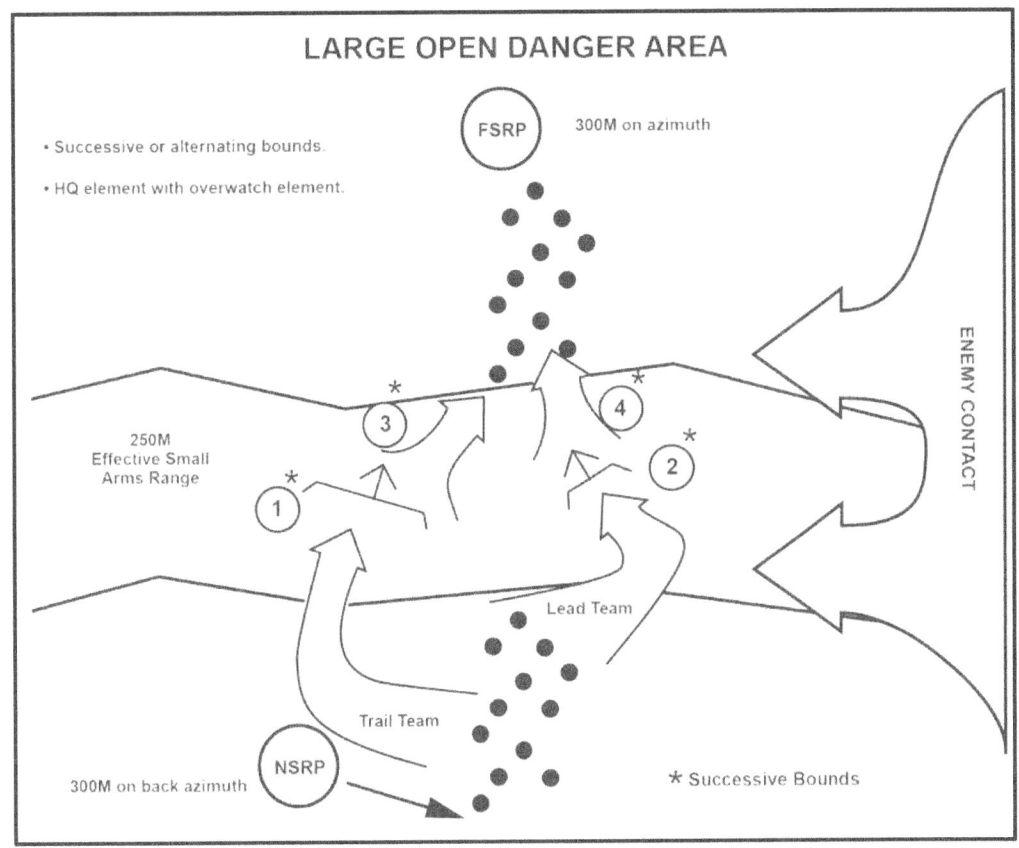

Figure A-8. Large open danger area

LEGEND
FSRP – farside rally point; HQ – headquarters; M – meter; NSRP – nearside rally point

Appendix A

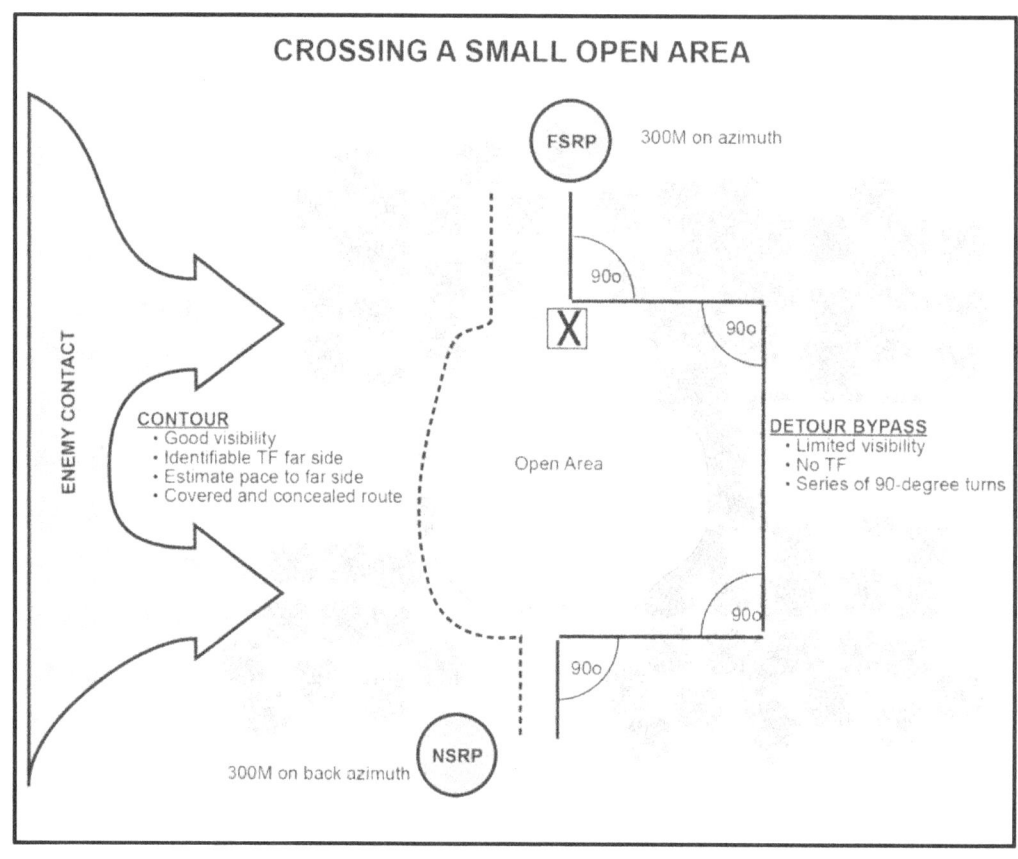

Figure A-9. Crossing a small open area

LEGEND
FSRP – farside rally point; M – meter; NSRP – nearside rally point; TF – terrain feature

Resources

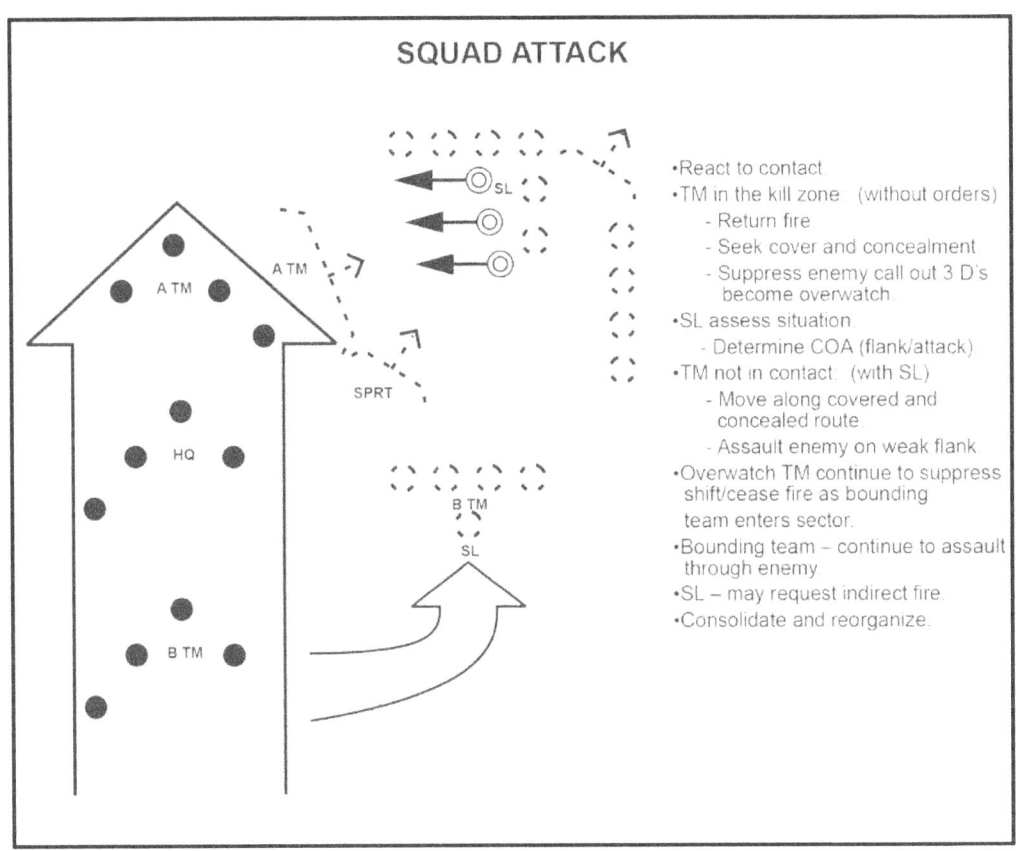

Figure A-10. Squad attack

LEGEND
3 D's – distance, direction, description; COA – course of action; HQ – headquarters; SL – squad leader; SPRT - support; TM - team

Appendix A

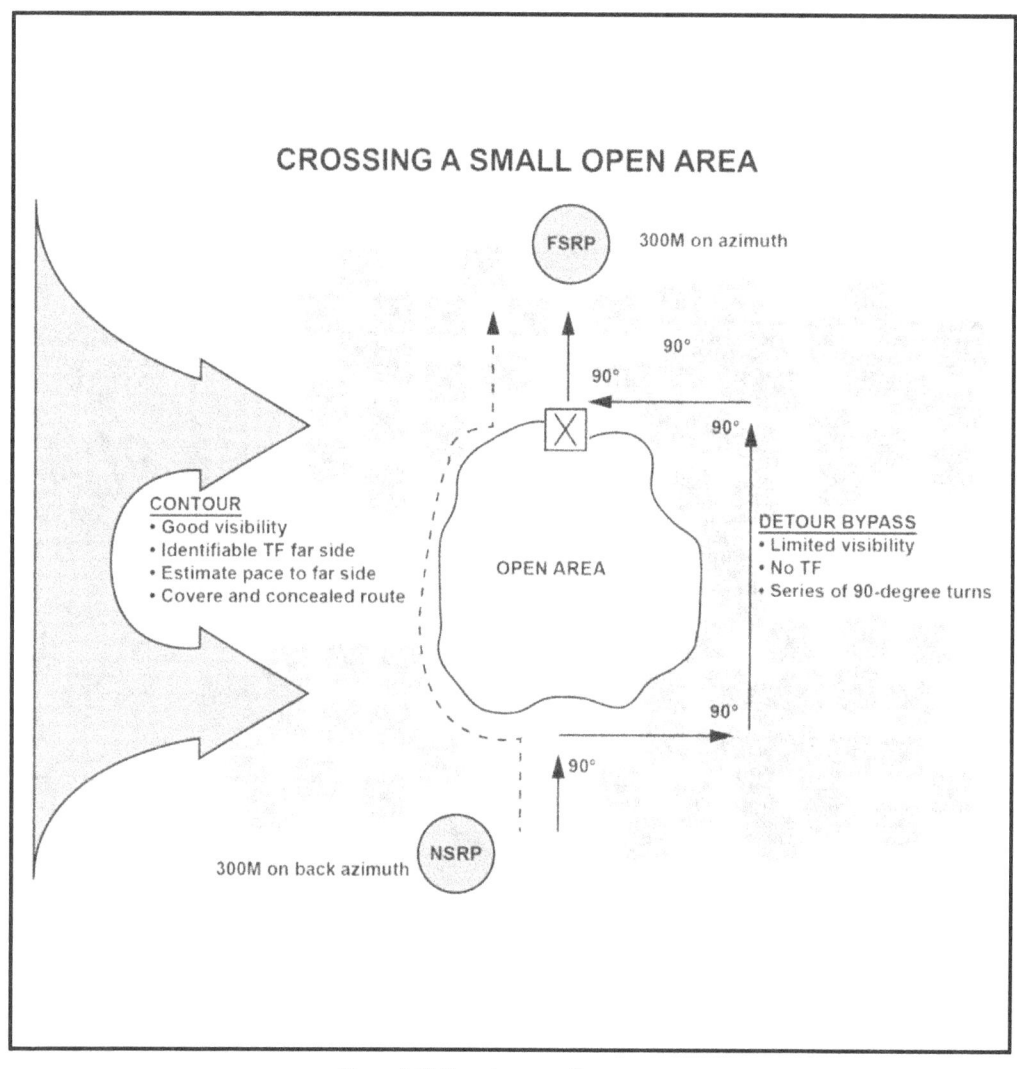

Figure A-11. Crossing a small open area

LEGEND
FSRP – farside rally point; M – meter; NSRP - nearside rally point; TF – terrain feature

Resources

Table A-1. Raid boards (left)

RAID BOARDS (LEFT)
GENERAL INFO

Raid References:
- RHB Chapter 7
- ATP 3-21.8

PURPOSES:
1. Destroy
2. Liberate
3. Collect Intelligence

Planning Considerations:
- Mission
- Enemy
- Troops
- Terrain/OAKOC
- Time
- Civilians

STUDENT INPUT WRITTEN HERE

Raid:
A surprise attack on a fixed position or installation ending in a planned withdrawal.

Characteristics:
1. Surprise - Gain
2. Coordinated fires - Maintain
3. Violence of actions - Retain
4. Planned withdrawal

INITIATIVE

LEGEND
ATP – Army Training Publication; intel – intelligence; OAKOC - observation and fields of fire, avenues of approach, key terrain, obstacles, and cover and concealment; RHB – Ranger Handbook

Appendix A

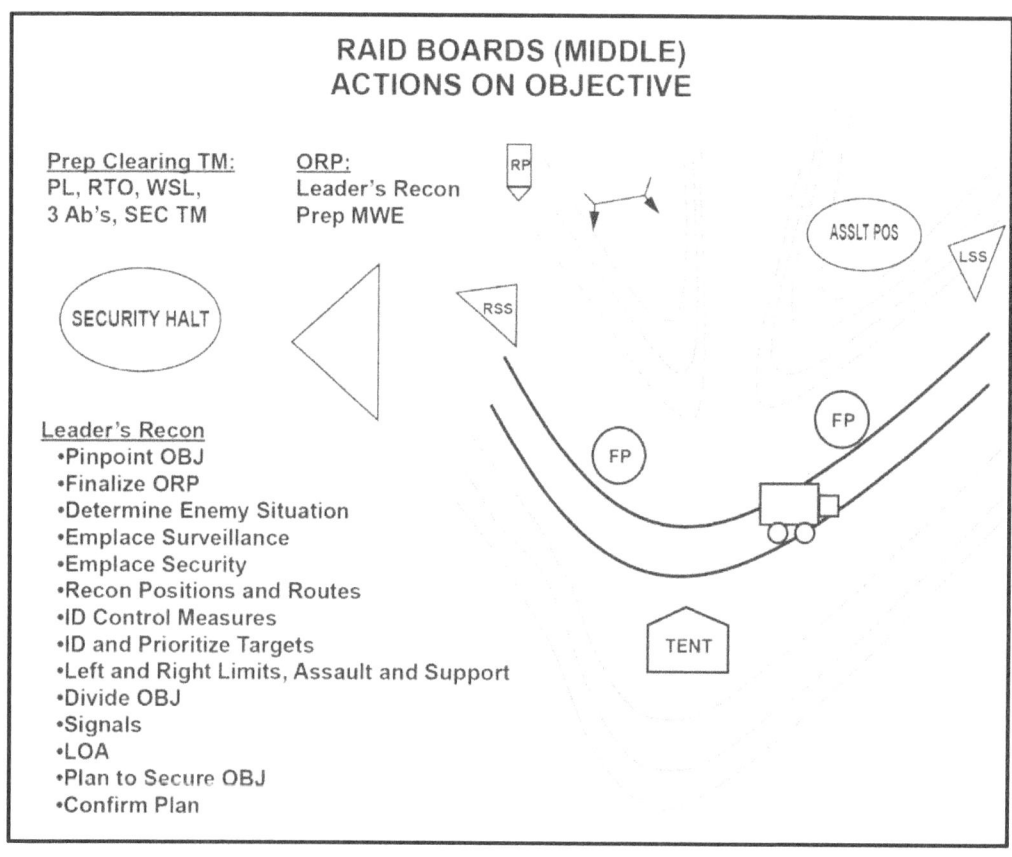

Figure A-12. Raid boards (middle)

LEGEND
Ab's – ammunition bearers; ASSLT POS – assault position; FP – firing point; ID – identify; LOA – limit of advance; LSS – left side security; MWE – men, weapons, and equipment; OBJ – objective; ORP – objective rally point; Prep - prepare; Recon – reconnaissance; RP – rally point; RSS – right side security; RTO – radiotelephone operator; SEC TM – security team; WSL – weapons squad leader

Resources

Table A-2. Raid boards (middle) task organization

ELEMENT	WHO	WHY	WHAT
HQ	PL, PSG, RTO, FO, Medic	C2, mission command	1 each AN/PRC 119F radio 2 each AN/PRC 148 radios 1 each AN/PSN 13 navigation set 1 each AN/PVS 14 NVD (for each individual) 1 each AN/PEQ 15 ATPIAL (For each WPN)
ASSLT 1	Squad (+) (-)	Destroy	3 each AN/PRC 148 radios 1 each AN/PSN 13 navigation set 2 each M18A1 Claymore mines 1 each AT4 antitank WPN 1 each AN/PVS 14 NVD (for each individual) 1 each AN/PEQ 15 ATPIAL (per WPN)
ASSLT 2	Squad (+) (-)	Destroy	3 each AN/PRC 148 radios 1 each AN/PSN 13 navigation set 2 each M18A1 Claymore mines 1 each AT4 antitank WPN 1 each AN/PVS 14 NVD (for each individual) 1 each AN/PEQ 15 AtPIAL (per WPN)
Support	Weapons Squad (+) (-)	Suppress	3 each M240B MG 1 each AN/PSN 13 navigation set 1 each AN/PVS 14 NVD (for each individual 3 each AN/PRC 148 radios 1 each AN/PEQ 15 ATPIAL (per WPN)
Security	Squad (+) (-)	Delay	1 each AN/PSN 13 navigation set 1 each AN/PVS 14 NVD (for each individual 2 each AN/PRC 148 radios 1 each AN/PEQ 15 ATPIAL (per WPN)

LEGEND
ASSLT – assault; ATPIAL – advanced target pointer/illuminator/aiming light; FO – forward observer; HQ – headquarters; NVD – night vision device; PL – platoon leader; PSG – platoon sergeant; RTO – radiotelephone operator; SQD – squad; WPN - weapon

Appendix A

Table A-3. Raid boards (right) SOP

RAID BOARDS (RIGHT) SOP	
CONTINGENCY	SOP

1. Contingency
   a. ORP
   b. Leader's recon
   c. Occupation
2. MASS CAL
3. Counterattack
	1. EPW search
	2. Aid litter
	3. MEDEVAC
	4. CCP
	5. Withdrawal plan

LEGEND
CCP – casualty collection point; EPW – enemy prisoner of war; MASS CAL – mass casualties; MEDEVAC – medical evacuation; ORP – objective rally point; recon - reconnaissance

Resources

Table A-4. Ambush SOP (left)

AMBUSH SOP (LEFT)

References:
- RHB Chapter 7
- ATP 3-21.8

Ambush:
A surprise attack from a covered and concealed position on a moving or temporarily halted target.

Types:
- Point
- Area

Fundamentals:
1. Surprise - Gain
2. Coordinated fires - Maintain
3. Violence of actions - Retain

PURPOSES:
1. Disrupt/Destroy
2. Collect Intelligence
3. Block or Deny Access
4. Canalize

Raid:
Deliberate: Platoon has specific target at a predetermined time and location.

Hasty: Platoon makes visual contact with the enemy and has time to establish an ambush without being detected.

Planning Considerations:
- Mission - Task/Purpose
- Enemy - MPCOA, MDCOA
- Terrain - Map Recon/Ldrs Recon
- Time - 1/3 - 2/3 Rule
- Troops - Task Org
- Civilians

LEGEND
ATP – Army Training Publication; Ldrs – leaders; MDCOA – most dangerous course of action; MPCOA – most probable course of action; Org – organization; Recon – reconnaissance; RHB – Ranger Handbook; SOP – standard operating procedures

Appendix A

Table A-5. Ambush boards (middle)

ELEMENT	WHO	WHY	WHAT
HQ	PL, PSG, RTO, FO, Medic	C2, Mission Command	1 each AN/PRC 119F radio, 2 each AN/PRC 148 radios, 1 each AN/PSN 13 navigation set, 1 each AN/PVS 14 NVD (for each individual), 1 each AN/PEQ 15 ATPIAL (for each WPN)
Security	Minimum of three 2-man teams	Early Warning Seals off OBJ LSS, RSS, ORP Security	3 each AN/PRC 148 radios, 1 each AN/PSN 13 navigation set, 2 each M18A1 Claymore Mines, 1 each AT4 Antitank Weapon, 1 each AN/PVS 14 NVD (for each individual), 1 each AN/PEQ 15 ATPIAL (for each WPN)
Support	WPNs squad, WSL	Suppressive Fire Security on OBJ during RECON/REORG	3 each M240B MG, 1 each AN/PSN 13 navigation set, 1 each AN/PVS 14 NVD (for each individual), 3 each AN/PRC 148 radios, 1 each AN/PEQ 15 ATPIAL (for each WPN)
Assault	Two squads (+) (-)	Main Effort Block, Destroy, Canalize, Capture PIR	3 each AN/PRC 148 radios, 1 each AN/PSN 13 navigation set, 2 each M18A1 Claymore mines, 1 each AT4 Antitank weapon, 1 each AN/PVS 14 NVD (for each individual), 1 each AN/PEQ 15 ATPIAL (for each WPN)

LEGEND
ATPIAL – advanced target pointer/illuminator/aiming light; FO – forward observer; HQ – headquarters; LSS – left side security; NVD – night vision device; OBJ – objective; ORP – objective rally point; PIR – priority intelligence requirement; PL – platoon leader; PSG – platoon sergeant; RECON/REORG – reconnaissance/reorganize; RSS – right side security; RTO – radiotelephone operator; SQD – squad; WPN – weapon; WSL – weapon squad leader

Resources

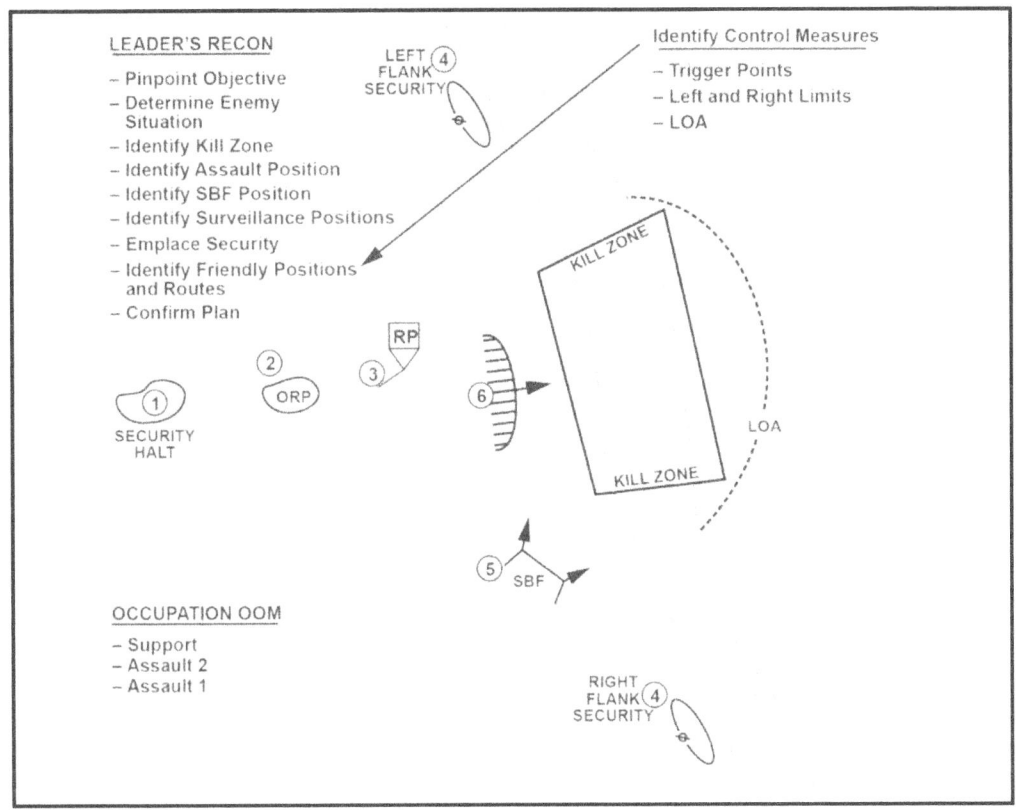

Figure A-13. Leader's reconnaissance

LEGEND
LOA – limit of advance; OOM – order of movement; ORP – objective rally point; Recon – reconnaissance; RP – release point; SBF – support by fire

BUILDING CLEARING TECHNIQUES

CLEAR A ROOM

A-3. On the signal, the team enters through the entry point (or breach). As the team members move to their points of domination, they engage all threats or hostile targets in sequence in their sector. The direction each Ranger moves should not be preplanned unless the exact room layout is known. However, each Ranger moves in a direction opposite the Ranger in front of him. For example, the—

Appendix A

- #1 Ranger enters the room and eliminates any immediate threat. Then moves left or right, moving along the path of least resistance to a point of domination—one of the two corners and continues down the room to gain depth.
- #2 Ranger enters almost simultaneously with the first and moves in the opposite direction, following the wall. The #2 Soldier must clear the entry point, clear the immediate threat area, and move to the point of domination.
- #3 Ranger simply moves in the opposite direction of the #2 Ranger inside the room, moves at least one meter from the entry point, and takes a position that dominates the sector.
- #4 Ranger moves in the opposite direction of the #3 Ranger, clears the doorway by at least one meter, and moves to a position that dominates the sector. (See Chapter 8, figures 8-9, 8-10, and 8-11 on pages 8-19 and 8-20.)

DIAMOND FORMATION (SERPENTINE TECHNIQUE)

A-4. The serpentine technique is a variation of a diamond formation that is used in a narrow hallway. The #1 Ranger provides security to the front. The sector of fire includes any enemy Soldiers who appear at the far end or along the hallway. The #2 and #3 Ranger's cover the left and right sides of the #1 Ranger.

A-5. Their sectors of fire include any enemy combatants who appear suddenly from either side of the hall. The #4 Ranger (normally carrying the M249 machine gun) provides rear protection against any enemy Soldiers suddenly appearing behind the team. (See figure A-14.)

VEE FORMATION (ROLLING-T TECHNIQUE)

A-6. The rolling-T technique is a variation of the vee formation and is used in wide hallways. The #1 and #2 Ranger's move abreast, covering the opposite side of the hallway from the one they are walking on. The #3 Ranger covers the far end of the hallway from a position behind the #1 and #2 Ranger's, firing between them. The #4 Ranger provides rear security. (See figure A-14.)

Resources

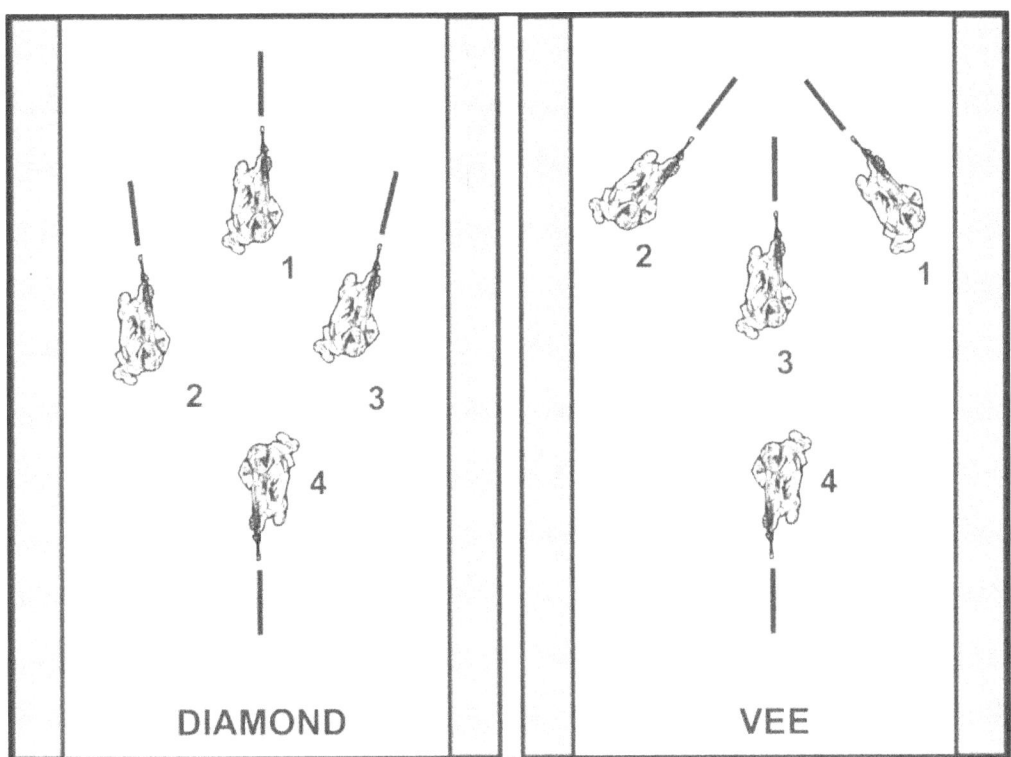

Figure A-14. Diamond and vee formations

CLEARING HALLWAY JUNCTIONS

A-7. Hallway intersections are danger areas and should be approached cautiously. Figure A-15 depicts the fire team's actions upon reaching a "T" intersection when approaching along the "cross" of the "T." The unit is using the diamond (serpentine) formation for movement. (See figure A-15A.) To clear a hallway—

- Team configures into a modified 2-by-2 (box) formation with the #1 and #3 Ranger's abreast and toward the right side of the hall. The #2 Ranger moves to the left side of the hall and orients to the front, and the #4 Ranger shifts to the right side (his left) and maintains rear security. When clearing a right-hand corner, use the left-handed firing method to minimize exposure. (See figure A-15B.)
- #1 and #3 Ranger's move to the edge of the corner. The #3 Ranger assumes a low crouch or kneeling position. On signal, the #3 Ranger, keeping low, turns right around the corner and the #1 Ranger, staying high, steps forward while turning to the right. (Sectors of fire interlock and the low/high positions prevent Soldiers from firing at one another. (See figure A-15C.)

Appendix A

- #2 and #4 Ranger's continue to move in the direction of travel. As the #2 Ranger passes behind the #1 Ranger, the #1 Ranger shifts laterally to the left until reaching the far corner. (See figure A-15D.)
- #2 and #4 Ranger's continue to move in the direction of travel. As the #4 Ranger passes behind the #3 Ranger, the #3 Ranger shifts laterally to the left until reaching the far corner. As the #3 Ranger begins to shift across the hall, the #1 Ranger turns into the direction of travel and moves to the original position in the diamond (serpentine) formation. (See figure A-15E.)
- As the #3 and #4 Ranger's reach the far side of the hallway, they, too, assume their original positions in the serpentine formation, and the fire team continues to move. (See figure A-15F.)

Resources

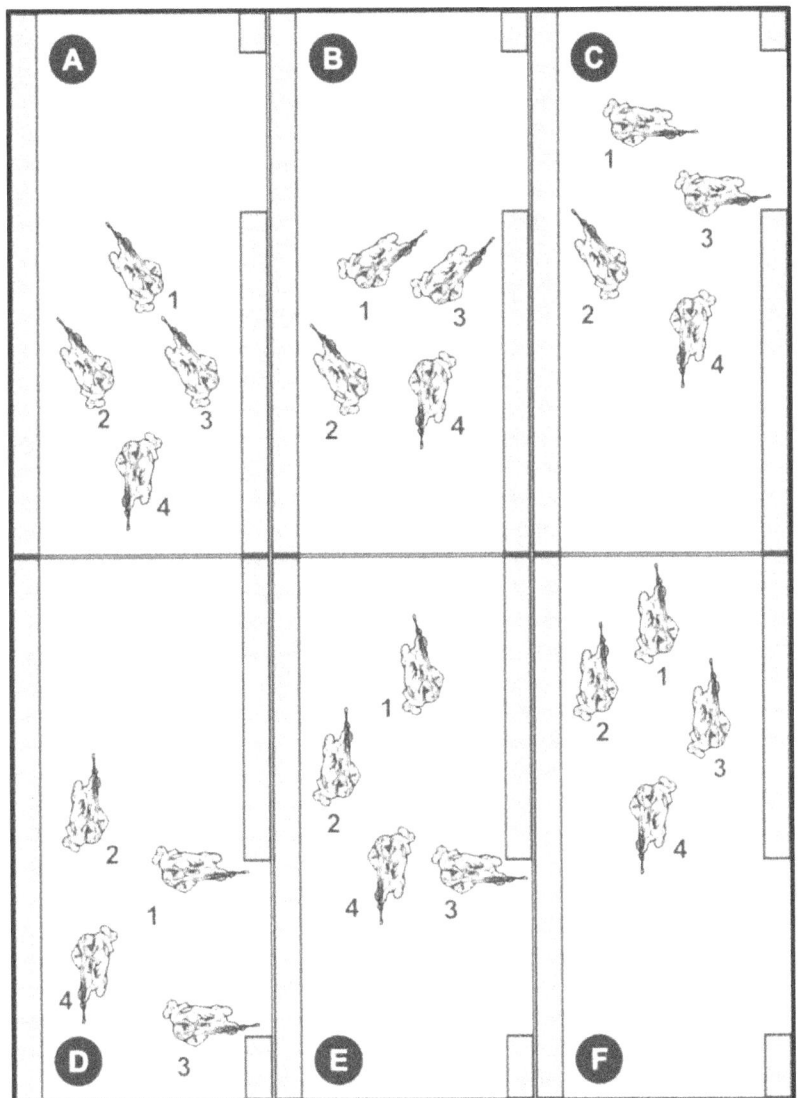

Figure A-15. Clearing hallway junctions

Appendix A

CLEARING A "T" INTERSECTION

A-8. Figure A-16 depicts the fire team's actions upon reaching a "T" intersection when approaching from the base of the "T." The fire team is using the diamond (serpentine) formation for movement. (See figure A-16A.) To clear a "T" intersection:

- The team configures into a 2-by-2 (box) formation with the #1 and #2 Ranger's left and the #3 and #4 Ranger's right. When clearing a right-hand corner, use the left-handed firing method to minimize exposure. See figure A-16B.)
- The #1 and #3 Ranger's move to the edge of the corner and assume a low crouch or kneeling position. On signal, the #1 and #3 Ranger's simultaneously turn left and right respectively. (See figure A-16C.)
- At the same time, the #2 and #4 Ranger's step forward and turn left and right respectively, while maintaining their (high) positions. Sectors of fire interlock and the low/high positions prevent Soldiers from firing at one another. See figure A-16D.)
- Once the left and right portions of the hallway are clear, the fire team resumes the movement formation. (See figure A-16E.) Unless security is left behind, the hallway will no longer remain clear once the fire team leaves the immediate area.

Resources

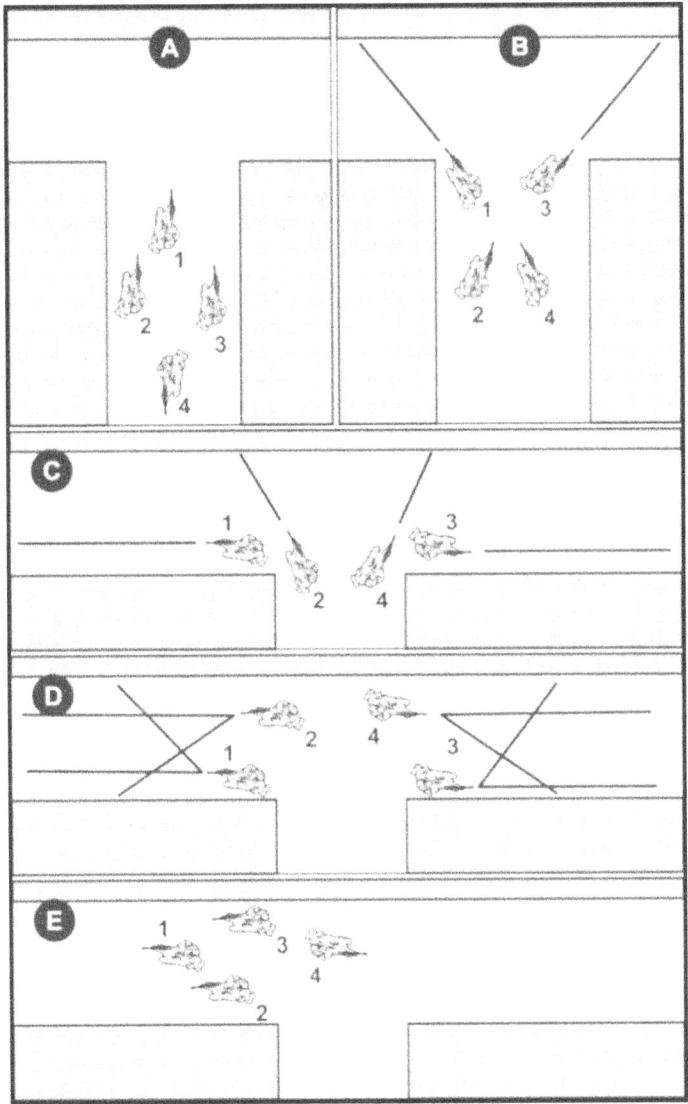

Figure A-16. Clearing a "T" intersection

CLEARING STAIRWELLS AND STAIRCASES

Appendix A

A-9. Stairwells and staircases are comparable to doorways because they create a fatal funnel. The danger is intensified by the three-dimensional aspect of additional landings. The ability of units to conduct the movement depends upon which direction they are traveling and the layout of the stairs. Regardless, the clearing technique follows a basic format:
- The leader designates an assault element to clear the stairs.
- The unit maintains 360-degree, three-dimensional security in the vicinity of the stairs.
- The leader then directs the assault element to locate, mark, bypass, or clear (or both) any obstacles or booby traps that may be blocking access to the stairs.
- The assault element moves up (or down) the stairway by using either the two-, three-, or four-man flow technique, providing overwatch up and down the stairs while moving. The three-man variation is preferred. (See figure A-17.)

Three-Man Flow Clearing Technique

A-10. The following figure best portrays the three-man flow clearing technique. All Ranger's work together, covering each other and focusing on their sector of fire.

Resources

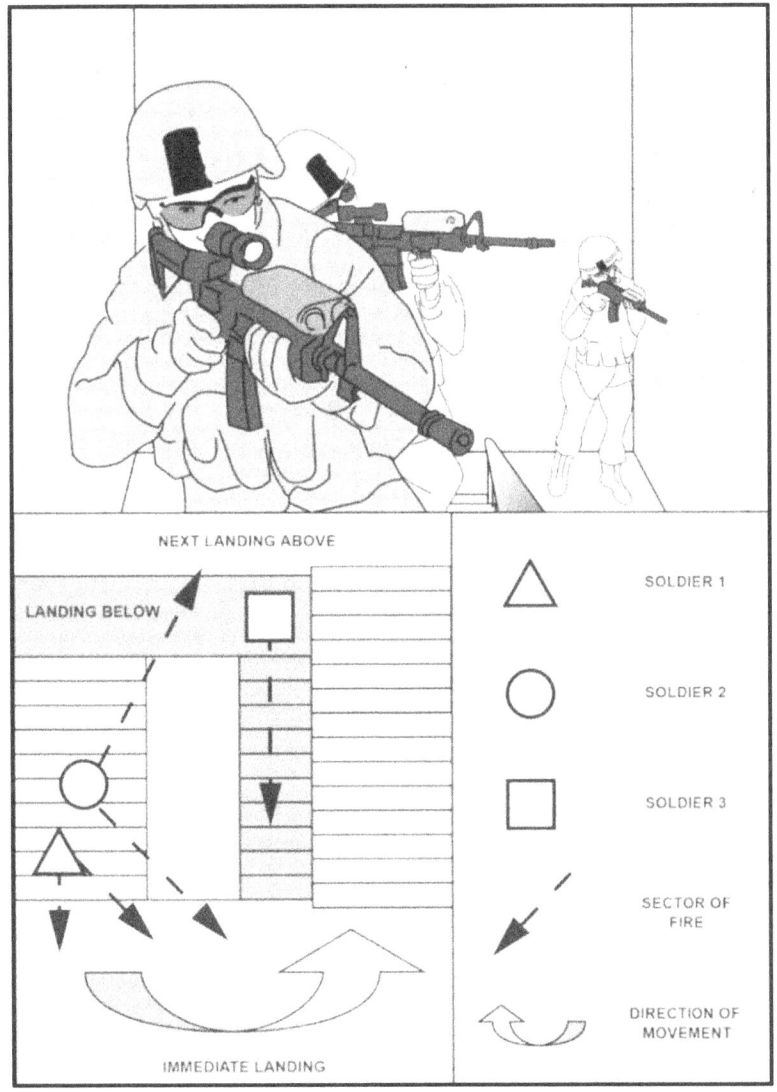

Figure A-17. Three-man flow clearing technique

FOLLOW THROUGH

Appendix A

A-11. After securing a floor (bottom, middle, or top), selected members of the unit are assigned to cover potential enemy counterattack routes to the building. Priority is given initially to securing the direction of attack. Security elements alert the unit and place a heavy volume of fire on enemy forces approaching the unit. Units must guard all avenues of approach leading into their area. These may include—
- Enemy mouse holes between adjacent buildings.
- Covered routes to the building.
- Underground routes into the basement.
- Approaches over adjoining roofs or from window to window.

A-12. Units that performed missions as assault elements should be prepared to assume an overwatch mission and to support another assault element. To continue the mission—
- Momentum must be maintained. This is a critical factor in clearing operations. The enemy cannot be allowed to move to its next set of prepared positions or to prepare new positions.
- The support element pushes replacements, ammunition, and supplies forward to the assault element.
- Casualties must be evacuated and replaced.
- Security for cleared areas must be established according to the OPORD or tactical standing operating procedure (TSOP).
- All cleared areas and rooms must be marked according to the unit SOP.
- The support element must displace forward to ensure that it is in place to provide support (such as isolation of the new objective) to the assault element.

SITE EXPLOITATION

A-13. Sensitive site exploitation (SSE) means to collect information, material, and persons from a designated location and analyze them to answer information requirements, facilitate subsequent operations, or support criminal prosecution. The duties and responsibilities of the site exploitation members are:
- Team leader (TL)—
    - Gathers mission planning products.
    - Briefs SSE during warning order and operations order.
    - Preplans objective search priorities.
    - Establishes premission checks and rehearsals.
    - Coordinates exploitation tasking and prioritization.
    - Establishes priority rooms.
    - Directs activities on target.
    - Orchestrates and leads back brief.
- Assistant team leader—
    - Conducts inventory of SSE kit.
    - Conducts precombat inspections.
    - Conducts rehearsals and rock drills with TL.
    - Determines marshalling area.
    - Determines evidence placement point (EPP).
    - Prepares and packages evidence for transfer.
    - Sketches the objective.
    - Records events.

Searchers (working in two-person teams, if feasible)—
- Train additional searchers (if necessary or feasible).
- Assist with SSE kit inventories.
- Conduct precombat inspections.

Resources

- Conduct rehearsals.
- Assist with packaging and labeling of evidence.
- Communicate all findings to TL, tactical questioner, or both.
- Marshalling officer (MO)—
    - Conducts precombat inspections on equipment.
    - Conducts a detailed search of marshalling area.
    - Collects biometrics.
- Tactical questioning—
    - Assists with collection of detainee biometrics.
    - Communicates collected PIR to the TL.
    - Helps searchers with intelligence gathering and organization.
    - Prioritizes all personnel under control.
    - Prepares questions during precombat inspections.

- A-14. During the initial assessment, team conducts 5/25 meter checks around the target looking toward and away from the target building. The TL determines how much time is left on target, what support is available, who is on target, and what has been found on target, while the assistant team leader determines the marshalling area. The TL and assistant team leader conduct the initial walk through of the target—TL has assistant team leader walk through the target building and discuss what was found on target.
- TL labels rooms while moving through the target (assistant team leader can label rooms if TL is busy).
- Assistant team leader sketches, the sketch includes:
    - Priority of the rooms based on the walk through and PIRs.
    - TL prioritizes rooms in order of importance to be searched.

A-15. During a room and building search, the TL determines the priority area. This is searched first, followed by a systematic 360-degree search of the entire room. This is conducted:

- Clockwise, low, medium, and high to ensure all areas are covered. Label all drop holes and false walls.
- Move all cleared items, such as tables, chairs, pots, to the center of the room.
- Place evidence bags in the EPP while searching the room. Items from the individual room EPP are later moved to the centralized EPP.
- Ensure that at least two people search each room.
    - Mark the label of the room with an "X" once the room is searched. The "X" informs team members that two people have searched that room.
    - Search all remaining rooms clockwise at low, medium, and high points.
- The assistant team leader continues to sketch, identifying dead space in the building EPP, and making sure the EPP is in an area that has already searched and cleared.

A-16. The MO and TQ clear the marshalling area, set up two pits, and then clear the area where the detainees were found on target. The two pits are called dirty and clean. The—

- Dirty pit is for detainees who have been searched once for weapons.
- Clean pit is for detainees who have had a detailed search for any items.

A-17. In the marshalling area, the MO and TQ search all detainees, collect biometrics, separate detainees, and relay all information to the TL.

- There are two searches:
    - The first search ensures the detainees do not have weapons and is conducted in the dirty pit. Eye, ears, flex cuffs, and detainee bag are put on the detainee at this time.

Appendix A

- The second search is a systematic search from top to bottom, front to back, and left to right in the clean pit that ensures every article of clothing has been examined.
- When collecting biometrics, the MO and TQ:
  - Collect fingerprints, iris scan, and a photo through the secure Biometric Live Scan device or ten-print card.
  - Collect all media devices for further exploitation.
- The TQ separates individuals from the rest of the detainees for further questioning.
- TQ and MO relay all pertinent information to TL to assist in decision making.

A-18. Once all objectives are met, the TL collapses the objective and conducts a final walk through to make sure no signs of U.S. TTPs were left on the target. The TL insures accountability of men, weapons, equipment items, and detainees found on the target and prepares for exfiltration.

Appendix B
Quick Reference Cards

This appendix is designed as a quick reference for Rangers to use when training or in the field. Several procedures are detailed in this chapter. Figures B-1 through B-4 on pages B-2 through B-5, and figure B-5 on pages B-9 and B10 depict critical reference cards that Rangers routinely use during their missions. Table B-1 on pages B-6 through B-8 explains in greater depth the nine lines of the MEDEVAC card. Although discussed in depth in the publication, these reference cards are designed to be a handy reminder for Rangers in the field. Check periodically for updated information on the Army Training Network (ATN).

Appendix B

IED / UXO

Procedures when IEDs are found

Security - Maintain 360 degrees. Scan close in and far out, up high and down low.

Always - Scan your immediate surroundings for more IEDs.

Move - Move away. Vary safe distances, but plan for 300m minimum safe distance and adapt to your METT-TC.

Attempt - To confirm suspected IEDs using optics while staying back as far as possible.

Cordon - Off the area. Direct people out of danger area. Do not allow anyone to enter except for EOD. Question, search, and detain suspects as defined by your Existing ROE.

Report - Your situation using the 9 line IED / UXO spot report format.

This could be your hand if you try to dispose of UXOs or IEDs. The enemy has developed anti-handling to catch you when you try defusing. Leave it to experts!

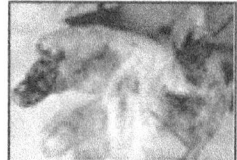

Call EOD - Don't be a Hero!

IED / UXO Report

LINE 1. DATE-TIME-GROUP: when the item was discovered
LINE 2. REPORT ACTIVITY AND LOCATION: Unit and grid location of the IED/UXO.
LINE 3. CONTACT METHOD: Radio frequency, call sign, POC, AND telephone number.
LINE 4. TYPE OF ORDNANCE: Dropped, projected, placed, of thrown. Give the number of items, if more than one.
LINE 5. NBC CONTAMINATIONS: Be as specific as possible.
LINE 6. RESOURCES THREATENED: Equipment, facilities, or other assets that are threatened.
LINE 7. IMPACT ON MISSION: Short description of current tactical situation and how the IED/UXO affecs the status of the mission.
LINE 8. PROTECTIVE MEASURES: Any measures taken to protect personnel and equipment.
LINE 9. RECOMMENDED PRIORITY: Immediate, Indirect, Minor, No Threat.

Priority

Immediate: Stops unit's maneuver and mission capability or threatens critical assets vital tothe mission.

Indirect: Stops the unit's maneuver and mission capability or threatens critical assets important to the mission.

Minor: Reduces the unit's maneuver and mission capability or threatens critical assets of value.

No Threat: Has little or no effect on the unit's capabilities or assets.

Figure B-1. IED/UXO card

LEGEND
EOD – explosive ordnance disposal; IED – improvised explosive device; m- meter; METT-TC - mission, enemy, terrain and weather, troops and support-time available, and civil considerations; NBC – nuclear, biological, and chemical; POC – point of contact; ROE – rules of engagement; UXO – unexploded ordnance

Quick Reference Cards

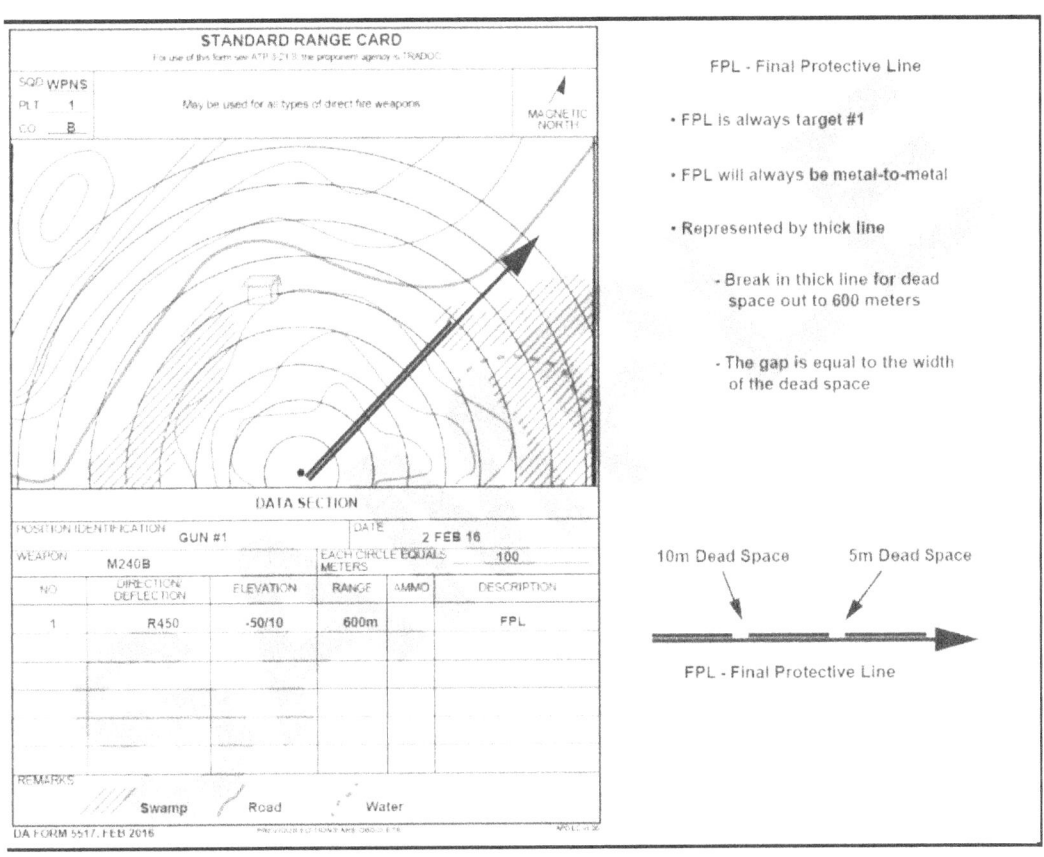

Figure B-2. Range card with final protective line

LEGEND
AMMO – ammunition; ATTP – Army Tactics, Techniques, and Procedures; CO – company; DA – Department of the Army; FEB – February; FPL – final protective line; m – meter; NO. – number; PLT – platoon; TRADOC – Training and Doctrine Command; WPNS – weapons

Appendix B

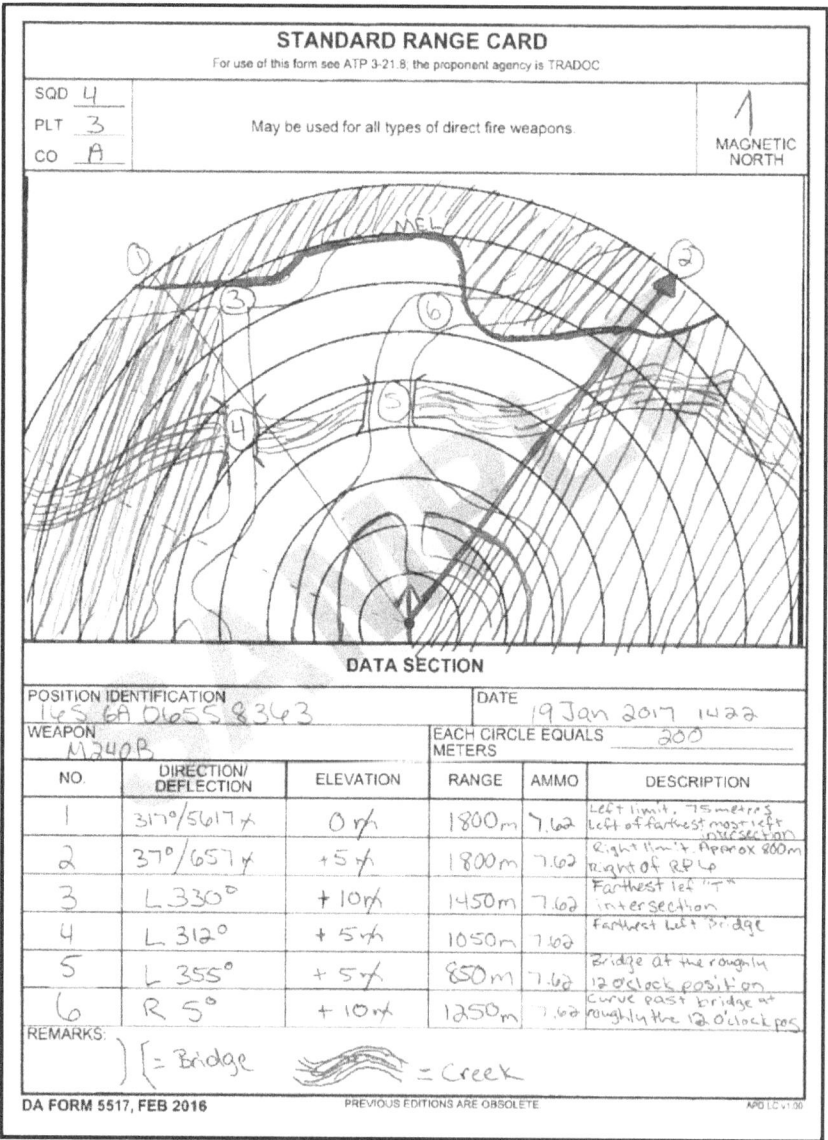

Figure B-3. Standard Range Card

Quick Reference Cards

NATO 9-Line MEDEVAC Request Format
(use brevity codes for non-secure communication or use full description for more clarity)

1. Location of Pick-up Site	AB 48734972			
2. Call Sign & Frequency of Requesting Unit	37500 RANGER 6			
3. # Patients by Precedence				
Urgent (1 hr):	Priority (4 hrs): X	Routine (24 hrs):		
4. Special Equipment Required				
A. None X	B. Hoist	C. Extrication Equip.	D. Ventilator	E. Other (describe)
5. # Patients by Type				
L. Litter: 1	A. Ambulatory:	E. Escort (women/children/HVT)		
6. Security at Pick-up Site				
N. No Enemy	N			
P. Possible enemy troops in area				
E. Enemy troops in area (approach with caution)				
X. Enemy troops in area (armed escort required)				
7. Method of Marking Pick-up Site				
A. Panel X	B. Pyrotechnic	C. Smoke X	D. None	E. Other (describe)
8. Patient Nationality & Status (# by type)				
A. US/Coalition Military, Nationality:	X			
B. US/Coalition Civilian, Nationality:				
C. Non-US/Coalition Military, Nationality:				
D. Non-US/Coalition Civilian, Nationality:				
E. Enemy Prisoner of War:				
F. High Value Target (escort required):				
9. Terrain Description	FLAT, NEAR LAKE			

MIST Report
(Required for Each Patient; Reference Patient's DD 1380 TCCC Card)

Patient ID (i.e. Battle Roster):	J05794	
M - Mechanism of Injury i.e. blast, gunshot wound (GSW), etc.; can be NONE if medical complaint	GSW	
I - Injuries Sustained i.e. penetrating wound, laceration, burn, amputation, etc.; include body location	PENETRATING WOUND	
S - Signs and Symptoms		
Pulse	68 R	69 R
Blood Pressure	120/80	118/80
Respiratory Rate	20	20
Level of Consciousness (AVPU)	A-ALERT	A-ALERT
Other		
T - Treatment Given i.e. tourniquet, NPA, needle-D, fluids, medications	FENTANYL 800 MG OTFC 1100	MORPHINE 20 MG IV 1115

Figure B-4. NATO 9-Line MEDEVAC Request card and MIST report

Appendix B

Table B-1. Explanation of the nine lines on a MEDEVAC Request card

LINE	ITEM	EXPLANATION	WHERE OR HOW OBTAINED	WHO NORMALLY PROVIDES	REASON
1	Location of pickup site.	Grid coordinates of the pickup site should be sent by secure communication. To prevent confusion, the grid zone letters are included in the message.	From map or navigational device, determine the military grid reference system six-digit grid coordinates of the pickup site.	Unit leader(s).	Required so evacuation vehicle knows where to pick up patient/casualty. Also, so that the unit coordinating the evacuation mission can plan the route for the evacuation vehicle (if the vehicle must pick up from more than one location).
2	Radio frequency call sign and suffix.	Frequency of the radio at the pickup site, not a relay frequency. The call sign (and suffix, if used) of person to be contacted at the pickup site may be transmitted in the clear.	From automated net control device or other approved means.	RTO.	Required so that evacuation vehicle can contact requesting unit while en route (obtain additional information or changes in situation, or directions).
3	Number of patients by precedence.	A: URGENT. B: URGENT – SURGERY. C: PRIORITY. D: ROUTINE. E: CONVENIENCE. If two or more categories must be reported in the same request, insert the word "BREAK" between each category.	From evaluation of patients.	Medic or senior person present.	Required by unit controlling vehicles to assist in prioritizing missions.

Quick Reference Cards

Table B-1. Explanation of the nine lines on a MEDEVAC Request card (continued)

LINE	ITEM	EXPLANATION	WHERE OR HOW OBTAINED	WHO NORMALLY PROVIDES	REASON
4	Special equipment required.	A: None. B: Hoist. C: Extraction equipment. D: Ventilator.	From evaluation of patient(s) and the situation.	Medic or senior person present.	Required so that the equipment can be placed onboard the evacuation vehicle prior to the start of the mission.
5	Number of patients by type.	Report only applicable information and encrypt the brevity code. If requesting medical evacuation for both types, insert the word "BREAK" between the litter entry and ambulatory entry: L + number of patients = litter. A + number of patients = ambulatory (sitting).	From evaluation of patient(s).	Medic or senior person present.	Required so that the appropriate number of evacuation vehicles may be dispatched to the pickup site. They should be configured to carry the patients requiring evacuation.
6	Security of pickup site **(wartime)**.	N: no enemy troops in area. P: Possible enemy troops in area (approach with caution). E: enemy troops in area (approach with caution). X: enemy troops in area (armed escort required).	From evaluation of situation.	Unit leader.	Required to assist the evacuation crew in assessing the situation and determining if assistance is required. More definitive guidance can be furnished to the evacuation vehicle as it is en route (specific location of enemy to assist an aircraft in planning its approach).
6	Number and type of wound, injury, or illness. **(peacetime)**.	Specific information regarding patient wounds by type (gunshot or shrapnel). Report serious bleeding and blood type (if known).	From evaluation of patient(s).	Medic or senior person present.	Required to assist evacuation personnel in determining treatment and special equipment needed.

Appendix B

Table B-1. Explanation of the nine lines on a MEDEVAC Request card (continued)

LINE	ITEM	EXPLANATION	WHERE OR HOW OBTAINED	WHO NORMALLY PROVIDES	REASON
7	Method of marking pickup site.	A: Panels. B: Pyrotechnic signal. C: Smoke signal. D: None. E: Other.	Based on situation and availability of materials.	Medic or senior person present.	Required to assist the evacuation crew in identifying the specific location of the pickup. (NOTE: the color of the panels or smoke should not be transmitted until the evacuation vehicle contacts the unit just prior to arrival.) For security, the crew identifies the color and the unit verifies it.
8	Patient nationality and status.	The number of patients in each category need not be transmitted: A: U.S. military. B: U.S. citizen. C: Non-U.S. military. D: Non-U.S. citizen. E: Enemy prisoner of war.	From evacuation platform.	Medic or senior person present.	Required to assist in planning for destination facilities and need for guards. Unit requesting support should ensure there is an English-speaking representative at the pickup site.
9	Chemical, Biological, Radiological, and Nuclear contamination **(wartime)**.	Include this line only when applicable: C: Chemical. B: Biological. R: Radiological. N: Nuclear.	From situation.	Medic or senior person present.	Required to assist in planning the mission. (Determine which evacuation vehicle accomplishes the mission, and when it is accomplished.)
9	Terrain description **(peacetime)**.	Include details of terrain features in and around proposed landing site. If possible, describe relationship of site to prominent terrain feature (lake, mountain, tower).	From area survey.	Personnel present.	Required to allow evacuation personnel to assess routes and avenues of approach into area. Of particular importance if hoist operation is required.

Quick Reference Cards

TACTICAL COMBAT CASUALTY CARE (TCCC) CARD

BATTLE ROSTER #: JD5794

EVAC: ☒ Urgent ☐ Priority ☐ Routine

NAME (Last, First): DOE, JOHN   LAST 4: 5794

GENDER: ☒ M ☐ F   DATE (DD-MMM-YY): 04 JUL 17   TIME: 0910 Z

SERVICE: ARMY   UNIT: RANGERS   ALLERGIES: NKDA

Mechanism of Injury: (X all that apply)
☐ Artillery ☐ Blunt ☐ Burn ☐ Fall ☐ Grenade ☒ GSW ☐ IED
☐ Landmine ☐ MVC ☐ RPG ☐ Other: _____

Injury: (Mark injuries with an X)

TQ: R Arm
TYPE:
TIME:

TQ: L Arm
TYPE:
TIME:

TQ: R Leg
TYPE: CAT
TIME: 0911

TQ: L Leg
TYPE:
TIME:

GSW (X) — right hip/groin area
GSW (X) — right foot

Signs & Symptoms: (Fill in the blank)

Time	0915	0950	1115	
Pulse (Rate & Location)	68 R	69 R	67 R	
Blood Pressure	120/80	118/80	120/80	/
Respiratory Rate	20	20	16	
Pulse Ox % O2 Sat	99	99	98	
AVPU	A	A	A	
Pain Scale (0-10)	8	5	3	

DD Form 1380, JUN 2014   TCCC CARD

Figure B-5A. DD Form 1380, Tactical Combat Casualty Care card (front)

Appendix B

```
BATTLE ROSTER #: JD 5794
     EVAC: ☒ Urgent  ☐ Priority  ☐ Routine
```

Treatments: (X all that apply and fill in the blank)   Type
C: TQ- ☒ Extremity ☐ Junctional ☐ Truncal   CAT
   Dressing- ☐ Hemostatic ☒ Pressure ☐ Other   ETD
A: ☒ Intact ☐ NPA ☐ CRIC ☐ ET-Tube ☐ SGA
B: ☐ O2 ☐ Needle-D ☐ Chest-Tube ☐ Chest-Seal
C:

	Name	Volume	Route	Time
Fluid				
Blood Product				

MEDS:

	Name	Dose	Route	Time
Analgesic (e.g., Ketamine, Fentanyl, Morphine)	Fentanyl	800 mcg	OTFC	0915
	Fentanyl	800 mcg	OTFC	0950
	morphine	20 mg	IV	1115
Antibiotic (e.g., Moxifloxacin, Ertapenem)				
Other (e.g., TXA)				

OTHER: ☒ Combat-Pill-Pack ☐ Eye-Shield (☐ R ☐ L) ☐ Splint
☒ Hypothermia-Prevention Type: HPMK

NOTES:

FIRST RESPONDER
NAME (Last, First): Monty, Harold   LAST 4: 0275
DD Form 1380, JUN 2014 (Back)   TCCC CARD

Figure B-6B. DD Form 1380, Tactical Combat Casualty Care card (back)

Glossary

TC 3-21.76 uses joint terms where applicable. Selected joint and Army terms and definitions appear in both the glossary and the text. Terms for which TC 3-21.76 is the proponent publication (the authority) are italicized in the text and marked with an asterisk (*) in the glossary. Terms and definitions for which TC 3-21.76 is the proponent publication are boldfaced in the text. For other definitions shown in the text, the term is italicized and the number of the proponent publication follows the definition.

SECTION I – ACRONYMS AND ABBREVIATION

AO	area of operations
APS	Anchor Prusik System
ATC	air traffic controller (belay device)
BCT	brigade combat team
BTC	bridge team commander
CAS	close air support
CASEVAC	casualty evacuation
CO_2	carbon dioxide
CRRC	combat rubber raiding craft
DA	Department of the Army
EPP	evidence placement point
EPW	enemy prisoner of war
FDC	fire direction center
FLC	field load carrier
FO	forward observer
FRAGORD	fragmentary order
FSO	fire support officer
HAMK	high-angle mountaineering kit
HE HQ	high explosive
IED	headquarters
KIA	improvised explosive device
LACE	killed in action
lb	liquid, ammunition, casualties, and equipment
LDA	pound
LOA	linear danger area
LOS	limit of advance
LZ	line of sight
m	landing zone
MB	meter
MDI	main body
MEDEVAC	modernized demolition initiator
	medical evacuation

Glossary

METT-TC	mission, enemy, terrain and weather, troops and support available-time available, and civil considerations
MO	marshalling officer
mm	millimeter
mph	mile per hour
MRE	meal, ready to eat
MTC	movement to contact
NLT	not later than
NVD	night vision device
OAKOC	observation and fields of fire, avenues of approach, key terrain, obstacles, and cover and concealment
OPORD	operation order
ORP	objective rally point
PB	patrol base
PIR	priority intelligence requirement
PL	platoon leader
PSG	platoon sergeant
PZ	pickup zone
QP	quartering party
R&S	reconnaissance and surveillance
RED	risk estimate distance
ROE	rules of engagement
RP	release point
SAW	squad automatic weapon
SBF	support by fire
SL	squad leader
SLLS	stop, look, listen, and smell
SOI	signal operating instructions
SOP	standard operating procedure
SSE	sensitive site exploitation
TC	training circular
TL	team leader
TLP	troop leading procedures
TM	technical manual
TQ	tactical questioning
UHF	ultrahigh frequency
UO	urban operation
U.S.	United States
VHF	very high frequency
WARNORD	warning order
WSL	weapons squad leader

Glossary

SECTION II - TERMS

5/25 meter check
Five meter and 25 meter search.

1/3 - 2/3 rule
Of the available time allotted for troop-leading procedures (planning), the leader uses 1/3 of the time, allowing the subordinate 2/3 of the time for their actions.

3-Ws
Who, what, and where.

4-W1s
Who, what, where, and why.

5-Cs
Confirm, clear, call, cordon, and control.

5-Ss
Secure, search, segregate, safeguard, and speed.

5-Ws
Who, what where, when, and why.

100 mph tape
Scrim-backed, pressure-sensitive tape.

Class I
Food, rations, and water.

Class III
Petroleum, oils, and lubricants.

Class V
Ammunition.

Class VII
Major end items.

Class VIII
Medical supplies, minimal amounts.

Class IX
Repair parts.

MIST
Mechanism of injury, injuries sustained, signs and symptoms, and treatment given

SALUTE
Size, activity, location, unit, time, and equipment.

STANO
Surveillance, target acquisition, and night observation.

This page intentionally left blank.

References

REQUIRED PUBLICATIONS

ADRP 1-02, *Terms and Military Symbols*, 16 November 2016.
DOD Dictionary of Military and Associated Terms, February 2017.

RELATED PUBLICATIONS

Most Army doctrinal publications and regulations are available at http://www.apd.army.mil.
Most joint publications are available online at http://www.dtic.mil/doctrine/doctrine/doctrine.htm.
Other publications are available on the Central Army Registry on the Army Training Network: https://atiam.train.army.mil.
Military Standards are available online at http://www.everyspec.com/MIL-STD/

ADP 5-0, *The Operations Process*, 17 May 2012.
ADRP 3-0, *Operations*, 11 November 2016.
ATP 3-09.32, *JFire Multi-Service Tactics, Techniques, and Procedures for the Joint Application of Firepower {MCRP 3-16.6A; NTTP 3-09.2; AFTTP 3-2.6}*, 21 January 2016.
ATP 3-21.8, *Infantry Platoon and Squad*, 23 August 2016.
ATP 4-01.45, *Multi-Service Tactics, Techniques, and Procedures for Tactical Convoy Operations*, 18 April 2014.
ATTP 3-06.11, *Combined Arms Operations in Urban Terrain*, 10 June 2011.
FM 3-22.27, *MK 19, 40-mm Grenade Machine Gun, Mod 3*, 28 November 2003.
FM 3-22.68, *Crew-Served Weapons*, 21 July 2006.
FM 27-10, *The Law of Land Warfare*, 18 July 1956.
STP 21-1-SMCT, *Soldier's Manual of Common Tasks, Warrior Skills Level 1*, 10 August 2015.
TC 3-97.61, *Military Mountaineering*, 26 July 2012.
TM 3-34.82, *Explosives and Demolitions*, 7 March 2016.
TM 9-1005-201-10, *Operator's Manual for Machine Gun, 5.56-MM, M249 W/Equip*, 26 July 1991.
TM 9-1005-213-10, *Operator Maintenance for Machine Gun, Caliber .50: M2A1, W/Fixed Headspace and Timing*, 11 January 2016.
TM 9-1005-313-10, *Operator's Manual for Machine Gun 7.62MM, M240*, 15 November 2002.
TM 9-1010-230-10, *Operator's Manual for Machine Gun, 40 MM, MK19 MOD 3*, 31 August, 2012.

RECOMMENDED READINGS

ADP 6-22, *Army Leadership*, 1 August 2012.
ADRP 6-22, *Army Leadership*, 1 August 2012.
AR 350-1, *Army Training and Leader Development*, 19 August 2014.
ATP 3-18.12, *Special Forces Waterborne Operations*, 14 July 2016
ATP 3-18.14, *Special Forces Vehicle-Mounted Operations Tactics, Techniques, and Procedures*, 12 September 2014.
ATP 3-75, *Ranger Operations*, 26 June 2015.
ATP 4-02.2, *Medical Evacuation*, 12 August 2014
ATP 6-02.53, *Techniques for Tactical Radio Operations*, 7 January 2016.

References

ATP 6-02.72, MCRP 3-40.3A, NTTP 6-02.2, AFTTP 3-2.18, *TAC Radios Multi-Service Tactics, Techniques, and Procedures for Tactical Radios,* 5 November 2013.
ATTP 3-21.50, *Infantry Small Unit Mountain Operations,* 28 February 2011.
FM 3-09, *Field Artillery Operations and Fire Support,* 4 April 2014.
FM 3-21.38, *Pathfinder Operations,* 25 April 2006.
FM 6-02, *Signal Support to Operations,* 22 January 2014.
FM 6-99, *U.S. Army Report and Message Formats,* 19 August 2013.
STP 8-68W13-SM-TG, *Soldier's Manual and Trainer's Guide MOS 68W, Health Care Specialist, Skill Level 1, 2 and 3,* 3 May 2013.
TC 3-25.26, *Map Reading and Land Navigation,* 15 November 2013.
TC 31-34-4, *Special Forces Tracking and Countertracking,* 30 September 2009.
TC 8-800, *Medical Education and Demonstration of Individual Competence,* 15 September 2014.
TM 3-22.31, *40-mm Grenade Launchers,* 17 November 2010.

OTHER PUBLICATIONS AND DOCUMENTS

PRESCRIBED FORMS

This section contains no entries.

REFERENCED FORMS

Unless otherwise indicated, DA forms are available on the Army Publishing Directorate (APD) web site (http://www.apd.army.mil). DD Forms are available on the Office of the Secretary of Defense (OSD) web site (www.dtic.mil/whs/directives/infomgt/forms/formsprogram.htm).
SF Forms are available on the GSA Forms Library web site (http://www.gsa.gov/portal/forms/type/SF).

DA Form 2028, *Recommended Changes to Publications and Blank Forms.*
DA Form 5517, *Standard Range Cards.*
DD Form 1380, *Tactical Combat Casualty Care Card.*

Index

Entries are listed by paragraph number.

5
5-S format. 2-20

A
ambush. 7-37
    deliberate. 7-41
    hasty. 7-39
AN/PRC-119F. 13-20, 4-3
    troubleshooting. 4-4
anchors. 9-30
    bowline. 9-44
    constant tension. 9-33
antenna
    construction. 4-11
    expedient 292-type. 4-15
    length planning. 4-20
    whip. 4-7
    wire. 4-8
avoiding detection. 7-20

B
belaying. 9-45
air traffic controller. 9-48
body belay. 9-46
command sequence. 9-49
    mechanical. 9-47
boat formations. 13-32

C
call-for-fire. 3-10
carabiners. 13-19, 9-27
casualty criterion. 3-6
charges. 5-25
    formula. 5-27
combat rubber raiding craft (CRRC). 13-24
curvature of the earth. 4-23

D
demolitions
    knots. 5-23

E
explosives. 5-1
    expedient. 5-13
    MDI blasting caps. 5-3
    safety. 5-9

F
fire support overlay. 3-8
first aid
    airway obstructions. 15-3
    bleeding. 15-9, 15-6
    burns. 15-10
    fractures. 15-8
    shock. 15-7
    weather and environment. 15-11
forward observer. 1-23

G
GOTWA. 7-13

Index

H

hitches
    clove. 9-35, 9-36
    Munter. 9-40
horizontal lift. 9-15

K

knots. 9-31
    bowline. 9-44
    figure-eight. 9-34,
    Munter mule. 9-41
    Prusik. 9-43
    slipknot. 9-39
    square. 9-32

L

limited visibility. 6-20

M

machine guns. 10-1
    ammunition planning. 10-28
    base of fire. 10-11
    controlled occupation. 10-17
    M2. 10-9
    M240B. 10-8
    M249. 10-8

maneuver element. 10-13
    MK19. 10-9
marches. 6-15
mark buildings and rooms. 12-29
medic. 6-17, 1-20
METT-TC. 11-1
mounted patrol
    five phases. 11-3
movement techniques
    bounding overwatch. 6-7
    traveling. 6-7

O

occupation of the ORP. 7-23

P

password systems. 7-10
platoon leader. 6-17, 1-4
platoon sergeant. 6-17, 1-6

R

R&S teams. 7-46, 7-33, 7-32, 7-29, 7-4
radio operator. 1-21
rally points. 7-14
rappelling. 9-58
    extended ATC. 9-61
    rappel seat. 9-37
    seat hip. 9-60
rope bridges. 9-53, 9-54, 9-56
    stream crossing. 13-1
ropes. 9-21
    bowline. 9-44
    care of. 9-25
    cordelette. 9-23
    cords. 9-24
    inspection. 9-29
    installations. 9-50
rotary aircraft formations
    diamond. 14-13
    echelon left or right. 14-15
    heavy left or right. 14-12
    staggered trail left or right. 14-17

Index

trail. 14-16
vee. 14-14

rotary aircraft landing space requirements. 14-21

rotary aircraft specifications. 14-25

S

selecting a PZ or LZ. 14-7
    ground slope. 14-7
    markings. 14-10

squad leader. 6-17, 1-11

T

team leader. 1-16

U

urban operations
    civilians. 12-18
    obstacles. 12-13
    task organization. 12-3
    terrain. 12-12
    troops and time. 12-17

V

vertical lift. 9-14

W

water navigation terminology. 13-29

weapons squad leader. 1-15

work, rest, and water consumption schedules. 15-19

Z

Z-Pulley System. 9-57

TC 3-21.76
26 April 2017

By Order of the Secretary of the Army:

MARK A. MILLEY
General, United States Army
Chief of Staff

Official:

GERALD B. O'KEEFE
Administrative Assistant to the
Secretary of the Army
1708907

DISTRIBUTION:

Active Army, Army National Guard, and United States Army Reserve: To be distributed in accordance with the initial distrubution number (IDN) 116076, requirements for TC 3-21.76.

PIN: 201554-000

www.ingramcontent.com/pod-product-compliance
Lightning Source LLC
Chambersburg PA
CBHW082201220526
45470CB00010B/3009